Plumbing

NVQ and Technical Certificate Level 2

Mike Phoenix

John Thompson

JTL
Connecting to the future

www.heinemann.co.uk

✓ Free online support
✓ Useful weblinks
✓ 24 hour online ordering

01865 888058

Heinemann

Inspiring generations

Heinemann Educational Publishers
Halley Court, Jordan Hill, Oxford OX2 8EJ
Part of Harcourt Education

Heinemann is the registered trademark of
Harcourt Education Limited

Text © JTL, 2005

First published 2005

10 09 08 07 06 05
10 9 8 7 6 5 4 3 2

British Library Cataloguing in Publication Data is available
from the British Library on request.

10-digit ISBN: 0 435401 94 7
13-digit ISBN: 978 0 435401 94 8

Typeset by HL Studios, Long Hanborough, Oxford

Original illustrations © Harcourt Education Limited, 2005

Cover design by GD Associates

Printed in the UK by Scotprint

Cover photo © Gareth Boden/Harcourt Education

Acknowledgements
Every effort has been made to contact copyright holders of material reproduced in
this book. Any errors or omissions will be rectified in subsequent printings if notice
is given to the publishers.

Contents

Acknowledgements

I would like to thank the following colleagues for contributing to this book:

John Thompson, who started the project and put in a good foundation, in particular the original training manuals from which this book has been developed.

Keith Arrell for some great proof-reading and observations, also Dave Bowers, Mark Brown, Karen Easton, Julian Hodgson, Keith Powell, and everyone else that lent a hand.

I would also like to thank the Plumbing staff at Building Services Training Caerphilly, JTL Training Centre, Malton and Oxford College for allowing the Heinemann photoshoot team to take over the centres for several days. The professional help given by so many individuals helped to make the photoshoots such a success.

At Heinemann I would like to thank Jenni Johns, Julia Bruce and Jane Hance, as well as Gareth Boden and his team.

Finally, but equally importantly, I would like to thank those many others who have been indirectly involved with this book. Their patience, understanding and support during the project have made this achievement possible. Thank you to all who knowingly or unwittingly provided support when required to do so.

Mike Phoenix

July 2005

The authors and publishers would like to thank the following for permission to reproduce copyright material:

Amicus, JTL Limited, WRAS

PICTURE ACKNOWLEDGEMENTS

The authors and publishers would like to thank the following for permission to reproduce photographs:

All pictures **Harcourt Education Ltd/Gareth Boden** apart from the following:

Alamy pages 216, 387, 390, 396; **Alamy/Brand X Pictures** page 82; **Alamy/Comstock Pictures** page 264 (middle); **Alamy/Brian Harris** page 120 (bottom); **Alamy/Ingram Publishing** page 254 (top); **Alamy/Niall McDiarmid** page 256 (middle); **Alamy/Photofusion/Steve Morgan** page 386; **Alamy/The Photolibrary Wales** page 120 (top); **Arco** page 36 (middle); **Art Directors and Trip/Chris Parker** page 72; **Art Directors and Trip/Helene Rogers** pages 36, 93, 263 (bottom); **Construction Images** pages 1, 19, 47, 70, 176, 357; **Fernox** page 280; **Jake Fitzjones/Redcover.com** page 257; **GAH Group** page 261; **Getty** page 2, 282; **Harcourt Education Ltd/Jules Selmes** page 61; **Harcourt Education Ltd/Ginny Stroud-Lewis** pages 35, 36 (bottom), 38, 87, 122, 259, 264 (bottom); **Institute of Heating and Plumbing Engineers** page 214; **Lead Sheet Association** pages 360, 361, 362, 366, 368, 369, 377, 378, 381, 382; **Science Photo Library/Sheila Terry** page 178

Introduction

How to use this book

This book has been designed with you in mind. It has a dual purpose:

1 To lead you through the Level 2 MES Plumbing qualifications, providing background information and technical guidance.

2 To provide a future reference book that you will find useful to dip into long after you have gained your qualification.

The book has been specially produced to assist you in completing your Plumbing 6129 Scheme and NVQ qualification. The schemes are applicable in England, Wales and Northern Ireland and are designed to set a quality standard for learning and training while attending college/a learning centre or as a mature experienced worker wishing to update qualifications.

The book details the various sections that you will undertake to complete the Level 2 Units from the Plumbing 6129 Technical Certificate Scheme, all of which support the full NVQ. Each chapter concludes with job knowledge checks which are essential parts of the overall qualification.

The book is also a key reference document for you to consult in support of your plumbing work. Your tutor will further explain the content of the various sections to you when you attend an approved assessment centre.

The chapters are in a similar format to the units of the City & Guilds assessment, making it easier to complete a section with a successful assessment before moving on to the next.

Qualifications

1 Technical Certificate: this is the job knowledge and training part of the qualification for new entrants.

2 NVQ: this in full means National Vocational Qualification, which is the collection of evidence that you have done the work in the real workplace.

Together they give the full qualification for an operative to be able to work with a company under supervision.

For those of you who are already engaged in the trade and have been for some time, you can simply do the tests of the Technical Certificate to accompany the NVQ workplace evidence.

Either way, this book has the content you require.

How this book can help you

There are other key features of this book which are designed to help you make progress and reinforce the learning that has taken place. Such features are:

- **Photographs:** easy-to-follow sequences of key operations

- **Illustrations:** clear drawings, many in colour, showing essential information about complex components and procedures

- **Margin notes:** short helpful hints to aid you to good practice

- **Tables, bullet points and flowcharts:** easy-to-follow features giving information at a glance

- **On-the-job scenarios:** typical things that happen on the job – what would you do?

- **Safety tips and Remember:** margin notes to emphasise important points

- **Did you know?:** useful information on things you always wondered about

- **End-of-chapter knowledge checks:** test yourself to see if you have absorbed all the information; are you ready for the real test?

- **Glossary:** clear definitions and explanations of those strange words and phrases.

Why choose plumbing?

Plumbing is a very satisfying and rewarding industry to join, with a variety of work on a range of systems. There is always a demand for workers who have the skills, knowledge and qualifications.

Why this book?

Because it is structured to give all the basic information required to gain the relevant qualification and set you on course for an exciting career in a buoyant industry. Well done for choosing such a good start!

There are another two similar qualifications at Level 3 and there's another book like this to match!

chapter 1

Industry and qualification overview

OVERVIEW

This chapter will introduce you to what you can expect when you start to work in the plumbing industry. This chapter includes:

- **An overview of the plumbing industry**
  - What is plumbing?

- **Key industry organisations and bodies**
  - Plumbing industry links with the construction industry
  - The size of the plumbing industry

- **The apprenticeship scheme**
  - The Advanced Apprenticeship (AA) in England and Wales
  - The MES (Mechanical Engineering Services) Plumbing National Vocational Qualification (NVQ)
  - The MES Certificate in Plumbing

- **Key Skills**
  - Training and assessment

- **Career opportunities**
  - Employment rights and statements of employment

An overview of the plumbing industry

This section will explore what the job of a plumber really involves and should prepare you for what to expect on site. It will also provide information about the structure of the plumbing industry and some of its key organisations. At the end of this section you should be able to:

- describe the type of work carried out by a plumber
- understand the nature of the industry, including the various types of plumbing businesses
- know who the key industry organisations are, what they do and how to obtain more information about them.

What is plumbing?

Generally, plumbing satisfies the basic needs of people to keep clean, warm and healthy by providing heating, hot and cold water systems and sanitation 365 days a year, 24/7.

A plumber's job usually includes the **installation**, **service** and **maintenance** of a wide range of systems such as:

- cold water, including underground services, to a dwelling
- hot water
- heating systems fuelled by gas, oil or solid fuel
- sanitation (or above-ground drainage), including the installation of baths, wash-hand basins, WCs and sinks
- sheet lead weathering
- rainwater runoff and drainage
- drainage
- associated electrical systems.

A qualified plumber could install, service and maintain all the various components contained within those systems, for example:

- pipe materials, fittings, fixings, controls
- heat exchangers, boilers, radiators
- pumps, accelerators and motorised/isolating valves
- storage vessels, cylinders, cisterns
- sanitary appliances, baths, WCs etc.

Some well-used tools of the trade!

- domestic appliances including washing machines and dishwashers

- cabling and electrical components

- sheet weathering, aprons, gutter backs, step flashings, soakers, lead slates

- gas appliances (natural or LPG), boilers, water heaters, cookers and fires.

Plumbers work in many and varied work locations. Plumbing companies tend to specialise in either industrial/commercial work – factories, office blocks – or domestic properties, such as houses and flats. Some companies specialise in certain types of system installation or maintenance, such as central heating systems, others undertake the full range of activities.

A competent plumber should be able to:

- work safely at all times

- plan the job and agree a schedule of work with the customer

- prepare the work location, including ensuring adequate access, protecting customer's property and making sure all the tools, materials and equipment are available

- mark out, measure and work out the installation requirements

- fabricate, position and fix system components

- pre-commission (including testing), commission and decommission systems

- service and maintain system components

- work effectively with customers, workmates and other site visitors

- work in an environmentally friendly manner

- promote the products and services of the plumbing business.

Find out

What range of plumbing activities does the company you work for provide?

What skills and knowledge do you think a plumber needs in order to be able to carry out the job competently?

Figure 1.1 Essential knowledge

- Principles of plumbing systems, including basic design
- Codes of Practice
- Regulations
- Essential knowledge for plumbing professionals
- Health and Safety – safe working practices
- Manufacturers' technical data
- Commercially agreed standards

A plumber also needs to be able to read and interpret drawings, specifications and manufacturers' catalogues – plumbing is not just about unblocking drains!

Key industry organisations and bodies

- SummitSkills
- Association of Plumbing and Heating Contractors (APHC)
- Institute of Plumbing (IOP)
- AMICUS
- JTL
- British Plumbing Employers' Council (BPEC) Training Limited
- BPEC Certification Limited
- Plumbing and Heating Industry Alliance (PHIA)
- National Association of Plumbing Teachers (NAPT)
- Joint Industry Board for Plumbing and Mechanical Engineering Services (JIB for PMES)
- Confederation of Registered Gas Installers (CORGI)
- Heating and Ventilating Contractors' Association (HVCA)
- Electrical Contractors' Association (ECA)
- Water Regulations Advisory Scheme (WRAS).

Plumbing links with the construction industry

Plumbing is seen as part of the construction industry. It also belongs to the building engineering services sector, which includes heating, ventilation, air conditioning, refrigeration and electrical installations. The gas supply from the consumer's meter to their appliances is also regarded as a plumber's job. The construction industry will be covered in more detail in Chapter 3 on effective working relationships.

The training and development needs of the construction sector are looked after by an organisation called the Construction Industry Training Board or CITB.

Plumbing links with the building engineering services sector

The plumbing industry has close links with the building engineering services sector, which was strengthened in 2003 by the introduction of **SummitSkills**. SummitSkills is the **Sector Skills Council** for plumbing, heating, ventilating, air conditioning, refrigeration and electrical installation.

SummitSkills looks after the training and development needs of the sector, as well as the National Standards used for NVQs. The Board of SummitSkills is made up of

Definition

SummitSkills – an industry organisation dedicated to improving skills in the construction industry by training and development senior representatives from industry, including employer representatives from trade associations such as the Association of Plumbing and Heating Contractors (APHC), the Heating and Ventilating Contractors Association (HVCA) and the Electrical Contractors Association (ECA), as well as Trade Union representatives from AMICUS and other industry sectors.

Trade associations provide a number of services to their members, including technical support, legal advice, representation to Government on industry-related matters, regular updates on industry developments through their own magazines, and work to raise the profile of their membership among potential customers. Membership includes all sizes of business from sole traders to larger organisations. Most trade associations operate rigid membership selection criteria to ensure a high standard of membership.

Trade unions represent the interests of the employees of businesses on such issues as pay and conditions, and offer free advice and guidance on many employment issues.

The principal trade union for the plumbing sector is AMICUS.

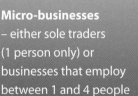

Definition

Association of Plumbing and Heating Contractors (APHC) – the principal trade association for the plumbing industry

The size of the plumbing industry

The plumbing industry contains approximately 25,000 businesses. About 80 per cent of these are classified as micro-businesses. Recent research indicates that the industry employs around 110,000 people.

You will find that smaller firms tend to concentrate on domestic plumbing, covering installation, servicing and maintenance. Some firms specialise in installation only, such as new house build. Larger firms tend to be more multi-disciplinary, providing a plumbing service to their customer, but also covering electrical installation and heating and ventilating in domestic, industrial and commercial premises. Some will also cover industrial and commercial gas work.

Plumbing has always been a mobile industry, moving from one site to another, and the trend these days is more and more towards working further afield, particularly if you work for a larger firm.

The structure of a business will vary depending on the size of the firm. In a micro-business, the owner/manager or sole trader will be responsible for every aspect of running the business, such as getting the work, pricing the job, ordering the materials, organising the job, doing the job and customer care, as well as managing the finances of the business. In a larger firm, some of the responsibilities of running the business are shared among the management team and technical support staff such as designers and estimators.

Definition

Micro-businesses – either sole traders (1 person only) or businesses that employ between 1 and 4 people

The apprenticeship scheme

At the end of this section you should be able to:

- describe what an Advanced Apprenticeship (AA) scheme is and what it covers
- describe what an MES (Mechanical Engineering Services) Plumbing National Vocational Qualification (NVQ) is and what it covers
- describe what an MES Certificate in Plumbing Level 3: Plumbing Studies is and what it covers
- distinguish between plumbing training and NVQ assessment.

You have chosen to study for a career as a plumber, which means you are now probably registered on an Apprenticeship in Plumbing, but if you are not, the following will give an insight into apprenticeships. Apprenticeships are managed and administered by a managing agent. There are a number of these in the UK, but at the current time the largest managing agent in plumbing is JTL.

JTL is a registered charity, jointly owned by the Electrical Contractors' Association (ECA) and AMICUS (a trade union). It is the leading training provider for the plumbing and electrical industries, and has trained a number of plumbing apprentices via the Advanced Apprenticeship scheme in England and Wales. JTL arranges and manages the training and funding for the apprentices by liaising with them and their employers. The company operates in 12 regions across England and Wales.

The role of a managing agent is to find suitable young people for apprenticeship employment with plumbing contractors; they will process all the applications for the apprenticeship scheme and conduct selection tests at various locations. Agents produce careers brochures that they send out to careers advisers. Agents also distribute information to employers, explaining their apprenticeship schemes and encouraging them to take on an apprentice.

The managing agent then oversees the arrangements in colleges or training centres that apprentices attend for the off-the-job learning part of their apprenticeship. JTL, for example, have training officers who review and monitor the progress of apprentices in the workplace and at college, and who are technically qualified to undertake the NVQ assessments in the workplace.

The Advanced Apprenticeship (AA) in England and Wales

If you are already following apprenticeship, your training officer should already have described the framework for you, so this part of the session will serve as a recap. Others will find this a useful insight into the apprenticeship system.

The MES Plumbing Advanced Apprenticeship is made up of three main parts:

1 An NVQ Level 3 in MES Plumbing

2 An MES Certificate in Plumbing Level 3: Plumbing Studies

3 Key Skills at Level 2 in Communication and Application of Number.

The scheme also requires the apprentice to successfully complete the following:

- Gas ACS assessments
- Water Regulations Certificate
- Domestic Unvented Hot Water Systems Certificate.

These will be explained in more detail later.

The plumbing industry has always operated an apprenticeship scheme, but the apprenticeship is based on a common framework for all industries and is regulated by the government.

The AA framework document

The actual framework document has a lot of detail, but the main thing to remember is that it contains:

- an Individual Apprenticeship Plan, which sets out what you will be doing throughout your apprenticeship
- a Training Agreement, which is signed by you, your employer, sometimes your parent or guardian (if you are under 18), and a Local Learning and Skills Council (LLSC) or Education and Learning Wales (ELWA) representative. The LLSC and ELWA are the bodies responsible for funding AAs. This money is used to pay for the running of the scheme and the cost of training in the centre.
- the Unit and Element Titles of MES Plumbing NVQ Levels 2 and 3, Key Skills and the technical certificate you are completing in the centre
- milestones and suggested timescales
- the partners involved in advanced apprenticeships
- progression routes – this shows how you could progress within the industry once your apprenticeship is completed.

What this means in practice

The AA framework sets the rules about what will happen during your apprenticeship. Because you are an apprentice, you will be looked after by a training officer. It is their job to ensure that you are properly trained and assessed, both at work and in your training centre. This involves a number of monitoring review visits for your training, as well as ongoing assessments in support of the achievement of your NVQ. Your training officer will explain this process to you.

Find out

What are the rules and procedures at your Training Centre regarding attendance, holidays, absence and disciplinary issues?

Definition

Milestones – key targets that you need to achieve at specified times within your apprenticeship

The MES (Mechanical Engineering Services) Plumbing National Vocational Qualification (NVQ)

MES Plumbing NVQs are the only vocational (work-based) qualifications that are recognised by the industry. Their development included in-depth discussions with everyone involved with plumbing training, such as employers, union representatives, trainers, professional bodies, and City & Guilds (C&G). The development of the NVQ is led by SummitSkills (see page 4).

City & Guilds

City & Guilds is known as the Awarding Body for the MES Plumbing NVQs. It is the organisation responsible for actually awarding the qualifications. City & Guilds works with two key organisations to offer the awards: the Association of Plumbing and Heating Contractors (**APHC**) – the plumbing employer body – and the Joint Industry Board for PMES (**JIB for PMES**) – an organisation that is responsible for determining such issues as the pay and conditions of plumbing operatives. Plumbing NVQ assessment centres and assessors have to be approved by C&G, and are strictly monitored to make sure they are meeting the quality standards laid down by C&G when assessing candidates.

NVQs are designed to show that someone can do a job competently in the workplace, so most of your assessment will be based on what you actually do while at work. This will be performed by your managing agent's training officer, who is a qualified NVQ assessor, and it is known as an **observed assessment**. Because it is not possible to observe all your work activities, due to time and cost, all other work carried out on-site will be recorded in a **workplace evidence record**. This will be used by your assessor as a basis for completing your site assessments.

To complete the award successfully you are also required for each evidence task to provide **supplementary evidence**.

Your employer or work supervisor is also required to confirm that you carried out the work for each activity to the company standard. The evidence record also has space for you to plan your future assessment activities, record details of any questions that your assessor asked to prove that you can do the work, and record details of the work observed in the workplace by your NVQ assessor.

The practical tasks related to the NVQ units of competence

There is also a range of practical work that is assessed in the centre and not in the workplace, and this is known as **simulated assessment**. Your assessor will go through the assessment documentation with you in detail prior to your assessment starting.

The NVQ also requires your job knowledge, or the theory aspect of the job, to be assessed. This is mainly done using multiple-choice questions, where you have to select the correct answer from four or five options. The same style multiple-choice questions will be used to assess the MES Certificate in Plumbing Level 2 and 3: Plumbing Studies.

What your MES NVQ in Plumbing contains

The AA includes the achievement of a Level 3 NVQ, but in practice you will also work through the NVQ at Level 2. NVQs are made up of a number of units; the unit titles describe what a candidate *must be able to do*.

Definition

Mandatory units – the ones candidates *must* do

Level 2

Level 2 mandatory units:

- contribute to the improvement of the plumbing work environment
- decommission non-complex plumbing systems
- maintain effective plumbing working relationships
- maintain a safe working environment when undertaking plumbing work activities
- install non-complex plumbing systems and components
- maintain non-complex plumbing systems and components.

Level 3

At the time of writing, all candidates at Level 3 have to do a mandatory gas option (gas or oil in Northern Ireland), but candidates will have more choice in the future, as they will be able to select optional units. There are also a number of additional units which can be selected outside the Advanced Apprenticeship; these will be available at a later date.

Mandatory units at Level 3:

- service and maintain complex domestic plumbing systems and components
- commission and decommission complex domestic plumbing systems
- maintain effective plumbing working relationships
- plan complex domestic plumbing work activities
- maintain the safe working environment when undertaking plumbing work activities
- contribute to the improvement of plumbing business products and services
- install complex domestic plumbing systems and components.

Optional units at Level 3

- Domestic gas heating systems (mandatory in England and Wales)
- Domestic oil heating systems

The following additional units are included in the qualification structure:

- Design of domestic plumbing systems
- Sheet lead weathering systems
- Fire control systems.

The MES Certificate in Plumbing

The MES Certificate in Plumbing Level 2: Basic Plumbing Skills and Level 3: Domestic Plumbing Studies forms part of the AA framework. A further element to be included at Level 3 on industrial and commercial installations is still being developed.

While the NVQ is concerned with proving competence in the workplace, the Certificate in Plumbing is a qualification based on what you are taught at the training centre. It covers both practical and theoretical aspects. The practical is assessed by a number of workshop assignments, and the theory combined and assessed in the NVQ knowledge assessments.

What the Certificate contains

Like the NVQ, you will be aiming for a Level 3 certificate, but will complete the Level 2 on the way. The technical certificate units are:

- Safety
- Key Plumbing Principles
- Common Plumbing Processes
- Cold-Water Systems
- Hot-Water Systems
- Sanitation Systems
- Central-Heating Systems (pipework only)
- Electrical Supply and Earth Continuity
- Sheet Lead Weathering
- Environment Awareness
- Effective Working Relationships
- Practical.

You will see that this book follows the units quite closely, but Effective Working Relationships are dealt with early in this book as they relate to all the other subject areas and will help you as you work through the other units.

Level 3

- Cold-Water Systems
- Hot-Water Systems (including unvented)
- Sanitation Systems
- Central-Heating Systems (including boilers and controls wiring)
- Gas Supply Systems
- Improvement of Business Products and Services
- Practical.

This book has been designed to cover all the required content of the Certificate in Plumbing at Level 2.

The practical assessments for the certificates require you to complete a full range of nationally set tasks, which are designed to exacting standards to ensure that you can demonstrate a high degree of competence in the basic plumbing skills.

Key Skills

At the end of this section, you should:

- know which Key Skills you require for your Plumbing NVQ
- understand the different ways to demonstrate you have these skills
- know how you will be assessed.

The AA requires that you obtain a Key Skills certificate in *Communication and Application of Number* at Level 2. Achieving Key Skills involves developing a portfolio of evidence as you work through a topic, and completing a final test. This will be incorporated into your course work. You will be exempt from the end test if you have a GCSE grade C or above in either English (Communication) or Maths (Application of Number). This GCSE must have been achieved within three years of starting the AA. You will be expected to achieve the Key Skills certificate within the first year of the scheme. The Key Skills are designed to ensure that you have adequate skills to communicate with people and to carry out essential calculations; they are therefore a very important part of the programme.

Training and assessment

You will undergo both practical and theory training at work and in the training centre. You will be given instruction on how to do something correctly, and taught the theory that underpins the job. Training also gives you the opportunity to practise things until you can do the job competently.

Assessment is a way of finding out whether you can do a job competently and if you understand the theory behind it. There are various forms of assessment. The NVQ includes assessment that is done at the workplace, either watching you do the job or assessing what you have included in your workplace record. It also includes knowledge assessment, which is the same for the NVQ and the Certificate in Plumbing. Finally, there are assessments that are linked to the NVQ simulations and the practical content of the Certificate in Plumbing, which will be carried out in the training centre.

Find out

What range of plumbing activities does the company you work for provide?

What skills and knowledge do you think a plumber needs in order to be able to carry out the job competently?

FAQs

Can I get my NVQ with just my workplace evidence record?

No, you will also need to have some observed assessments done by your Assessor. All new entrants must, of course, complete the 6129 Technical Certificate.

Do I have to take a gas option to gain my Level 3 NVQ?

Yes, the gas option is mandatory except in Northern Ireland where you can take gas or oil. In the future there will be more optional units available.

Career opportunities

At the end of this section, you should have a better idea of the range of job opportunities in the plumbing industry.

We appreciate that you have only just started your apprenticeship, and you are focusing on getting to know your new workmates, working environments and the job itself. It is worth taking a few moments, however, just to have a think about what your future in plumbing could hold. The opportunities are many and varied: you could consider running your own business, designing systems that are installed in large buildings, or training future plumbers.

The chart on p.13 gives some idea of the career paths available.

Employment rights and statements of employment

When your company takes you on, it must comply with a number of legal requirements. Many of these are there to protect your rights, while others relate to aspects of employment such as your tax, National Insurance, working time and health and safety.

There are a number of laws affecting employment, the main ones being:

- The Employment Rights Act 1996
- The Employment Relations Act 1999
- The Employment Act 2002.

The Department of Trade and Industry (DTI) publishes most employment legislation.

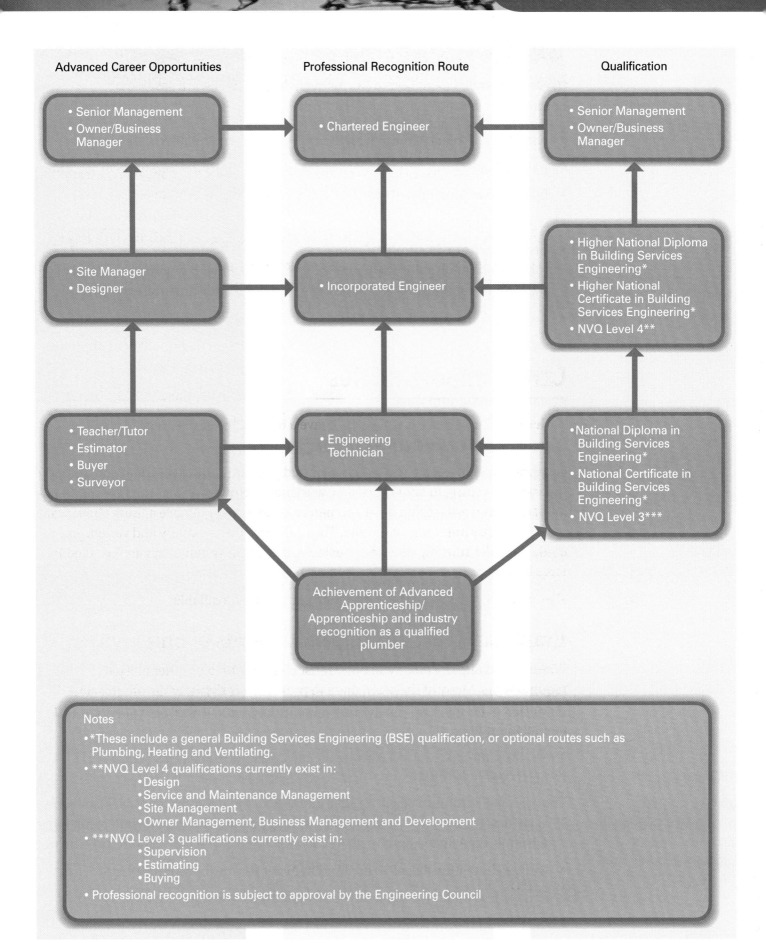

Figure 1.2 Career development paths

What are your basic rights?

In general, you'll be entitled to the following rights, which are covered by employment law:

- written particulars of employment
- paid annual leave
- limits on working time
- equal pay for equal work
- protection from discrimination on the grounds of disability, religion, race, gender and sexual orientation
- protection against unfair dismissal
- sick pay
- maternity pay
- national minimum wage

- a safe and healthy work place
- paternity leave and time off for family emergencies
- notice of termination of employment
- redundancy pay
- guaranteed payments in respect of lay-off and short-time working
- time off to fulfil trade union duties
- protection of employment upon the transfer of a business
- no unlawful deductions from wages.

Terms and conditions of employment

When you are offered a job, you should receive confirmation in writing, though the offer can also be made and accepted verbally. This should be followed by a statement of employment containing the terms and conditions of employment (see later).

Leave

You qualify immediately for holiday leave and pay, and you don't have to be employed for a minimum period first. You have a statutory entitlement to four weeks of paid holiday each year, and entitlement builds up through the holiday year; these four weeks may include payment for bank holidays. However, your employer is not required to allow you to carry over holidays not taken during one holiday year into the next year.

Your employer should inform you in writing of the start date of the holiday year. If they fail to do this, you can ask for your entitlement to start from your starting employment date.

Minimum wage

Find out

What is the current minimum wage?

The National Minimum Wage Act 1998 and the National Minimum Wage Regulations 1999 are designed to cover workers in all industry areas. There are certain regulations which apply directly to apprentices, the main points of which are summarised below:

- workers under the age of 18 do not qualify for the minimum wage
- apprentices under the age of 19 do not qualify for the minimum wage
- the minimum 'development rate' wage for 18–21 year olds is £4.10* per hour

- the minimum wage for adult workers over the age of 22 is £4.85* per hour.

* at the time of writing

Any worker who feels they have been unfairly dismissed as a result of becoming eligible for the minimum wage has a right to claim unfair dismissal, no matter how long they have worked with the employing organisation.

Wages and Statutory Sick Pay (SSP)

When you are paid during your apprenticeship, you should receive a payslip. Your payslip must show the following information:

- the gross wages (amount before tax) earned in a given time (week or month)
- the amounts of (and reasons for) any fixed deductions – e.g. National Insurance
- the amounts of (and reasons for) any variable deductions – e.g. Income Tax or PAYE (Pay As You Earn)
- the net pay, or amount of pay you actually 'take home'.

It is possible that on occasion your wages will be wrong, so it is important always to check your payslip. Your employer also has the right to make additional deductions from your wages if you have been wrongly overpaid at a previous date.

If you are off sick for four consecutive days or more you are entitled to statutory sick pay (SSP). Your employer pays SSP in the same way as wages and must keep records of payments made and dates of any absence lasting at least four consecutive days. Your employer is responsible for the payment of SSP for periods of up to 28 weeks.

Termination of employment

There are a number of regulations that must be adhered to if your employer wishes to terminate your employment. The procedures that have to be followed are usually linked to the amount of time you have been employed by the company. The periods of notice that must be given are:

- one week if employed for between one month and two years
- an additional week's notice for every continuous year of employment between two and twelve years
- minimum of twelve weeks' notice if the employee has been in continuous employment with the company for 12 years or more.

If you have been employed by a company for one year or more, your employer must give written reasons for your dismissal. After a year you are also entitled to make a claim for unfair dismissal if there is just cause. Advice on any employment disputes can be found at the Advisory, Conciliation and Arbitration Service (Acas).

Discrimination

Equality in employment is required by both UK and European law. UK employment law currently recognises a number of types of unlawful discrimination – sex, race, disability, religion and sexual orientation. This requires that jobs, training and promotion must be open to all regardless of colour, race, nationality, ethnic or national origin, sex or marital status, and regardless of whether someone is undergoing or has undergone gender reassignment. Men and women must get the same pay for the same or like work and for work of equal value.

The main pieces of legislation that cover discrimination are:

- Sex Discrimination Acts 1975 and 1986
- Race Relations Act 1976
- Disability Discrimination Act 1995 (this Act excludes prison officers, fire-fighters and police officers).

Statements of employment

To present you with a statement of employment it will be necessary for your company to hold records of your personal details. To do this your employer must adhere to the Data Protection Act 1984 and register with the Data Protection Registrar. Statements of employment were previously known as Contracts and Conditions of Employment. Employment law requires that you receive a statement of employment within *two months* of starting work. Your statement of employment should contain:

- the name of your employer and your name
- the date your employment started
- your job title and a summary of your duties
- the period of employment, stating whether it is a permanent position
- the place of work
- how much you will be paid, how often, and the method of payment; it should also include information such as travel allowances and any deductions from pay
- hours of work
- holiday entitlement
- procedures for dealing with absence from work through illness, or for other reasons, and how to notify the employer if absent
- details of pension scheme if applicable
- details of how to terminate employment (for example, length of notice required by both you and your employer)
- disciplinary rules and procedures (these are usually contained in a separate document such as a staff handbook)
- grievance procedures (which again could be contained in a staff handbook).

Did you know?

In most circumstances, the first action to take with any grievance is to notify your immediate supervisor, or employer if employed by a small company

Information on the following items could also be included in the statement, or alternatively will be found in a staff handbook:

- appraisal arrangements
- training and development
- trade union membership
- health and safety matters
- maternity rights
- redundancy policy
- company vehicles
- smoking policy.

Your organisation should also have a company policy statement containing information on:

- safety policy
- customer relations initiatives
- the company's responsibilities under the Consumer Protection Act (1987)
- service agreement procedures.

FAQs

How long do I have to work before I can take some holiday?

You qualify for holiday immediately and are entitled to four weeks holiday every year, which accrues during the year.

On what grounds can I claim unfair dismissal?

Check your rights with your union, but you can claim unfair dismissal if you can prove it was on the grounds of race, gender, religion, disability or sexual orientation, or if your employer did not go through the correct procedures for dismissal.

If I am overpaid, can I keep the money?

No, unfortunately not. If you think you have been overpaid you must inform your employer. They have the right to deduct the overpayment from subsequent wages.

Knowledge check

1   State one of the laws related to employment.

2   List five employment rights that are covered under employment law.

3   List six items that you would expect to see on a statement of employment.

4   Which one of the following statements do you think is correct:

a. The main method of assessing the job knowledge aspect of the NVQ is by using a written examination.

b. The main method of assessing the job knowledge aspect of the NVQ is by using multiple-choice questions.

c. The main method of assessing the job knowledge aspect of the NVQ is by using simulations.

d. The main method of assessing the job knowledge aspect of the NVQ is by using practical tests.

5   Summarise in not more than 30 words the type of work carried out by a plumber.

6   Is the statement below true or false?
About 80% of plumbing businesses are classified as micro-businesses.

7   What are the main parts that make up an MES Plumbing Advanced Apprenticeship?

chapter 2

Health and Safety

OVERVIEW

Health and Safety forms a very important part of your everyday working life. This chapter provides the knowledge and learning to meet the requirements of Level 2 – Health and Safety – a unit required to complete the City & Guilds Certificate: Domestic Plumbing Studies.

It will include sections on the following topics:

- Personal safety responsibilities and the safety of others
- Regulations
- Risk assessments
- Personal safety
- Access to work
- Fire safety and emergencies
- Electric shock and first aid
- Control of substances hazardous to health (COSHH)
- Safe use of LPG and lead working.

Health and Safety is a very important aspect of your job. There are a number of legal Health and Safety requirements that must be met by you and your employer. These have been put in place to keep you, your fellow workers and all site visitors free from accidents. Having a good grasp of Health and Safety requirements will also help you to be aware of the potential hazards when working on site and in the plumbing workshop.

Every year in the construction industry tens of thousands of building-site accidents are reported to the Health and Safety Executive (HSE), some of which will be fatal. These figures are not intended to put you off your new career but to make sure you understand that accidents do happen if you do not follow Health and Safety guidance.

There are two main bodies that establish and implement Health and Safety laws:

1 The Health and Safety Commission (HSC)

2 The Health and Safety Executive (HSE)

During this chapter, you will learn more about these and about other regulations designed to keep you safe at work. You will also learn about the various hazards you may face, safe methods of working and actions to be taken in the event of accidents.

Personal safety responsibilities and the safety of others

At the end of this section you should be able to:

- describe your Health and Safety responsibilities

- state the requirements of your employer's Health and Safety responsibilities

- explain the importance of personal hygiene

- explain the importance of having a positive personal attitude towards training

- state the main methods of accident prevention.

Health and Safety responsibilities

The rules that govern your own and your employer's Health and Safety responsibilities come directly from the Health and Safety at Work Act (HASAWA) 1974. This crucial piece of legislation will be referred to frequently throughout this chapter. Under the HASAWA, you, as an employee, can be prosecuted for breaking safety laws. You are legally bound to cooperate with your employer to ensure your company complies with the requirements of the Act. It is the legal responsibility of all workers to take reasonable care of their own Health and Safety and to ensure that they act in a responsible manner so as not to endanger other workers or members of the public. Some of the most important aspects of the HASAWA are detailed below:

- In domestic premises, the occupier is normally covered by insurance (Occupier's Liability) for visitors such as friends. When you are working on domestic premises, however, you as the plumber become responsible for the health, safety and welfare of the occupant, because it is your place of work.

- It is an offence under the Act to misuse or interfere with equipment provided for your health and safety or that of others. Employers provide you with safety equipment and should instruct you in the use of it. Your employer also has a duty to ensure that the safety equipment provided is kept in good condition, but they cannot cater for human error or negligence. You should always use Health and Safety equipment correctly, for the purpose it is provided for.

- Substances such as oil, grease, cutting compounds, paints and solvents are hazardous if spilled on the floor. Items such as off-cuts of pipe, cables, tools and even food are also dangerous if left underfoot. You have a responsibility both to yourself and to others to keep the workplace hazard free.

What do you need to do?

You should be beginning to get an idea of some of your Health and Safety responsibilities. Your precise legal requirements are listed below. The information may seem obvious, but it is very important that you stick to these requirements throughout your apprenticeship and working career.

Your legal responsibilities

1. Take reasonable care at work of your own Health and Safety and that of others who may be affected by what you do or do not do.

2. Do not intentionally or recklessly interfere with or misuse anything provided for your Health and Safety.

3. Cooperate with your employer on Health and Safety matters. Assist your employer in meeting their statutory obligations.

4. Bring to your employer's attention any situation you think presents a serious and imminent danger.

5. Bring to your employer's attention any weakness you might spot in their Health and Safety arrangements.

What does your employer need to do?

Your employer also has important Health and Safety responsibilities. You should know what your employer's legal responsibilities are for keeping you safe at work. The HASAWA requires your employer to ensure, so far as is reasonably practicable, your health, safety and welfare at work. The matters to which this duty extends include:

1 The provision and maintenance of plant and systems of work that are safe and without risk to health.

2 Safety in the use, handling, storage and transport of articles and substances.

3 The provision of information, instruction, training and supervision as necessary to ensure the Health and Safety at work of employees.

4 The provision of access to and exit from the workplace that is safe and without risk.

5 The provision of adequate facilities and arrangements for welfare at work.

The HASAWA also says that employers have other duties towards employees in companies that employ five or more members of staff. Such companies must have a Health and Safety policy statement. These will be examined in more detail during the session on Regulations.

Ensuring personal safety and the safety of others
Personal hygiene

It is in your own interests to keep yourself as hygienic as possible while at work, but this may be hard at times, given the nature of some plumbing jobs! To ensure good personal hygiene you should:

- keep your overalls as clean as possible and ensure they are washed regularly
- wash your hands thoroughly before contact with food
- avoid washing with solvents (e.g. white spirits), which can cause dermatitis.

Get into the habit of using barrier cream before starting a job. Barrier cream fills the pores of the skin with a water-soluble antiseptic cream, so that when you wash your hands the dirt and germs are removed with the cream. This should be reapplied each time you wash your hands.

Accident prevention

The best way to reduce the risk of accidents is to try and remove the cause. It is important that, wherever possible, the workplace has clearly defined passageways, good lighting, ventilation, reduced noise levels and non-slip floorings.

Another potential hazard is removed by locking dangerous substances in approved locations so that people will not be tempted to interfere with them, thus causing danger to themselves and others.

Storing materials correctly will also reduce the risk of accidents. Lengths of tube and pipe should be stored horizontally to minimise the risk of them falling.

Hazards that cannot be removed or reduced can be guarded against. For example:

- safety guards and fences can be put on or around machines
- safe systems of work can be introduced
- wearing safety goggles, safety helmets and safety shoes can be made standard practice
- other protective clothing can be supplied for your personal protection such as ear defenders, respirators, eye protection and overalls.

Personal attitudes to safety training

It is essential that you adopt a positive attitude towards safety training, not only for your own welfare but also to ensure the safety of your work mates and the general

Remember

Your employer must provide protective clothing, free of charge, when the work process requires its use public. To do this, you must make a positive decision to act and work in a careful and responsible manner and to put into practice what you have learned. You must make yourself aware of your company's Health and Safety policies/procedures and act accordingly. You need to know what dangers can occur, what protection is available, how to use it and how to prevent accidents.

Personal habits such as drinking alcohol and drug abuse can render a worker a hazard not only to himself but also to others. It stands to reason that evidence of alcohol and drug abuse will not indicate a positive personal attitude towards safety training.

Regulations

At the end of this section you should be able to:

- state the main points and priorities of the Workplace Regulations
- explain the main requirements of the HASAWA
- state the main aspects of the Construction (Design and Management) Regulations
- state the main aspects of the Construction (Health, Safety and Welfare) Regulations
- state the main requirements of the Manual Handling Operations Regulations
- state the main requirements of the Fire Precautions Act
- list the main requirements of the Electricity at Work Regulations.

The Workplace Regulations (1992)

These Regulations apply to *all* workplaces. Their requirements are imposed upon every employer or any person who, has to any extent, control of a workplace. They do not, however, cover domestic premises. The Regulations impose requirements that affect the working environment, safety and facilities.

1 Working environment	2 Safety	3 Facilities
Regulations 5 to 11 require that: • effective and suitable provision is made to ensure that every enclosed workplace is ventilated by a sufficient quantity of fresh air. • The temperature in all workplaces inside buildings is reasonable, i.e. a minimum of 16°C for offices and 13°C in areas where manual work is carried out.	Regulations 12 to 19 require that: • Every floor and surface area is suitable for the purpose for which it is used. • No floor is slippery, uneven, has holes or a slope, which may expose any person to a risk to their Health and Safety. • Where necessary, every floor must have a means of drainage.	Regulations 20 to 25 require that: • Suitable toilets are provided at easily accessible places. • A supply of hot and cold water, soap and towels are provided. • A supply of clean drinking water is accessible to all persons within the workplace. (continued over)

Figure 2.1 Workplace regulations

1 Working environment	2 Safety	3 Facilities
• There is a sufficient number of thermometers provided for monitoring. • Every workplace has suitable and sufficient lighting, and if possible this should be natural lighting. • All furniture, furnishings, fittings, surfaces of floors, walls and ceilings of all workplaces must be kept clean. • Every room where persons work has sufficient floor area, height and unoccupied space for the purposes of health, safety and welfare. • No room is so overcrowded as to cause risk to the Health and Safety of persons working in it. • The number of persons at work at any one time is such that the amount of space allowed for each is not less than 11 cubic metres. • Every workstation is arranged so that it is suitable for both any person at that workstation and any person who is likely to use that workstation. • In the case of outdoor workstations, protection is provided from adverse weather.	• Measures are taken to ensure persons are protected from falling objects and from falling from height. • Workplaces are organised in such a way that pedestrians and vehicles are separated and can circulate in a safe manner. • Escalators and moving walkways are equipped with any necessary safety devices and are fitted with one or more emergency stop controls, which are easily identifiable and readily accessible.	

Figure 2.1 Workplace regulations (continued)

The Health and Safety at Work Act (HASAWA – 1974)

This Act was a milestone in bringing in laws dealing with the health, safety and welfare of persons at work. Duties are placed on all persons connected with Health and Safety at work, whether as employers, employees, self-employed workers, manufacturers or suppliers of plant and materials. Protection is also given to members of the public affected by the activities of persons at work.

In addition to setting out the basic Health and Safety requirements of employers and employees, the HASAWA also gives more detailed guidance to employers on aspects such as Health and Safety policy statements, which are required for companies employing five or more members of staff.

Additional HASAWA employer requirements

The Health and Safety policy statement must be brought to the attention of all members of staff; arrangements for ensuring this must be in place. Under the HASAWA employers also have a duty to:

- carry out an assessment of risks associated with all the company's work activities
- identify and implement control measures
- inform employees of the risks and control measures
- periodically review the assessments
- record the assessment if more than five persons are employed.

Employers must be prepared to consult a safety representative, if one is appointed by a recognised trade union, about matters affecting their employees' health and safety. Employers must also (if requested in writing by any two safety representatives) establish a safety committee within three months of the request being made.

Construction (Design and Management) Regulations (1994) (CDM)

The aim of the CDM regulations is to improve the overall management and cooperation of health, safety and welfare throughout all stages of a construction project. These Regulations were designed to reduce the large number of serious and fatal accidents and cases of ill health that arise every year as a result of work relating to the construction industry.

The CDM Regulations place duties on all those who can contribute to the Health and Safety of a construction project. Duties are placed upon clients, designers and contractors. The Regulations create a new duty holder – the planning supervisor. They also introduce new documents – Health and Safety plans and the Health and Safety file.

The CDM Regulations apply to most construction projects. However, there are a number of situations where the Regulations do not apply. These include:

- construction work other than demolition that does not last longer than 30 days and does not involve more than four people
- construction work for a domestic client
- construction work carried out inside offices and shops or similar premises without interrupting the normal activities in the premises and without separating the construction activities from the other activities
- the maintenance or removal of insulation on pipes, boilers or other parts of heating or water systems.

Construction (Health, Safety and Welfare) Regulations (1996)

These Regulations are aimed at protecting the health, safety and welfare of everyone who carries out construction work. They also give protection to other people who may be affected by the work.

The main duty-holders under these Regulations are employers, the self-employed and those who control the way in which construction work is carried out. Employees have duties to carry out their own work in a safe way. Also, anyone doing construction work has a duty to cooperate with others on matters of Health and Safety and to report any defects to those in control.

These Regulations also contain guidance on legal provisions relating to working at height, the use of scaffolds and the means of accessing them. The guidance given for this regulation will be relevant to much of the work you undertake during your career as a plumber. Summary details of the most important aspects of this Regulation follow.

Duty to take precautions against falls

- Prevent falls from height by taking physical precautions or, where this is not possible, provide equipment that will check falls.

- Ensure there are physical precautions to prevent falls through fragile materials, e.g. roofs.

- Erect scaffolding, access equipment, harnesses and nets; this should be done under the supervision of a competent person.

- Ensure criteria for using ladders are provided.

Falls account for more than half of the fatal accidents in construction. The aim of the Regulations is to prevent falls from any height, but there are specific steps to be taken for work over 2 metres high:

1 Above 2 m, where work cannot be done safely from the ground, the first objective is to provide physical safeguards to prevent falls. Where possible, means of access and working places should be of sound construction and capable of safely supporting both people and the materials needed for the work. Guardrails and toe boards or an equivalent standard of protection should be provided at any edge from which people could fall.

2 Sometimes it is either not possible to provide the above safeguards, or the work is of such short duration or difficulty that it would not be reasonably practicable to do so. If so, consider using properly installed personal equipment such as rope access or boatswain's chairs.

3 If, for the same reasons, these methods of work cannot be used, it will be necessary to consider using equipment that will check falls, e.g. safety harnesses or nets. Scaffolds, personal harnesses and net equipment have to be erected or installed under the supervision of a competent person.

Manual Handling Operations Regulations (1992)

Manual Handling Operations Regulations refer to the human effort involved in handling loads. This includes effort applied directly or through straining on a rope or lever. You will already be aware that your job as a plumber involves the manual handling of tools, equipment and materials. The principles of safe manual handling will be covered in more detail later.

The Manual Handling Operations Regulations themselves present guidance to employers and employees.

The employer

Employers:

- should endeavour to avoid the need for employees to undertake manual handling operations at work which involve a risk of their being injured

- where it is not reasonably practicable to avoid the need for employees to undertake any manual handling operations at work which involve a risk of their being injured, the employer should:
  - assess all such manual handling operations to be undertaken by the employees and reduce the risk of injury to those employees arising out of them undertaking any such manual handling operations
  - provide any of those employees who are undertaking any such manual handling operations with certain information about the loads to be carried by them – to include the supply of relevant training and safety equipment.

The employee

Employees:

- will, while at work, make full and proper use of systems of work provided for his or her use by the employer, in compliance with the employer's duty under the Manual Handling Operation Regulations.

Fire Precautions Act (1971)

Many larger or high-risk premises require a fire certificate. This includes factory premises employing more than 20 people, or buildings that store/use highly flammable materials. Employers in all premises are required to carry out a fire risk assessment to identify the hazards from fire. Once identified, precautions must be implemented to minimise the risks, including:

- reduction/elimination of ignition sources
- elimination/isolation of materials likely to assist in the spread of fire
- means of giving warning of fire
- provision of means of escape in case of fire
- provision of fire-fighting equipment

- appropriate signing of fire exits and fire-fighting equipment

- training for employees in what to do in case of fire

- periodic revision and maintenance of above.

It is the duty of the Fire Authority (or the HSE in Special Premises) to enforce the regulations through the appointment of inspectors. A person may be guilty of an offence if, as a result of their failure to comply with the Regulations, one or more employees are put at risk. If a breach of the Regulations occurs, the authority can issue either an improvement or prohibition notice requiring the items to be put right.

The Electricity at Work Regulations (1989)

The Electricity at Work Regulations, which came into force on 1 April 1990, lay down general Health and Safety requirements regarding electricity at work for employers, self-employed persons and employees.

The Regulations are made under the HASAWA 1974 and, broadly speaking, impose a duty upon every employer and self-employed person to comply with the provisions of the Regulations. They also impose a duty on every employee to cooperate with their employer so far as is necessary to enable the Regulations to be complied with.

As a plumber you will need to be especially careful when dealing with electricity. The Electricity at Work Regulations provide the full safety standards requirement; more will be said about the practical use of electricity on site in a later section.

The Regulations refer to a person as a 'duty holder' in respect of systems, equipment and conductors. These Regulations are therefore statutory, and consequently penalties can be imposed on those people found guilty of malpractice or misconduct.

Definitions relevant to persons who have responsibilities under these Regulations

1. Employer

For the purpose of the Regulations, an employer is any person or company whom (a) employs one or more individuals under a contract of employment or apprenticeship, or (b) provides training under the schemes to which the HASAWA applies.

2. Self-employed

A self-employed person is an individual who works for gain or reward other than under contract of employment, whether or not he or she employs others.

3. Employee

Regulation 3(2) (b) reiterates the duty placed on employees by the HASAWA Act.

Regulation 3(2) (b) places duties on employees equivalent to those placed on employers and self-employed persons where these matters are within their control.

This will include apprentices – like you – who will be considered as employees under these Regulations.

This arrangement recognises the level of responsibility that many employees in the plumbing trade are expected to take on as part of their job. The 'control' that they exercise over the electrical safety in any particular circumstances will determine to what extent they hold responsibilities under the Regulations to ensure that the Regulations are complied with.

A person may find that they are responsible for causing danger to arise elsewhere in an electrical system, at a point beyond their own area of work. This situation may arise, for example, if you are working on a circuit while somebody else is working in a different room on that same circuit. This is obviously a dangerous situation. Because such circumstances are 'within [your] control', the effect of Regulation 3 is to bring responsibilities for compliance with the rest of the Regulations to you, thus making you a duty holder.

Remember

Can you remember your responsibilities under the HASAWA?

4. Absolute/reasonably practicable

Duties in some of the Regulations have a qualifying term called 'reasonably practicable'. Where qualifying terms are absent, the requirement in the Regulation is said to be absolute. The meaning of 'reasonably practicable' has been well established in law. The interpretations below are given only as a guide.

5. Absolute

If the requirement in a Regulation is 'absolute' – for example, if the requirement is not qualified by the words 'so far as is reasonably practicable' – the requirement must be met regardless of cost or any other consideration.

6. Reasonably practicable

Someone who is required to do something 'so far as is reasonably practicable' must think about the amount of risk of a particular work activity or site and, on the other hand, the costs in terms of the physical difficulty, time, trouble and expense which would be involved in taking steps to reduce the risks to Health and Safety of a particular work process. For example, in your own home you would expect to find a fireguard in front of a fire to prevent young children from touching the fire and being injured. This is a cheap and effective way of preventing accidents and would be a reasonably practicable situation.

If the cost or technical difficulties of taking certain steps to prevent those risks are very high, it might not be reasonably practicable to take those steps.

In the context of the Regulations, where the risk is often that of death from electrocution, and where the necessary precautions are often very simple and cheap – for example, insulating surrounding cables – the level of duty to prevent that danger approaches that of an absolute duty.

FAQ

Do I really need to know all these Health and Safety Regulations?

Yes. You have a responsibility to work within the bounds of these Regulations. They are there to ensure your safety and the safety of others.

Risk assessments

At the end of this section you should be able to:

- state the purpose of risk assessments
- understand what is meant by 'hazard ranking'
- work in accordance with risk-assessment forms.

Risk assessments are a kind of checking system designed to keep you safe at work. The aim of risk assessments is to make you think about the possible hazards you may face while performing a specific working task. At this stage of your apprenticeship you will not be required to write risk assessments, but you must have a working knowledge of them and understand their purpose.

What is risk assessment?

A risk assessment is nothing more than a careful examination of what, in your work, could cause harm to you or other people. It enables you to weigh up whether you have taken enough precautions or could do more to prevent any harm. The main aim of a risk assessment is to ensure that no one gets hurt or becomes ill.

On the job: Risk assessment

Jane, a new apprentice, was carrying out a risk assessment with the senior plumber. Jane spotted some obvious hazards where a number of nails were sticking out from upturned floorboards. The plumber told Jane not to bother about the nails because no one would be silly enough to tread on them. Jane knew that loads of casual labourers would be clearing the building in the morning.

1   What should Jane do in this situation?

2   Has the senior plumber contravened any Health and Safety legislation by not completing the risk assessment correctly?

3   If so, what legislation has been contravened?

Your employer needs to decide whether a hazard is significant, and whether it is covered by satisfactory precautions so that the risk is small. Your employer is also legally bound to assess the risks in your workplace. Legislation that governs the requirements of risk assessment is contained within the Management of Health and Safety at Work Regulations (1999).

Generic risk assessment

Hazard and risk identification

In order to assess the potential of a risk, processes are usually undertaken to 'rank' hazardous operations. The example that follows is from a model developed by the British Plumbing Employers' Council (BPEC) specifically for the plumbing industry. It is based on a number of factors, which are presented on a generic risk-assessment form. These are used to assess any risk associated with the plumbing industry.

Here is an example of a generic risk-assessment form for working with specialist tools.

TASK		
MANUAL HANDLING OF LOADS SPECIALIST TOOLS		

APPLICATION OF EQUIPMENT	APPLICATION OF SUBSTANCE
PIPE-BENDING MACHINES, STILSONS, ROPES, LEAD DRESSERS, BENDING SPRINGS, BLOCK AND TACKLE, SPANNERS ETC.	N/A

ASSOCIATED HAZARDS
RISK OF MUSCLE STRAINS
RISK OF SPRAINS
RISK OF MUSCULO-SKELETAL INJURY

LIKELIHOOD	CONSEQUENCE	RISK FACTOR
3	3	9

RISK EXPOSURE	SAFEGUARDS HARDWARE
EMPLOYEES	NIL

CONTROL MEASURES
1 SPECIFIC TRAINING AND INSTRUCTION TO EMPLOYEES – KINETIC LIFTING
2 INDIVIDUAL ASSESSMENT TO BE PERFORMED FOR ALL TASKS
3 WORKPLACE INSPECTIONS CONDUCTED AT 3 MONTHLY INTERVALS
4 RANDOM SAFETY INSPECTIONS
5 SUITABLE AND SUFFICIENT PERSONAL PROTECTIVE EQUIPMENT
6 MEDICAL SCREENING FOR STAFF AT RISK

Figure 2.2 Risk-assessment form

Most of the form is self-explanatory, but we want to focus here on the following items from the form:

- likelihood of accident occurring
- maximum consequences (severity) of an accident
- risk factor (likelihood of accident occurring × maximum consequences (severity) of an accident)

Likelihood of accident occurring

This is assessed from the following table:

Likelihood	Scale value
No likelihood	0
Very unlikely	1
Unlikely	2
Likely	3
Very likely	4
Certainty	5

Figure 2.3

Maximum consequences of an accident

This is assessed from the following table:

Injury or loss	Scale value
No injury or loss	0
Treated by first aid	1
Up to 3 days off work	2
Over 3 days off work	3
Specified major injury	4
Fatality	5

Figure 2.4

Risk factor

Risk factors are then calculated using a simple formula:

Likelihood × Consequence = Risk

The outcome of the likelihood of an accident occurring and the maximum consequences should it happen, will reveal a risk factor of between 1 and 25.

- a figure between 1 and 6 = minor risk, can be disregarded but closely monitored
- a figure between 8 and 15 = significant risk, requires immediate control measures
- a figure between 16 and 25 = critical risk, activity must cease until risk is reduced.

Other factors contained on the assessment form are discussed below.

Risk exposure

Risk exposure describes the individuals or groups of people that may be affected by the work activity or process. Control measures must take account of all those people.

Safeguards hardware

Safeguards hardware describes the in-built safety features of work equipment. For example, on powered machines this would include machine guards or trip switches. In this particular example safeguards hardware is nil.

Control measures

These describe the additional safeguards that underpin your arrangements. Where these are identified they must be followed through, and a record kept of any outcomes etc.

The list below shows some of the most common risks you will face during your plumbing career.

Checklist of possible hazardous operations

- Working with electrical-powered plant:
  - power transformers
  - extension cables
  - plugs and sockets
  - portable tools
  - electric arc welders
  - threading machines
  - large hammer drills
  - specialist equipment
  - fixed equipment
- Working with non electrical-powered plant:
  - hand tools
  - specialist tools
  - pneumatic tools
  - hydraulic tools
- Working with non-powered tools:
  - specialist equipment
- Manual handling of loads:
  - specialist tools
  - general lifting

- Working with hazardous substances
- Working in excavations:
  - physical conditions
  - interruption of gas, water or electrical services
- Working with powered industrial trucks:
  - fork-lift trucks
  - dump trucks
  - JCBs/tractors
  - tail-lift vehicles
  - road vehicles
- Working with highly flammable liquids
- Working with liquefied petroleum gases
- Working with lead:
  - manual handling
  - lead hygiene

- Working at heights:
  - ladders
  - scaffolds
  - ropes/harnesses
- Working with demolitions
- Working with electrical installations
- Controlling work with fumes
- Controlling work with noise
- Controlling work with dust
- Working with asbestos
- Working within vessels
- Working in confined spaces
- Clearing of hazardous waste
- Cartridge-fixing devices
- Working on suspended timber floors.

Personal safety

At the end of this section you should be able to:

- name the various types of safety signs and state their purpose
- explain the need for Personal Protective Equipment (PPE)
- describe the types of PPE you should expect to use as a plumber
- state the basic safety requirements of tools and equipment used in the plumbing trade
- state the main requirements to ensure safe working with electricity and electrical equipment on site
- explain the correct lifting and carrying procedures that must be used to ensure the safe movement of tools, equipment and materials.

Safety signs

Find out

Find out what the common warning, information and prohibition signs found in construction environments are. You will need to remember these. When looking at the signs, try to make a mental note of which group they fit into

During your plumbing career you will spend time on various building/construction and other work sites. All sites are littered with potential dangers. Any site you enter will have safety signs. Safety signs are designed to keep visitors and workers safe, pass on useful information (regarding exits, fire extinguishers etc.) and to warn personnel about the possible dangers they may be exposed to during their time on site.

There are four categories of safety sign:

1 mandatory signs
2 prohibition signs
3 warning signs
4 information signs.

	Prohibition signs	Mandatory signs	Warning signs	Information or safe condition signs
Shape:	Circular	Circular	Triangular	Square or rectangular
Colour:	Red borders and cross bar. Black symbols on white background	White symbol on blue background	Yellow background with black border and symbol	White symbols on green background
Meaning:	Shows what must NOT be done	Shows what must be done	Warns of hazard or danger	Indicates or gives information on safety provision
Example:	No smoking	Wear eye protection	Danger electric shock risk	Danger electric shock risk

Figure 2.5 Safety signs

General introduction to 'PPE'

The guidelines that govern protective clothing are set out in the Personal Protective Equipment at Work Regulations (1992). You will usually hear colleagues refer to personal protective equipment as 'PPE', which is defined as all equipment designed to be worn or held to protect against a risk to health or safety, and includes most types of protective clothing and equipment such as eye, hand, foot and head protection.

Your employer has a duty to assess and, where possible, minimise the risks you face at work, although some jobs you will be asked to complete will carry a degree of risk. In these cases, PPE will be used as a last resort to keep you as safe as possible. Your employer should ensure:

- all PPE is provided free of charge
- PPE is suitable for its purpose
- information on procedures to maintain, clean and replace damaged PPE is freely available
- storage for PPE is provided
- PPE is maintained in an efficient state and is in good repair
- that PPE is properly used.

Eye protection

Your eyes are two of your most precious possessions. You need them for learning and for earning, because without your eyesight you probably could not perform your present job! Your eyes are the parts of your body that are most vulnerable to injury at work.

Every year, thousands of workers suffer eye injuries, which result in pain, discomfort, lost income and even blindness. Following safety procedures correctly and wearing eye protection can prevent these injuries. There are many types of eye protection equipment available – for example:

- safety glasses
- safety goggles
- welding goggles.

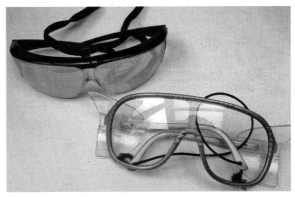

Safety glasses

Foot protection

Although this refers to items to protect your toes, ankles, and feet from injury, wearing the correct footwear can also protect your whole body from injury, for example from electric shock.

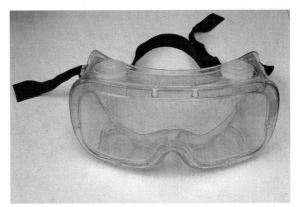

Safety goggles

Accidents to the feet have serious consequences and can result in pain and suffering, disability and loss of work and income.

Protective footwear can help prevent injury and reduce the severity of the injuries that do occur. The basic universal form of foot protection is the safety shoe, which will usually include some kind of metal toe protection, rubber soles and sturdy leather uppers. The photograph shows typical foot protection.

Safety boots

Hand protection

This involves the protection of two irreplaceable tools – your hands, which you use for almost everything: working, playing, driving, eating etc. Unfortunately hands are often injured. Almost 1 in 4 work-related injuries happen to hands and fingers.

One of the most common problems – other than cutting, crushing or puncture wounds – is dermatitis. Dermatitis is an inflammation of the skin normally caused by the hands coming into contact with irritating substances.

Gloves used for general work

This skin irritation may be indicated by sores, blisters, redness or dry, cracked skin that can easily become infected.

To protect your hands from irritating substances, you need to keep them clean by regular washing using approved cleaners. Make good use of barrier creams where provided, and wear appropriate personal protection (usually a strong pair of gloves) when required. The type of glove used depends on the work being carried out:

Specialist gloves used for working with chemicals

- prevention of cuts: general purpose gloves

- working with hazardous substances (an example could be the contents of a toilet bowl): rubber-type gloves.

Head protection

Head protection is important because it guards your most vital organ: your brain. A head injury can disable a person for life or

Safety helmet

even be fatal. One study revealed that over 80 per cent of industrial head injuries occurred as a result of people not wearing 'hard hat' protection.

On building sites the risk from falling objects is very real. There are also other potential traps, such as objects (e.g. scaffolding poles) at head height, just waiting for you to walk into them!

On the job: Head protection

Fazal is a first-year apprentice who is starting his first day on a construction site. The plumber in charge gives Fazal a hard hat to wear. Fazal notices that the hat has a crack down one side and reports this to his supervisor. The supervisor tells Fazal that he must wear this hat because there are no more in the store and he needs to get to work.

1   What should Fazal do in this situation?

2   If an object hit Fazal on the head and he was injured because of the crack in the hat, who would be at fault for this injury?

You need to wear your safety helmet whenever it is required, and you should ensure it is worn correctly.

Safety tip

Always wear your safety helmet the correct way round – despite what your workmates might be doing with theirs!

CHECKLIST FOR SAFETY HELMETS

✓ Adjust the fit of your safety helmet so that it is comfortable

✓ All straps should be snug but not too tight

✓ Do not wear your helmet tilted or back to front

✓ Never carry anything inside the clearance space of a hard hat e.g. cigarettes, playing cards

✓ Never wear an ordinary hat under a safety helmet

✓ Do not paint your safety helmet as this could interfere with electrical protection or soften the shell

✓ Handle it with care: do not throw it or drop it etc.

✓ Regularly inspect and check the helmet for cracks, dents or signs of wear.

✓ Check the strap for looseness or worn stitching and also check your safety helmet is within its specified 'use-by' date.

To keep safe, use your head:

- Know the potential hazards of your job and what protective gear to use.
- Follow safe working procedures.
- Take care of your protective headgear.
- Notify your supervisor of unsafe conditions and equipment.
- Get medical help promptly in the case of head injury.

Ear protection

Ear defenders

You may have to wear ear protection when you are carrying out a noisy job, for example using a large hammer drill to bore a hole through an external wall for an overflow pipe. Ear protection ranges from simple earplugs to earmuffs.

Earplugs

Respiratory protection

Respirator

There are many types of respiratory protection device: each is used to guard against a particular circumstance. In plumbing, respiratory protective devices are used primarily to guard against the following:

- Breathing in dusts, mists or fibre particles, for example in a loft space where loft insulation particles can cause irritation or during the removal of a solid-fuel fireback boiler where the dust and soot created can require the use of a respirator.

- Breathing in dangerous gases, for example in situations where you have to undertake lead welding indoors or where there is no mechanical extraction system. In these situations, a respirator would typically be used.

- Occasionally a plumber may have to enter a sewer system, which may contain harmful sewer gases – this is a dangerous situation where a plumber may be required to wear full breathing apparatus.

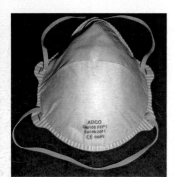

Dust mask

The consequence of not protecting yourself is the strong likelihood of lung problems in later life which could include potentially fatal conditions such as lung cancer and emphysema.

Safety tip

Remember that the filters on respirators require regular replacement.

Full breathing apparatus requires specialist training in its use

Tools and equipment

You will use a variety of tools and pieces of equipment as a plumber. All of them are potentially dangerous if misused or neglected. Practical instruction in the proper use of tools and equipment will form part of your training on-site and in your training centre. Tools and equipment are covered in greater detail in Chapter 4, Key plumbing principles.

Hand tools and manually operated equipment are often misused. You should always use the right tool for the job. Here are some guidelines to follow:

- Never use a hammer on a tool with a wooden handle (e.g. wood chisel) as you may damage the wooden handle and release dangerous splinters. If the handle splits as you hit it, the hammer could slip off and damage your hand.

- Cutting tools, saws, drills etc. must be kept sharp and in good condition. As a plumbing apprentice, you will frequently be asked to use cutting tools such as hacksaws and wood saws: you should ensure that the blades are always fitted properly and are sharp. Hacksaw teeth should be pointing in the forward direction of cut. After use, guards should be fitted where possible.

- Handles should be properly fitted to tools such as hammers and files, and should be free from splinters. Hammerheads should be secured correctly using metal or wooden wedges. 'Mushroom heading' of chisels is also a dangerous condition which can lead to serious eye injury. Unprotected file tangs present a serious danger of cuts and puncture wounds.

- It is particularly important to check that the plugs and cables of hand-held electrically operated power tools are in good condition. Frayed cables and broken plugs should be replaced. Electric power tools of 110 or 230 volts must be P.A.T. tested in accordance with your employer's procedures. It is good practice to check all electrical equipment and test labels to ensure they are in safe working order.

- Other common items of equipment, e.g. barrows, trucks, buckets, ropes and tackle etc. are all likely to deteriorate with use. If they are damaged or broken they will sooner or later fail in use and may cause an accident. Unserviceable tools and equipment should not be used. They should be repaired or replaced and the unsafe equipment removed from the site.

- You may have to use cartridge-operated tools during your career. If you do, you will be given the necessary instructions on the safe and correct methods for using them. They can be dangerous, especially if they are operated by accident or used as toys. This could cause ricochets, which could lead to a serious injury.

Trips and fire hazards

Tools, equipment and materials left lying about, trailing cables and welding hoses, spilled oil and so on cause people to trip, slip or fall. Clutter and debris, oily rags, paper etc. should be cleared away to prevent fire hazards. As an individual you may have no control over the general state of the work place, but you should see that your own work area is kept clear and tidy, as this is the mark of a skilled and conscientious plumbing tradesperson.

Electricity on site

The safe use of electricity on site is covered by the Electricity at Work Regulations 1989, which in turn are covered by the Health and Safety at Work Act 1974 (HASAWA). These Regulations impose specific duties on employers to put into place measures to protect their employees against death or personal injury from the use of electricity at work.

Safety tip

Never be tempted to just 'make do' with whatever tool you may have to hand

Definition

P.A.T. tests – Checking for safety and keeping maintenance records of all portable electrical equipment to ensure it is in safe working order

Find out

What is the minimum age required to use cartridge-operated tools?

Employers are required to have specific codes of practice for their employees. These practices must include keeping maintenance records for all portable equipment (P.A.T. tests). The HSE suggest that portable electrical equipment should be tested every 3 months for construction-site applications. These records must show that the equipment is tested regularly by a competent person using suitable test equipment. You must, however, make a visual inspection of a power tool to establish whether it is safe to use EVERY TIME YOU USE IT!

The supply to a work site may be from a generator or from the local public supply. Care should be taken to site a generator so that noise and fumes are reduced to a minimum. Whatever the source of supply, it must be routed to where it is needed on the site. This may involve electric cables being buried underground or, more often, suspended overhead on poles.

Both methods of distribution present hazards. Sites will often have **residual current devices** (RCDs) in place for added protection. These are intended to 'trip out' the electrical supply if there is any current leakage.

You may work on a site where a site distribution system is needed for the temporary supply of electrical power to the various locations around the site. A number of different units are used for this purpose, all of which should comply with the relevant British Standard 4363.

All the units will be clearly marked with details of the output voltage and the 'Danger–electricity' symbol. The installation of this system must conform to the requirements of the current edition of BS 7671. You should never interfere with or alter any installation. Only an authorised person may carry out any alterations.

Electricity on-site safety checklist

✓ Do not use lighting circuits for power tools

✓ Power tools should be double insulated

✓ Never carry a portable electric tool by its cable

✓ Ensure that equipment is not damaged before you plug it in

✓ Always have enough light for the job

✓ Keep lights clean

✓ Check that all cables are correctly insulated and not damaged or frayed

✓ Check that plugs and sockets are clean and in sound condition

✓ Check for current P.A.T. labels

✓ Check that RCD protection is provided where necessary.

Lifting and carrying (manual handling)

The manual handling or lifting of objects is the cause of more injuries on worksites than any other factor. Back strains and associated injuries are the main source of lost hours in the building services industries. Manual handling can involve: pushing, pulling, and the lifting and lowering of loads (tools, cylinders boilers, radiators, etc.).

The movement of loads requires careful planning in order to identify potential hazards before they cause injuries. You should follow the safety precautions and codes of practice at all times.

Before we consider the risks and procedures of manual handling, get into the habit of asking yourself whether the load can be moved another way with less risk of personal injury, for example using a sack trolley or a cart.

Before moving a heavy load, do the following basic risk assessment:

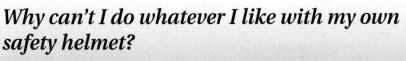

Safety tip

It takes time to become competent. Practise first with light weights until you become proficient.

If possible, use a mechanical means of lifting in preference to manual handling – it reduces the risk of personal injury

1. **The task – does it involve:**	2. **The load – is it:**
stooping?	heavy?
twisting?	bulky or unwieldy?
excessive lifting or lowering distances?	difficult to grasp?
excessive carrying distances?	unstable or with contents that are likely to shift?
excessive pushing or pulling distances?	
frequent or prolonged physical effort?	sharp, hot or otherwise potentially damaging?
the sudden risk of the load moving?	
3. **The working environment – does it have:**	4. **The individual – does he or she have:**
space constraints?	any restriction on their physical capability?
slippery or unstable floors?	the knowledge and training for manual handling?
variation in levels?	
poor lighting?	
hot/cold/humid conditions?	

These are factors that must be taken into account when carrying out manual handling operations. Take a few moments to consider the information in the above lists and the impact each factor could have on the safe movement of tools and materials.

Extreme care must be taken when lifting or moving heavy or awkward objects manually. The load generally accepted as maximum for a fit person to lift is 20 kg. The rules for correctly lifting a load are shown in detail on the next page.

FAQ

Why can't I do whatever I like with my own safety helmet?

It is an offence under HASAWA to misuse or interfere with equipment provided for your Health and Safety or the Health and Safety of others.

Testing weight with foot

- Ensure that the path where you want to move the load is clear from obstructions, that any doors you have to pass through are opened and that you have a clear area for placing the load.

- Test the load by gently applying force with your foot – if it feels heavy or difficult, you may need to seek help for a double lift.

Getting hold of load

- If it feels comfortable to move, start from a good base and stand with the feet hip-width apart.

- Grip the load firmly. (Use gloves to avoid injuries if the load has sharp or rough edges.)

- Balance the load, using both hands if possible.

- Take account of the position of the centre of gravity of the load when lifting.

Picking up with legs bent

- Maintain a straight back; bend your knees and let the strong muscles of your legs and thighs do the work.

- Keep your arms straight and close to the body.

- Avoid sudden movements and twisting of the spine.

Moving slowly

- Move slowly and evenly

- Never obstruct your vision with the load that you are carrying

Placing carefully

- Place load carefully at the end of the move. Do not drop or slam the load down.

Remember

A safe lifting technique requires thinking before acting, total concentration, correct use of body power and a smooth, rhythmical sequence of activity. Safe handling is a skill.

In cases where team lifting (two or more people) is required, the following points must be remembered:

- team members should ideally be of similar height and build
- all team members must know the lifting sequence
- one member must be nominated to act as coordinator
- good communication when lifting should reduce the risk of accidents happening.

Movement of loads and methods of transport

Making use of various methods of transport can make the movement of loads much simpler and safer. Flat trailers, sack trolleys and fork-lift trucks are some of the moving aids you will come across on-site.

Untrained people must NEVER use equipment such as dumper trucks and similar mechanical devices on construction sites to move equipment; these should only be used by people who are properly trained. It is also vital to ensure that in the use of the equipment and the item of mechanical equipment is safe to use for the task to be carried out.

Remember

Fork-lift trucks and certain other methods of transport can only be operated by a qualified person

Access to work

At the end of this section you should be able to:

- state the common defects found on ladders
- explain how to transport ladders of all sizes in the correct way
- state the correct requirements for safely positioning:
    - ladders
    - stepladders
    - trestle scaffolds.
- identify potential hazards on an independent and putlog scaffold
- explain the safety precautions when using roof ladders
- state the main requirements to ensure safe working in excavations.

Access equipment is a term that refers to items such as ladders, roof ladders, trestle scaffolds, independent and putlog scaffolds, and tower scaffolds. Access also includes working in excavations.

Access equipment forms a very important part of site work. It is vital that it is kept in good order. For this reason all access equipment is regularly checked and recorded. We will cover its correct use and limitations.

Ladders

As a plumber you will need to use ladders frequently, either working from the ladder directly or using it to gain access to the place of work or scaffold.

A ladder should be used only for:

- gaining access to a work platform, e.g. a scaffold, or
- relatively short-term working (duration 30 minutes or less) such as repairs to an external waste pipe.

A ladder is not suitable for long-term working; for this type of work another form of access equipment should be used.

Ladders can be manufactured from wood, aluminium or fibre glass. Because ladders are used frequently, their condition tends to be neglected, which can lead to defects. It is advisable to inspect any ladder before use.

You should check the following to ensure the ladder is safe for you to use:

- the **stiles**/strings are not cracked or warped
- the **rungs** are not split or dirty
- tie-rods are not missing and are not damaged
- there is no wood rot
- there are no temporary repairs
- the ladder should not be painted as the paint may be hiding defects.

Short ladders can be carried by one person, on the shoulder in either the horizontal or vertical position. Longer ladders should be carried by two people horizontally on the shoulders, one at either end holding the upper stile. When carrying ladders you should take care in rounding corners or passing between/under obstacles.

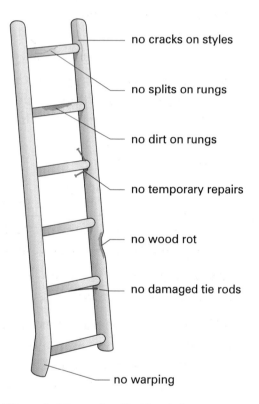

no cracks on styles

no splits on rungs

no dirt on rungs

no temporary repairs

no wood rot

no damaged tie rods

no warping

Figure 2.6 Example of ladder defects

There are certain rules for erecting ladders, which must be followed to ensure safe working:

- The ladder should be placed on firm, level ground. Bricks or blocks should not be used to 'pack up' under the stiles to compensate for uneven ground.
- If using extension ladders, these should be erected in the closed position and extended one section at a time. When extended, there must be at least three rungs overlap on each extension.
- If the ladder is placed in an exposed position it should be guarded by barriers.
- The angle of the ladder to the building should be in the proportion of 4 up to 1 out, or 75°.

- The ladder should be secured at the top and, as necessary, at the bottom to prevent unwanted movement. Alternatively the ladder may be 'steadied' by someone holding the stiles, and placing one foot on the bottom rung; this is commonly known as 'footing' the ladder. This person must not, under any circumstances, move away while someone is on the ladder.

- When the ladder provides access to a roof or working platform, the ladder must extend at least 1 m or 5 rungs above the access point.

- Ensure that the ladder is not resting against any fragile surface (e.g. a glass window) or against fittings such as gutters or drainpipes – these could easily give way, resulting in an accident.

Did you know?

There are three classes of ladder: 1, 2 and 3. A class 1 ladder is intended for industrial use

Safety tip

When accessing scaffolding on a construction site where heavy materials/equipment are moved or there is regular foot access, only a Class 1 ladder may be used

1m

Figure 2.7 Ladder attached to platform

- When climbing up ladders you must use both hands to grip the rungs. This will give you better protection if you slip.

- All ladders, stepladders and mobile tower scaffolds should be tested and examined on an annual basis. The results of the tests should be recorded. Ideally the item tested should be marked to show it has been tested. A competent person must carry out this test.

The diagram shows a correctly positioned ladder, securely lashed to the scaffold with sufficient extension past the access point (5 rungs).

Stepladders

Plumbers use stepladders extensively. The first essential check before using a stepladder is to make sure the ground is level and firm. If it is not, you should not use the stepladder on it. All four legs of a stepladder should rest firmly and squarely on the ground. They will do this provided that the floor or ground is level and that the steps themselves are not worn or damaged.

When using the steps, ensure that your knees remain below the top of the steps. The top of the steps should not be used unless the ladder is constructed as a platform.

On wooden stepladders, check that the hinge is in good condition and that the ropes are of equal length and not frayed.

Fibre-glass stepladder

Trestles

Some jobs cannot be done safely from a pair of steps. In such cases a working platform known as a 'trestle scaffold' should be used. This consists of two pairs of trestles or 'A' frames spanned by scaffolding boards which provide a simple working platform.

When erecting trestle scaffolds, the following rules should be observed:

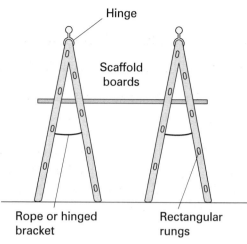

Figure 2.8 Trestles

- As with ladders. they should be erected on a firm, level base with the trestles fully opened.

- The platform must be at least two boards or 450mm wide.

- The platform should be no higher than two-thirds of the way up the trestle: this ensures that there is at least one-third of the trestle above the working surface.

- The scaffold boards must be of equal length and should not overhang the trestle by more than four times their own thickness, for example a 40 mm board must not overhang by more than 160 mm.

- The maximum span for boards is 1.3 m for 40 mm thick and 2.5 m for 50 mm thick boards.

- If the platform is more than 2 m above the ground, toe boards and guardrails must be fitted and a separate ladder provided for access.

- Trestles must not be used where anyone can fall more than 4.5 m.

Trestle scaffold boards

Scaffold boards are made to satisfy the requirements of BS2482/70 and are the only boards that should be used. The maximum length is usually no greater than 4 m. If a greater length is required, then special staging is used. Scaffold boards should be:

- clean, free from grease and dirt
- straight
- free from decay, damage or any splits
- unpainted.

Scaffolds
Independent and putlog scaffolds

As a plumber you will not be expected to erect either an independent or putlog scaffold. However, you will have to work from one at some stage, maybe to install guttering, or to access roof weatherings. It is important, therefore, that you are happy that the scaffold has been erected correctly and is safe for you to work from. As a rule of thumb, stand back and look at the scaffold and ask yourself the following questions:

- Does it look safe?

- Are the scaffold tubes plumb and level?
- Are there sufficient braces, guardrails and scaffold boards?
- Is it free from excessive loads such as bricks?
- Is there proper access from a ladder?

Make sure there are ledger-to-ledger braces on each lift for independent scaffolds, and that putlog scaffolds are tied to the building. Look for any gaps between the boards of the platform from which you will be working, and make sure a toe board and guardrail are fitted.

Mobile scaffold towers

Mobile scaffold towers may be constructed of basic scaffold components or may be specially designed 'proprietary' towers made from light alloy tube. The tower is built by slotting the sections together until the required height is reached. Mobile towers are fitted with wheels, static towers are fitted with base plates.

When working with mobile tower scaffolds the following points must be followed:

Mobile scaffold tower

- If the working platform of a tower scaffold is over 2 m from the ground, it must be fitted with guardrails and toe boards. The guardrail, also known as a 'knee rail', must be between 0.4 m and 0.7 m from the working platform and must consist of two horizontal bars. There must also be a hand rail which must be no more than 910 mm from the working platform.
- When the platform is being used, all four wheels must be locked.
- The platform must never be moved unless it is clear of tools, equipment and people. It should be pushed at the bottom of the base and not at the top.
- The stability of a tower depends upon the ratio of the base width to height. A ratio of base to height of 1:3 gives good stability.
- Outriggers can increase stability by effectively increasing the area of the base but, if used, must be fitted diagonally across all four corners of the tower and not on one side only. When outriggers are used they should be clearly marked (e.g. with hazard marking tape) to indicate that a trip hazard is present.
- Towers higher than 9 m should be secured to the structure of the building.
- Towers must not be built taller than 12 m unless they have been specially designed for that purpose.
- Access to the working platform of the tower should be by a ladder securely fastened inside the tower. Ladders must never be leaned against a tower as this may push the tower over.

Using roof ladders

The illustration shows what a roof ladder looks like. It is used by plumbers to gain access to chimneys in order to carry out work such as small repairs to sheet weathering. For a full sheet weathering installation, a scaffold would be needed. Roof ladders are also used to gain access to a roof when installing chimney flue liners for gas fire or gas boiler installations in existing flues. They are made of aluminium, which is extremely light, so that they can easily be manoeuvred up the roof. A roof ladder is positioned by turning it on its wheels, pushing it up the roof, and then turning it through 180 degrees to hook it over the ridge tiles.

Figure 2.9 Roof ladder

The key safety points are:

- the supports under the ladder must all fully rest on the roof surface
- the ladder must be long enough for the roof on which it is being used
- the ladder used to gain access from the ground to the roof must be securely fastened at the top and must be next to the roof ladder
- gaining access to a roof and completing the work using a roof ladder usually requires two people: one to access the roof, the other to assist with providing equipment.

Safety checklist for the roof ladder:

- stiles must be straight and in sound condition
- rungs must be sound
- the ridge hook must be firmly fixed
- wheels must be firmly fixed and running freely
- pressure plates must be sound (these are the parts that rest on the roof surface).

Working in excavations

As a plumber you may occasionally be required to work in trenches or holes below ground. We call these excavations. When working in or around excavations, it is necessary to work in accordance with the specific Health and Safety guidelines that apply.

The Health and Safety legislation is, in the main, designed to prevent the following occurrences:

- persons falling into the excavation
- collapse of the excavation walls.

Fall prevention

Fencing must be erected around any excavation into which someone could fall more than 2 m, but it is fairly obvious that precautions should also be taken for holes/trenches less than this depth in order to protect the safety of workers and the public. Similarly, if you are working in a block of flats under construction, there should be fencing around stairwells and lift shafts if there is any danger of people falling.

Wall collapse prevention

Material that is excavated must be piled a safe distance away to ensure it does not begin to spill back in, jeopardising the safety of those who may be working in the excavation.

When an excavation is deeper than 1.2 m, the sides should be either sloped or shored up (see figure 2.10). Excavation walls should always be shored up with strong wooden timbers or steel plates. The amount of these materials required will normally depend on the soil/ground type in a particular area or the depth of the excavation. Deep excavations should be inspected every 7 days by a competent person.

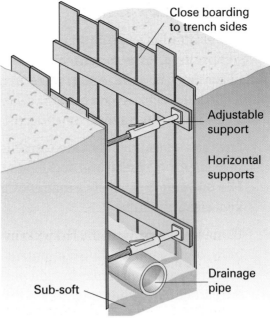

Figure 2.10 Close boarding of trenches

The example illustrated shows a method called 'close boarding of trenches', used to secure trenches that have a significant chance of collapse.

Note on Liquefied Petroleum Gas (LPG)

LPG is heavier than air, and therefore poses a particular danger when working in excavations. If LPG is being used in an excavation, any leakage could result in the formation of pockets of LPG. This is a potentially dangerous situation, as LPG is a highly explosive gas.

Fire safety and emergencies

At the end of this section you should be able to:

- explain the basic principles of fire safety and the procedures you should follow in the event of fire
- identify different types of fire extinguisher
- state the correct fire extinguishers to use on given types of fire
- explain what to do in case of an accident at work
- explain the use of accident report forms and how to complete them correctly
- state the basic requirements of RIDDOR and explain what RIDDOR stands for
- explain the correct action to take in the event of other emergencies.

All of the topics covered within this chapter are designed to keep you safe, but this is a very important section. Fire is a constant risk in any industry, but even more so for plumbers, who can come into contact with electricity, gas and heating equipment.

Safety tip

On no account should you ever move trench supports to locate pipework in excavations: the trench may collapse and the results could be deadly!

Figure 2.11 The Fire Triangle

Fire safety

Fire or burning is the rapid combination of a fuel with oxygen (air) at high temperature. A fire can reach temperatures of up to 1000°C within a few minutes of starting. For a fire to start there are three requirements: combustible substance (called the fuel), oxygen (usually as air) and a source of heat (spark, friction, match). When these three items come together in the correct combination, a fire occurs.

Fires can spread rapidly. Once established, even a small fire can generate sufficient heat energy to spread and accelerate the fire to surrounding combustible materials. Fire prevention is largely a matter of 'good housekeeping'. The workplace should be kept clean and tidy.

Plumbing is a particularly risky job in terms of potential fire hazards. The use of blow torches and welding equipment – which are often deployed near combustible materials, and sometimes in tight positions or difficult to access areas – means you are particularly vulnerable.

Because 'hot working' is such a risky job, your employer is required to provide strict working methods. These procedures constitute basic common sense, but are also a requirement of insurance companies. They usually include:

- providing a fire extinguisher in the immediate working area
- completing work with a blow or welding torch a minimum of an hour before leaving a site.

If you are working close to a combustible material (e.g. timber skirting, joists etc.), always make sure you protect the area around the fitting you are soldering with a heat-resistant mat. You can get these from your local plumbers' merchant. If the pipe is insulated, remove the insulation for about 300 mm on each side of the fitting from the area where you are using the torch.

If you think you may have caught any material with the flame, wet the area and check again after a few minutes to make sure it is all right. If you are working in a customer's home, make sure you pull back carpets or place curtains away from the area you are going to be working in.

Finally, building sites can be dirty paces. Timber shavings and other combustible materials find their way under floors, baths etc. Make sure the area where you are working is clean before you start as well as after you finish.

Another major cause of fires is electrical faults. All alterations and repairs in electrical installations must only be carried out by a qualified person, and must be to the standards laid down in the IEE Regulations.

You may sometimes work in an occupied building, such as an office block. You must be aware of the building's fire safety procedures, and find out about normal and alternative escape routes. Know where your assembly point is located and report your presence to your supervisor.

If you discover a fire, follow the procedures below:

- raise the alarm immediately
- leave by the nearest exit
- call the fire service.

In addition, windows should be closed to help starve the fire of oxygen, but they should not be locked.

Classes of fire

Fires are commonly classified into four groups, according to fuel type and how the fires are extinguished:

- **Class A** – fires involving solid materials, extinguished by water
- **Class B** – fires involving flammable liquids, extinguished by foam or carbon dioxide
- **Class C** – fires involving flammable gases, extinguished by dry powder
- **Class D** – fires involving flammable metals, extinguished by dry powder.

Fire-fighting equipment

If a fire is small, it may be possible to put it out quickly and safely. However, if your efforts to contain a fire prove fruitless and it starts to get out of control, when you finally decide to make your escape you could have difficulty finding your way through the smoke and the fumes. Smoke and fumes are just as lethal as the fire itself.

Fire-fighting equipment, including extinguishers, buckets of sand or water and fire-resistant blankets, should be readily available in buildings. In larger premises, you will find automatic sprinklers, hose reels and hydrant systems. The table below shows each type of fire extinguisher and the use to which they should be put.

Type of extinguisher	Colour code	Main use
Water	Red	Wood, paper or fabrics
Foam	Cream	Petrol, oil, fats and paints
Carbon dioxide	Black	Electrical equipment
Dry powder	Blue	Liquids, gases, electrical equipment

Figure 2.12 Fire extinguishers

FAQ

Do I really have a legal Health and Safety duty?

Yes you do! You *must* take resonable care of your own Health and Safety and that of others affected by you.

Water extinguisher

Powder extinguisher

Carbon dioxide extinguisher

The colour coding on extinguishers may not seem immediately obvious. Older extinguishers may have their whole body painted in the appropriate colour (e.g. black ones are filled with carbon dioxide). Recent European legislation dictates that new extinguishers must be coloured red whatever substance they contain, but will carry 5 per cent of the colour the extinguisher would have been under the original system: for example, a carbon dioxide extinguisher will be red with a black stripe, triangle or lettering etc. Care should therefore be taken when choosing which extinguisher to use to extinguish a fire that you choose the correct type.

The decision to have all extinguishers coloured red was made because red is recognisable by people with colour blindness. Previously, it was possible for a colour-blind person to use the wrong extinguisher for a specific type of fire. The fact that they are now all the same colour means you have to read the labels to discover their intended use.

Fire extinguisher safety checklist

- Never use a fire extinguisher unless you have been trained to do so.
- Do not use water extinguishers on electrical fires because of the risk of electric shock and explosion.
- Do not use water extinguishers on oils and fats as this too can cause an explosion.
- Do not handle the nozzle on carbon dioxide extinguishers as this can cause a freezer burn to the hands.
- Do not use carbon dioxide extinguishers or halon types in a small room as this could cause suffocation.
- Always read the operating instructions on the extinguisher before use.

Accidents

It is recommended that your Plumbing Apprenticeship training course includes a certificated first-aid course, but in case it does not, here are a few pointers about dealing with accidents.

First actions

No matter how many precautions are taken, sooner or later you are likely to witness an accident. Whatever the accident – a cut finger or a fall from a ladder – you must know what to do.

Consider a situation where you are working with another plumber who has just cut her hand. There is a lot of bleeding and your workmate is in distress. What are you going to do? More importantly, *what are you going to do first*? There are a number of things that need to be done after an accident:

1. Seek or administer immediate first aid

2. Get help if necessary, i.e. phone for an ambulance

3. Report the accident to the site supervisor

Which one of these things you carry out first will depend on several things:

- Are you a qualified first aider?
- How severe is the injury?
- Is your supervisor or a qualified first aider immediately available?

Later you will need to:

1 Write down the details in an accident report book

2 Complete a company accident report form.

There are several important things to remember. In this scenario, we are dealing with a cut hand; the patient is bleeding quite heavily so it would be necessary to administer first aid as soon as possible.

However, if the scenario were, for example, an electric-shock casualty, you must first make sure that the area is safe – to ensure that whatever caused the accident is not going to injure you. In this case, you should isolate the supply first and call for an ambulance if necessary; then someone can attend to the casualty and inform the supervisor.

If someone has fallen from a height, do not move them: if they have a serious back injury, your actions could make things worse. If they are conscious, make them as comfortable as possible until help arrives; this may include treating minor wounds. If the person is unconscious, they may be put in the recovery position, but only if you are confident that there is no back injury.

If someone is injured in an accident, the details have to be entered in the accident book. You may have to enter details on a casualty's behalf if they are unable to do so themselves. If the accident involved a piece of faulty equipment, do not tamper with it as it may be subject to an investigation by the Health and Safety Inspectorate.

Accidents that result in three or more days' absence from work must be reported to the Health and Safety Executive. The next page shows a typical example of an accident report form.

Did you know?

There are many providers of first-aid courses, and many courses are specific to the workplace. You can find plenty of information about them on the Internet. Some providers will run a course at your place of work

Remember

People with back injuries should only be moved by trained personnel

Remember

After an accident – seek first aid, inform your supervisor, complete the accident report form.

On the job: Accident report form

You are working with two other plumbers – Alan, a second-year apprentice, and Ishmail, who is the senior tradesperson in charge of the site. Alan was balancing on the top of a stepladder trying to secure some guttering when he overbalanced and fell to the ground. The ground was hard concrete and as a result of this fall Alan has sustained a broken wrist and bleeding to his hand and forehead. Ishmail accompanies Alan in the ambulance. You are left to complete the accident report form.

1 What information do you need to complete the form as well as you can?

2 Complete the accident report form on the next page, using the information contained in the scenario above and making assumptions about the exact time and necessary addresses.

Full name of injured person:	
Home address:	Sex: Male/Female
	Age:

Status: Employee
 	
          Contractor
 	
          Visitor	

Date of accident:	
Time of accident:	
Precise location:	
What was the accident and its cause? (You may have to give a detailed written description.)	
Name and address of witness if any:	
Details of apparent injuries:	
Summary:	

Figure 2.13 Typical accident report form

Reporting of Injuries, Diseases and Dangerous Occurrences Regulations (RIDDOR)

The RIDDOR Regulations came into effect in April 1996. The Regulations require injuries, diseases and occurrences in specified categories to be notified to the relevant enforcement authority. Failure to comply is a criminal offence.

Your company must notify the enforcing authority without delay, normally by a telephone call, of the following:

- any fatal injuries to employees or other people in an accident connected with your business

- any major injuries to employees or other people in an accident connected with your business

- any dangerous occurrences:
  - the collapse, overturning or failure of a load-bearing part of a lift, hoist, crane or mobile platform
  - the explosion or collapse of a closed vessel
  - electrical short circuit or overload causing fire or explosion
  - any explosion or fire resulting in the suspension of normal work for more than 24 hours
  - any unintended collapse of any building or structure under construction
  - any uncontrolled or accidental release or escape of any pathogen (virus)
  - any unintentional explosion
  - any incident in which plant or equipment comes into contact with overhead power lines exceeding 200 volts.

Did you know?

Any act of aggression by a colleague while at college or at work that causes a situation of conflict is also considered a dangerous occurrence

There are also prescribed diseases, certain poisonings, skin diseases, lung diseases and other conditions such as occupational cancer that have to be reported. Normally, an employer will carry out the reporting of these occurrences; however, it would be your responsibility to bring a matter to the attention of your employer or indeed the enforcing authority if the situation required you to do so. Remember: it is a criminal offence not to report such an occurrence.

Other emergencies

Emergencies other than a fire or an accident could include a bomb threat or chemical spillage. These events are pretty rare and it is unlikely that you will be involved in one of them. However, you still need to be aware of the procedure for dealing with emergencies:

1 Find a telephone in a safe environment well away from the emergency. Make sure you are not going to be trapped while making the call.

2 Dial the emergency service number, which is normally 999. However, this may not always be the case, especially if you are on an internal exchange. You need to familiarise yourself with any special procedures that may apply.

3 Keep calm and listen to what the emergency-service operator has to say.

4 When asked, give your name and the name of the service you require.

5 You will then be connected to the emergency service you require; again, stay calm, listen to what the operator has to say and answer the questions they ask.

6 You will be required to explain the nature of the emergency and where it occurred. Try to give the operator the exact location of the incident, such as the name of the company, its address etc.

7 When you have completed the call, arrange for someone to meet the emergency services to show them where the incident has occurred.

If you are working in an environment where there is an increased risk of an emergency situation occurring, your employer should provide you with sufficient information on what to do in an emergency. Some of these procedures will be similar to the fire evacuation procedures and should all be about getting outside to a place of safety. There may be other more specific procedures, and it is your duty to familiarise yourself with them and follow them when necessary.

Electric shock and first aid

At the end of this section you should be able to:

- explain the actions to take in the event of someone suffering electric shock

- describe the process of cardio-pulmonary resuscitation (CPR)

- explain how, once CPR has been administered, to place a person in the recovery position

- explain the basic actions that should be taken to treat people who have suffered burns, shocks and broken bones.

In this section you will be looking at the treatment of an unconscious casualty suffering from an electric shock. We will also be looking at the basic first-aid procedures for the treatment of burns, shock and breaks. You should be doing a purpose-made first-aid course; this section can offer only the basic theory.

Electric shock and its treatment

Electric shock occurs when a person becomes part of the electrical circuit. The severity of the shock will depend on the level of current and the length of time it is in contact with the body.

The lethal level is approximately 50 mA, above which muscles contract, the heart fibrillates and breathing stops (in other words, you die!). A shock current above 50 mA could be fatal unless the person is quickly separated from the supply. Below 50 mA, only an unpleasant tingling sensation may be experienced. However, this may cause someone to fall from a roof or ladder and the resulting fall may lead to a serious injury.

Action in the event of an electric shock

- First of all, check for your own personal safety to ensure that you will not be putting yourself at risk by touching the casualty – they are part of the circuit and they are live too.

- Break the electrical contact to the casualty by switching off the supply, removing the plug, or wrenching the cable free. If this is not possible, break the contact by pushing or pulling the casualty free using a piece of non-conductive material, e.g. a piece of wood. Guide the casualty to the ground, making sure that further injuries are not sustained, e.g. banging their head on the way down.

Remember

It is not the volts that give you the jolts; it's the amps that put out your lamps!

- Check the casualty's response, to assess their level of consciousness.

- Talk to them and gently shake them to gauge their level of response.

- If the casualty appears unharmed, they may only be shaken and should be advised to rest.

- If there is no movement or any sign of breathing summon help immediately. If there is someone with you, tell them to get help, i.e. ring 999. If you are on your own with the casualty, you will have to leave them for a moment while you get help.

- As soon as you return to the casualty you need to start cardio-pulmonary resuscitation (CPR).

To assess someone's consciousness level, ask a question or give a command – for example, 'what happened?' or 'open your eyes'. Speak loudly and clearly, close to the casualty's ear. Carefully shake the casualty's shoulders. A 'slightly conscious' casualty may mumble, groan or make slight movements. A fully unconscious casualty will not respond at all.

Cardio-pulmonary resuscitation

1 *Open the airway.* An unconscious casualty's airway may be narrowed or blocked, making breathing difficult and noisy or impossible. The main reason for this is that muscular control in the throat is lost, which allows the tongue to sag back and block the throat. Lifting the chin and tilting the head back lifts the tongue away from the entrance to the air passage.

2 *Remove any obvious obstruction from the mouth.* Place two fingers under the point of the casualty's chin and lift the jaw. At the same time, place your other hand on the casualty's forehead and tilt the head well back.

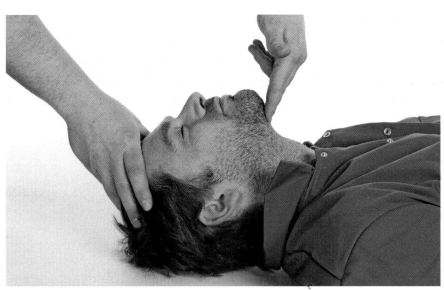

Remove obvious obstructions from the mouth

3 *To check for breathing,* put your face close to the casualty's mouth, look and feel for breathing as well as looking for chest movements. Listen for the sound of breathing. Feel for breath on your cheek. Look, listen and feel for 10 seconds before deciding that breathing is absent.

4 *Check for a pulse.* If the heart is beating adequately, it will generate a pulse in the neck (the carotid pulse) – this is where the main carotid arteries pass up to the head. These arteries lie on either side of the larynx (throat), between the Adam's apple and the strap muscle that runs from behind the ear across the neck to the top of the breastbone.

5 To check for the pulse ensure the head is tilted back and feel for the Adam's apple with two fingers. Slide your fingers back into the gap between the Adam's apple and the strap muscle, and feel for the carotid pulse. Feel for 10 seconds before deciding that the pulse is absent.

Mouth-to-mouth ventilation is given with the casualty lying flat on their back:

1 Open the airway by tilting the head and lifting the chin.

2 Remove any obvious obstruction including broken or displaced dentures from the mouth.

3 Close the casualty's nose by pinching it with your finger and thumb.

4 Take a full breath, and place your lips around the casualty's mouth, making a good seal.

5 Blow into the casualty's mouth until you see the chest rise. Blow for about 2 – 5 seconds for full inflation.

6 Remove your lips and allow the chest to fall fully.

7 Deliver subsequent breaths in the same manner.

In this situation, you need to deliver two complete breaths.

Fifteen chest compressions are now given:

1 With the casualty still lying flat on their back on a firm surface, kneel beside them, and find one of the lowest ribs using your index and middle fingers.

2 Slide your fingers upward to the point in the middle where the rib margins meet the breastbone.

3 Place your middle finger over this point and your index finger on the breastbone above.

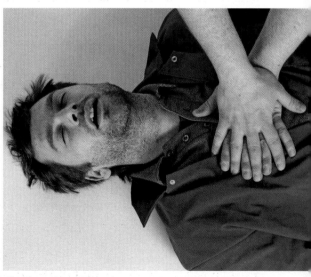

Chest compressions

4 Place the heel of your hand on the breastbone, and slide it down until it reaches your index finger. This is the point at which you will apply pressure.

5 Place the heel of your first hand on top of the other hand, and interlock fingers.

6 Leaning well over the casualty, with your arms straight, press down vertically on the breastbone to depress it approximately 4–5 cm, then release the pressure without removing your hands.

7 Repeat the compressions, aiming for a rate of approximately 15 compressions per 15 seconds (4 cycles per minute).

Return to the head and give two more ventilations, then a 15 further compressions. Continue to give two ventilations to every 15 compressions until help arrives. Do not interrupt CPR to make pulse checks unless there is any sign of a returning circulation. If a pulse is confirmed, check breathing, and if it is still absent continue with ventilation. Check the pulse after every 10 breaths, and be prepared to re-start chest compressions if it disappears. If the casualty starts to breathe unaided place him or her in the recovery position. Re-check breathing and pulse every two minutes.

The recovery position

Any unconscious casualty should be placed in the recovery position, so long as they have no obvious back injury. This position prevents the tongue from blocking the throat, and, because the head is slightly lower than the rest of the body, it allows liquids to drain from the mouth, reducing the risk of the casualty inhaling stomach contents should they vomit.

The head, neck and back are kept in a straight line, while the bent limbs keep the body propped in a secure and comfortable position. If you must leave an unconscious casualty unattended, they can be left safely in the recovery position while you get help.

The technique for turning is described here and assumes that the casualty is found lying on their back from the start.

1 Kneeling beside the casualty, open the airway by tilting the head and lifting the chin. Straighten the legs. Place the arm nearest you out at right angles to the body, elbow bent and with the palm uppermost.

2 Bring the arm furthest from you across the chest, and hold the hand, palm outwards, against the casualty's nearer cheek.

3 With your other hand, grasp the knee furthest from you and pull the knee up, keeping the foot flat on the ground.

4 Keeping the casualty's hand pressed against their cheek, pull the knee towards you to roll the casualty towards you and on to their side.

5 Tilt the head back to make sure the airway remains open. Adjust the hand under the cheek if necessary so that the head stays in this tilted position and blood flow to the hand is not restricted.

6   Adjust the upper leg, if necessary, so that both the hip and the knee are bent at right angles.

The recovery position

A casualty who has received an electric shock, or who has been unconscious for any reason, should be sent to hospital in an ambulance.

Treatment for burns, shocks and breaks

An electric shock may result not only in unconsiousness but also in other injuries; there may be burns at both the entry and exit points of the current. The treatment for these burns is to flood the site of the injury with cold water for at least 10 minutes. This will halt the burning process, relieve pain and minimise the risk of infection.

You would then need to treat for shock, which is the medical condition where the circulatory system fails and insufficient oxygen reaches the tissues. If shock is not treated quickly, the vital organs can fail leading ultimately to death.

To treat shock, you need to stop any external bleeding. Lay the casualty down, keeping the head low; raise and support the legs – but be careful if you suspect a fracture; loosen tight clothing, braces, straps or belts to reduce constriction at the neck, chest and waist.

Keep the casualty warm by wrapping them in a blanket or coat and continue to check and record breathing, pulse and level of response every 10 minutes. Be prepared to resuscitate if necessary.

If the casualty has sustained a broken bone, your first aim will be to prevent movement at the site of the injury. Do not move the casualty until the injured part is secured and supported – unless they are in danger, perhaps from electric shock.

You must arrange for immediate removal to hospital, maintaining comfortable support during transport. You would also treat the casualty for shock as mentioned earlier.

First-aid kits

Employers should provide first-aid kits. The HSE suggests that a basic first-aid kit should contain the following items:

- twenty adhesive dressings (assorted sizes)
- two sterile eye pads
- six triangular bandages
- six unmedicated wound dressings, 10cm × 8cm
- two unmedicated wound dressings, 13cm × 9cm
- six safety pins
- three extra sterile unmedicated wound dressings 28cm × 17.5cm.

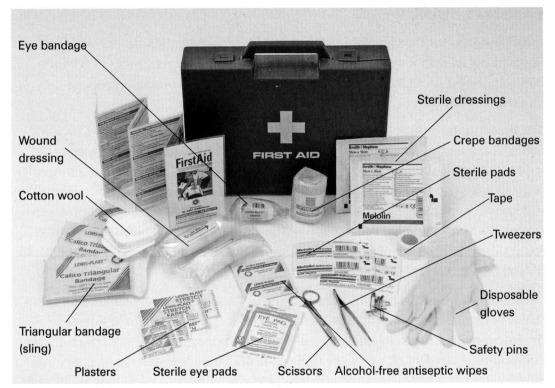

Figure 2.14 First-aid kit

The kit should only contain basic/first-aid items. Additional items such as headache tablets should never be stored in first-aid kits.

In circumstances where plumbers work in domestic properties, it is usual practice for a first-aid kit to be provided with every company vehicle.

FAQ

What is dermatitis?

Dermatitis is an inflammation of the skin normally caused by the hands coming into contact with irritating substances. Skin can become sore, cracked and easily infected. Protect your hands by washing with approved cleaners and use barrier creams.

Control of Substances Hazardous to Health

At the end of this section you should be able to:

- explain what is meant by COSHH and what aspects of work the Regulations refer to
- describe some of the protective measures that can be used to prevent accidents involving COSHH-registered substances
- clearly explain what the different types of chemicals are and the effects of chemical spillages
- explain the different types of asbestos and their effects.

COSHH 2002

COSHH stands for the Control of Substances Hazardous to Health. These Regulations were first introduced in 1988 and provide a legal framework for the control of substances hazardous to health in all types of business, including factories, farms, offices, shops – and plumbing!

The Regulations require that employers make an assessment of all work that is liable to expose any employee to hazardous solids, liquids, dusts, fumes, vapours, gases or micro-organisms. Any risks to health must be evaluated and a decision must be taken on the action to remove or reduce those risks. For example, if you were replacing waste pipework in a college laboratory, you would expect protective clothing to be provided to prevent injury to yourself while carrying out your work.

What is a substance hazardous to health?

Substances that are 'hazardous to health' include substances labelled as dangerous (i.e. very toxic, toxic, harmful, irritant or corrosive). They also include micro-organisms and substantial quantities of dust and indeed any material, mixture or compound used at work, or arising from work activities, that can harm a person's health. There is a wide variety of substances that fall under COSHH that are used in plumbing, including items ranging from fluxes to working with irritants such as loft insulation and working with human waste such as urine: all require specific actions to enable plumbers to work safely with them.

What does COSHH require?

- The risk to health arising from working with a substance must be assessed.
- The necessary measures to prevent exposure or control the risk must be identified.
- There must be control measures to ensure that equipment is properly maintained and procedures observed.
- Employees must be informed, instructed and where necessary receive the relevant training about the risks and the precautions that need to be followed.

Remember

COSHH stands for Control of Substances Hazardous to Health

Prevention will usually involve the use of protective clothing and equipment – respiratory equipment masks, dust masks, goggles, gloves and so on.

Chemical safety

Chemicals are substances that form the basis of many common items. However, they can be harmful to the people working with them if they come into contact with the skin by accident. When not contained or handled properly chemicals can be:

- inhaled as a dust or gas
- swallowed in small doses over a long period
- absorbed through skin or clothing
- touched by or spilled on unprotected skin.

Some chemicals can cause:

- injury to eyes, skin, organs – from fires and burns etc.
- silent illnesses – effects are not noticeable immediately after exposure, but can cause medical problems after many months or years of exposure (e.g. asbestosis)
- allergy to the skin as a rash, coughing and breathing problems
- death – some poisonous chemicals can kill outright.

Why do chemical accidents happen?

- Hurrying, overconfidence, fooling around or not adhering to instructions can lead to accidents.
- Spills and leaks can be dangerous if not wiped up.
- Vapours may build up where there is no proper ventilation.
- The exposure of some chemicals to heat or sunlight can cause explosion, fire and poisonous reactions.
- Contact between a chemical and the wrong material can cause harmful reactions.
- Neglect or failure to dispose safely of certain old chemicals is dangerous, as chemical changes can happen over time.

There are four main types of chemicals. You need to know what they are and how to guard against their hazards.

Safety tip

More haste, less speed! Try not to rush, that's when accidents happen

Toxic agents

Poisons such as hydrogen sulphide and cyanides can cause injury, disease and death. To protect yourself:

- close containers tightly when not in use
- be sure that the work area is well-ventilated
- wear personal protective equipment

- wash hands often
- carry cigarettes in a protective packet
- safely dispose of contaminated clothing
- keep any antidotes handy.

Corrosives

These are irritants such as acids and alkaline, which are especially dangerous to the eyes and respiratory tract. To protect yourself:

- wear personal protective equipment – goggles, breathing devices, protective gloves
- make sure ventilation is good
- run for the nearest water if corrosives come into contact with you; use a safety shower if one is available
- if your eyes are affected, flush with water for 20 minutes and get medical aid. Many first-aid kits contain eye-irrigation kits.

Flammables

These are liquids and gases that burn readily such as methanol, ethanol, ether, LPG and petrol.

To protect yourself:

- make sure no flames, sparks, or cigarette lighters are allowed near flammables
- keep only a small quantity of flammables in the work area
- store and dispose of flammables safely.

In an emergency:

- evacuate the area, if possible
- turn off all flames and sparking equipment
- clean up flammables and ventilate the area thoroughly
- call the emergency services.

Reactives

These are substances that can explode, such as nitrogen compounds. To protect yourself:

- Know your chemicals before working with them: read about them and test them for stability.
- Handle reactives with great care – for example certain proprietory cleaners mixed with bleach can give off hydrogen peroxide.
- At the first sign of trouble, close the doors and evacuate the room through doors that do not lead to the danger area.

Remember

Burning requires three things: the right amount of flammable fuel, oxygen, and a spark or other source of ignition

When using chemicals of any sort, you must understand the dangers involved. This information will usually be communicated to you using a risk assessment – a typical example produced by the British Plumbing Employers Council (BPEC) is shown below. The packaging of hazardous substances usually provides essential safety information – you must follow all recommended safety rules and procedures and you must know what to do in an emergency.

The example shows an assessment sheet for using sulphuric acid – a particularly nasty product. You should follow all the guidance detailed on the sheet to use the substance safely, paying particular attention to the control measures you must use.

SULPHURIC ACID (H_2SO_4) LOCATED: DRAINAGE WASTE SOIL-PIPE DISCHARGES MAIN SEWER SYSTEMS	CORROSIVE	HAZARDS FUMES CONTACT

HAZARDOUS OPERATIONS

FUMES – INHALATION OF FUMES WHERE INSUFFICIENT VENTILATION

CONTACT – INGESTION VIA EXPOSED SKIN OR INTO THE EYE

HEALTH RISKS

SULPHURIC ACID IS A STRONGLY CORROSIVE LIQUID WHICH WILL CAUSE SEVERE BURNS TO SKIN AND COULD ON ENTRY TO THE EYE CAUSE PERMANENT DAMAGE.

RESPIRATORY IRRITANT CAUSING SHORTNESS OF BREATH AND AS A RESULT OF LONG-TERM CONTACT, DELAYED PULMONARY OEDEMA

SPILLAGES COVER WITH SODA ASH AND LIFT. RINSE WITH COPIOUS AMOUNTS OF WATER	FIRST AID SEEK MEDICAL ADVICE IF INGESTED FROM SKIN CONTACT OR ENTRY INTO THE EYE. WASH WITH PLENTY OF WATER

FIRE   **STORAGE**

N/A   NEVER ADD WATER TO SULPHURIC ACID. STORE IN A SECURE LOCATION

RISK FACTORS DURATION	PROCESS	OUTDOORS
0–1 MINUTE	UNBLOCKING DRAINS	CRITICAL
1–10 MINUTES	UNBLOCKING DRAINS	CRITICAL
10 + MINUTES	UNBLOCKING DRAINS	CRITICAL

CONTROL MEASURES

CRITICAL RISKS   EYE PROTECTORS, IMPERVIOUS GLOVES, IMPERVIOUS OVERALLS AND CARTRIDGE-TYPE RESPIRATOR CAPABLE OF REMOVING FUMES

Figure 2.15 Risk assessment sheet for working with sulphuric acid

Asbestos dust safety

Asbestos is a particularly nasty substance and you must take great care if you encounter it on site. Asbestos is the cause of a sizeable number of deaths in the plumbing and heating industry – so be warned!

Asbestos is a mineral found in many rock formations. When separated from rock it becomes a fluffy, fibrous material that has many uses. About two-thirds of all the asbestos produced is used in the construction industry. It is used in cement production, roofing, plastics, insulation, floor and ceiling tiling, and fire-resistant boards for doors and partitions.

Asbestos becomes a health hazard if the fibres are inhaled through the nose or mouth; they cannot be absorbed through your skin. Some of the fine rod-like fibres may work their way into the lung tissue and remain embedded for life. This will become a constant source of irritation and can cause chronic illness.

Breathing in asbestos fibres can lead to you developing one of three fatal diseases:

- asbestosis – a scarring of the lung leading to shortness of breath.
- lung cancer
- mesothelioma – a cancer of the lining around the lungs and stomach.

There are three main types of asbestos:

- chrysotile – white (accounts for about 90 per cent of asbestos in use)
- amosite – brown
- crocidolite – blue.

They cannot be identified by colour alone; laboratory analysis is required.

There is no cure for asbestos-related diseases.

It is a legal requirement to know the hazards of your job; any asbestos on site should have been identified before you start work.

You may come across asbestos in the following:

- asbestos cement flue pipes
- asbestos gutter systems
- asbestos soil pipes
- artex ceilings containing asbestos.

If so, you need to work to the following guidelines:

- Where possible, keep asbestos materials damp when working on them.
- Never use power tools on asbestos materials as they create dust; use hand tools instead.
- Wear and maintain any personal protective equipment provided – and use it properly.

Remember

Identifying any asbestos on site should have been part of the risk assessment. Risk assessments must be documented by law

- Practice good housekeeping – only use special vacuums and dust-collecting equipment.
- Keep asbestos waste in suitably sealed containers and ensure that they are labelled to show that they contain asbestos.
- Report any hazardous occurrences, e.g. unusually high dust levels, to your supervisor.

The removal of asbestos, or working in an area of asbestos other than those in the examples listed above, must only be carried out by specialist contractors. Consider your health at all times.

Safe use of LPG, and lead working

At the end of this section you should be able to:

- explain what is meant by LPG
- describe some of the protective measures that can be used to prevent accidents involving LPG
- explain the effects of lead and the potential hazards involved when working with lead.

LPG

LPG stands for Liquefied Petroleum Gas and is any concentration of commercial propane or butane used in heating processes. Even small quantities of LPG, when mixed with air, create an explosive mixture. If 1 litre of LPG is boiled or evaporated it becomes 250 litres of gas (LPG is a gas when its temperature rises above –42°C). This is enough to make an explosive mixture in a large shed, room, store or office.

LPG is widely used in construction and building work as a fuel for burners, heaters and gas torches. The liquid, which comes in cylinders and containers, is highly flammable and needs careful handling and storage.

Whether in use or stored, cylinders must be kept upright in open-air secure compounds. When not in use cylinder valves should be closed and protective dust caps should be in place. When handling cylinders do not drop them or allow them to come into heavy contact with other cylinders.

When using a cylinder with an appliance, including a blowtorch, ensure it is connected properly, in accordance with the instructions you have been given. The amount of draw-off from LPG cylinders MUST be limited because, if this rate is too high, the control valve may freeze; this could potentially lead to leakage.

It is essential to make sure the gas does not leak. LPG is heavier than air; if it leaks, it will not disperse into the air, but will sink to the lowest point and form an explosive concentration, which could be ignited by a spark. Leakages are especially dangerous in trenches and excavations because the gas cannot flow out of these areas.

Safety tip

If you suspect the presence of asbestos seek advice from those in charge before you start work, and avoid exposure; always follow recommended controls

Remember

It is your Health and Safety you are looking after – and nothing is more important

If you are to assemble blowtorch equipment or if you suspect a leak from gas heating equipment, the equipment should be checked using a soap solution. If there is a leak, the soap solution will bubble and identify the leak source for you to repair.

Transportation of LPG

The transportation of LPG is covered by the Packaged Goods Regulations. When using LPG, everyone should understand the procedures to be adopted in case of an emergency. The appropriate fire extinguishers (dry powder) should always be available. When LPG is being transported in an enclosed van, only small quantities should be carried, and two dry-powder fire extinguishers must be available at all times. (It is also preferable that the vehicle is ventilated and there must be an LPG hazard warning sign displayed on the vehicle.) If significant quantities are being transported, an open vehicle should be used, and the driver will need to undertake special training in order to receive a TREMCARD – a requirement for moving significant quantities of LPG.

Lead safety

The vast majority of pipework and tube that you come into contact with will be copper, steel or plastic, but during your time as a plumber you will probably come across lead. Most dwellings are now free of lead pipework, but lead is still used on roof weatherings, and plumbers are still considered as lead-work specialists. Therefore, it is necessary for you to be aware of the hazards involved when working with lead. Extended unprotected exposure to lead working can lead to chronic illness.

The Control of Lead at Work Regulations (2002) set out measures similar to those concerning safe work with asbestos. Working with lead takes two forms:

* handling tasks – controlled by simple protective methods
* heating-process tasks – which require much more stringent control methods.

Handling lead materials can effectively be controlled by blocking the likelihood of lead entering the body through the skin by the use of PPE such as barrier cream, gloves, eye protection and overalls.

When lead is heated, as it would be during lead-welding processes, poisonous fumes are given off. It is therefore necessary to take precautions to guard against breathing these fumes in. This is usually achieved through using Respiratory Protective Equipment or, preferably, Local Extraction Ventilation systems.

Harmful lead particles can be inhaled, ingested (swallowed) or absorbed through the skin. To ensure you are kept safe from the adverse effects of lead, your employer should provide areas to eat, drink and smoke well away from the worksite.

And remember: wash your hands, arms and face after a lead-working session. Lead safety is revisited in Chapter 12, Sheet lead working.

FAQs

Why do I need to know all that stuff about COSHH and toxic substances? I don't even work with any of them.

It's a legal requirement to know the hazards of your job. Any dangerous substance on site should have been identified before you start work; this is called a risk assessment. This is not just desirable, it's the LAW.

Knowledge check

1 When working in domestic premises, who is responsible for the Health and Safety of the occupant?

2 Who ensures that the Fire Precautions Act is enforced in normal premises?

3 What type of safety sign shows a black symbol on a yellow background bordered by a black triangle?

4 What's the correct angle that a ladder should be placed at?

5 When dealing with a fire in an open oil-storage tank, what is the most suitable type of extinguisher?

6 Factories need a fire certificate when employing more than how many staff?

7 List five potentially hazardous operations that you may be involved in during your work as a plumber.

8 What type of respiratory protection would be most suitable while using an electric circular saw to remove softwood floorboards?

9 Does a risk assessment look at the level of risk to operatives working on a site or assess the value of property for insurance purposes?

10 When a ladder provides access to a working platform, how much of the ladder should there be above the access point?

11 What should you do if you suspect the presence of asbestos?

12 Is this statement true or false? An independent scaffold uses putlogs to fix it to the building.

13 What are the three elements that all fires require?

14 State three precautions a plumber should take when using a blowtorch.

chapter 3

Effective working relationships

OVERVIEW

As well as excellent technical skills – and you should always aim to be a top-class plumber – you need different skills when you have to deal with people. These are known as 'personal skills'. They are taught in the Key Skills part of the plumbing certificate programme.

Plumbing is part of the construction industry. In this chapter, we will look at the construction industry and some of the people who work in it, so that you will know what they do.

In this section we will look at:

- **The Construction Industry**
  - company structure

- **The construction team**
  - non-craft jobs
  - other occupations
  - how plumbing relates to the construction industry

The Construction Industry

At the end of this section you should be able to:

- explain how companies are structured
- describe the general structure of the construction industry
- explain how the plumbing industry fits in with the construction industry.

During your time as a plumber you are likely to meet a wide range of people. It is important that you know how to conduct yourself in all sorts of situations and that you can communicate clearly with other people.

Ninety-five per cent of employers in the construction industry employ between 1 and 13 people, 4 per cent 14–299, and 1 per cent 300+.

Small companies will tend to concentrate on one-off house-building, extensions, maintenance and repair work. They may also be subcontracted by larger construction companies, which usually means providing a specific service.

Larger contractors will develop construction projects from start to finish. These may be private or public housing developments or major construction projects such as the new Wembley Stadium. To do this type of work, companies need to employ a range of managers, office-based staff, designers etc. as well as the staff who will run the job on site.

Larger contractors are also involved in facilities management. This is the ongoing management of all aspects of running a large building (e.g. a hospital) or an estate of houses (local authority, housing association or private). We will look at what a facilities manager does later.

Many of the larger contractors carry out work on behalf of insurance companies. At one time, individual householders arranged for a contractor to effect repairs; now, most insurers place this work directly in the hands of the contractor, who handles everything from receiving the emergency call or claim to doing the job.

Company structure

You will appreciate by now that the construction industry is diverse both in terms of company size and the range of work undertaken. Whatever the size of company, it has to have a 'legal status'. This means it will be one of the following:

- sole trader
- partnership
- limited company (private)
- public limited company (plc).

Sole trader

One person will own the business and look after everything: quoting for work, arranging the jobs, organising materials, doing the job and then sorting out all the paperwork, invoices, tax and so on.

Did you know?

The Construction Industry Training Board (CITB) estimates that there are approximately 165,500 companies working in the industry

Did you know?

Roughly 80% of plumbing businesses employ between 1 and 4 people

Despite the name, a sole trader may employ one or two people, and maybe also an apprentice

A sole trader

Advantages	Disadvantages
Owner entitled to all profits	Owner liable for all business's debts
Minimal paperwork	Owner risks personal assets if business fails
Do not need to register with Companies House	No job security, pension, annual leave or sick leave benefits
Freedom to make own decisions	Owner entirely responsible for business

Figure 3.1 Advantages and disadvantages of being a sole trader

Partnerships

This is also a popular format for small construction and plumbing businesses. Often a partnership develops from trades people working together and deciding to go into business; a joiner and bricklayer setting up a partnership is a typical example, although partners do not have to have separate trades, and often two, or even more, plumbers set up partnerships.

Construction involves collaboration between trades

Advantages	Disadvantages
Partners bring additional expertise to the business	Partners are liable for all the business's debts
Usually all partners will invest some money in the business	Partners risk personal assets if business fails
Do not need to register with Companies House	If one partner wants to leave they have to be bought out
Responsibility for business is shared	Profits are shared

Figure 3.2 Advantages and disadvantages of partnerships

Limited companies (private)

Most businesses, if they begin to expand, tend to become limited companies. This means the company is registered with Companies House and has a legal identity. Its owners have shares in the company.

Advantages	Disadvantages
The owners are not personally liable for business debts, only for an amount up to their investment in the company	Limited companies have to comply with more regulations
There are tax advantages	Companies have to register with Companies House
Money can be raised for investment by selling shares	Lots of paperwork and accurate accounts and records must be kept

Figure 3.3 Advantages and disadvantages of private limited companies

Public limited companies (plcs)

These tend to be the largest companies. They are allowed to trade their shares on the stock market provided they have a share capital of at least £50,000. Some of the major contractors in the construction industry are plcs.

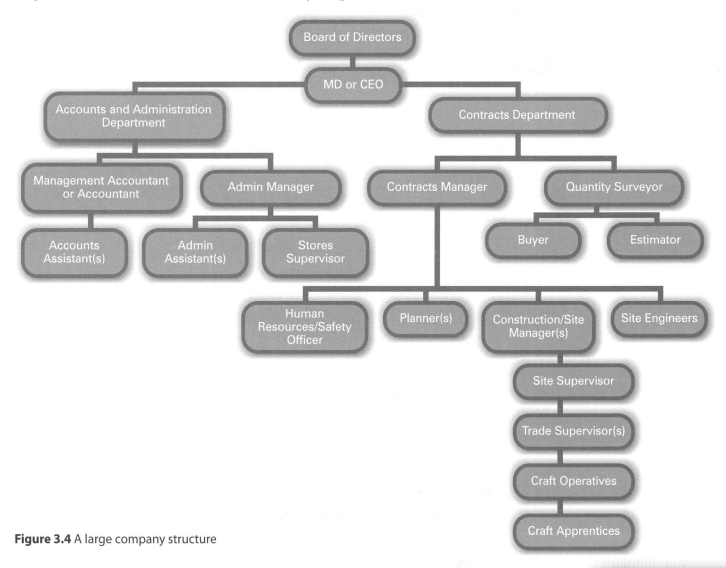

Figure 3.4 A large company structure

The structure of sole trader and partnership businesses is relatively simple. Limited companies and PLCs usually have a Board of Directors, and, depending on their size, will have a Chairman, a Managing Director (MD) or Chief Executive Officer (CEO), who are usually responsible for overseeing the management of the business and planning the business strategy. Companies' structures vary but the illustration on the previous page shows what the structure might look like for a company employing around 80–100 craft operatives.

The construction team

At the end of this section you should be able to:

- list the various members of the construction team
- explain briefly what each team member does.

The main non-craft job and department roles in the construction industry.

Job	Main responsibilities	Other aspects
Contracts manager	• Responsible for running several contracts • Works closely with construction management team • Link between the other sections of the business and the MD/CEO • Makes sure the job is running to cost and programme	• Office-based • Some site visits
Construction manager (Site manager, site agent or building manager)	• Responsible for running construction site or a section of a large project • Develops strategy for the project • Plans ahead to solve problems before they happen • Makes sure site and construction processes are carried out safely • Communicates with clients to report progress and seek further information • Motivates workforce	• On larger contracts, the construction manager could have the support of supervisory staff
Site supervisor (Foreman)	• Responsible for the trade supervisor and trade operatives • Supervises the day-to-day running of the job	• Reports to the construction manager
Trade supervisor (Charge hand)	• Supervises operatives from a specific trade (e.g. bricklayers)	• Reports to the site supervisor • This role is usually only needed on very large jobs

Figure 3.5 Non-craft jobs in construction (*continues overleaf*)

Job	Main responsibilities	Other aspects
Planner	• Ensures project completed on time and within budget • Reschedules projects as necessary • Works closely with estimators to establish working methods and costs • Plans the most effective use of time, people, plant and equipment • Schedules events in a logical sequence • Visits sites to monitor progress	• Works with construction manager
Site engineer	• Ensures the technical aspects of the construction projects are correct • Sets out the site so that things are in the right place • Interprets original plans, documents and drawings • Liaises with workforce and subcontractors on practical matters • Checks quality • Refers queries to the relevant people • Provides 'as built' details	• Key role in ensuring things are built correctly and to the right quality • May supervise parts of the construction
Quantity surveyor – contractor (Commercial manager, Cost consultant)	• Advises on and monitors the costs of a project • Allocates cost effectively work to specialist subcontractors • Manages costs • Negotiates with client's quantity surveyor on payments and final account • Arranges payments to subcontractors	
Buyer (Procurement officer)	• Identifies suppliers of materials • Obtains quotations from suppliers • Purchases all the construction materials needed for a job • Negotiates on prices and delivery • Resolves quality or delivery problems	• Liaises with other members of the construction team
Estimator	• Calculates how much a project will cost including plant, materials and labour • Identifies the most cost-effective construction methods • Establishes costs for labour, plant, equipment and materials • Calculates cash flows and margins • Seeks clarification on contract issues affecting costs	• This information is used as the basis of the tender the contractor submits to a client prior to getting the contract

Figure 3.5 Non-craft jobs in construction (cont'd)

(continues overleaf)

Job	Main responsibilities	Other aspects
Accounts department	• Invoices for work carried out • Pays staff, suppliers and subcontractors • Produces financial reports and budget forecasts • Deals with Inland Revenue and Customs and Excise (VAT)	• Works with Construction manager
Administration department	• Works closely with accounts to keep financial records • Deals with customers and other enquiries • Records and files work records, time sheets etc	• Most systems are electronic, but companies often keep hard-copy records
Human resources department	• Oversees employees' training and development needs • Assists with the recruitment and selection of new staff • Deals with industrial relations matters.	• Some businesses will also employ a human resources officer, who looks after Health and Safety matters. (Otherwise, Health and Safety would be the responsibility of the Construction or Contracts manager)

Figure 3.5 Non-craft jobs in construction (cont'd)

Any construction job is likely to involve several craft trades. The following table outlines the major trades and the tasks they are generally responsible for.

Craft or trade	Main aspects of job	Typical job details
Carpenter and joiner	Positions and fixes timber materials and components from roofs and floors to doors, kitchens and stairs.	Work on site is usually divided into two phases or 'fixes'. The **first fix** is usually completed before the building has been made watertight and includes fixing floor joists, boards and sheets, stud partitions, and roof trusses and timbers. The **second fix** takes place when the building is watertight. Staircases, doors, kitchen units, architraves and skirtings are common second-fix items.
Roof slater and tiler	Creates a waterproof covering for a building by applying individual slates and tiles to a basic framework.	Roofing felt is laid over the roof timbers and tacked down. Timber battens are then fixed horizontally at centres to suit the roof dimensions and type of slate or tile. The roof is loaded with slates or tiles and then they are laid. Tiles have to be cut to fit at valleys, hips and gable ends. Ridge tiles are bedded on mortar. On re-roofing or maintenance work, tiles and slates are removed and timber checked and replaced where necessary. Reclaimed tiles and slates must be checked and sorted before reuse.

Figure 3.6 Main tasks of construction trades (*continues overleaf*)

Craft or trade	Main aspects of job	Typical job details
Built-up felt roofer	Built-up felt roofing is used mainly on flat roofs and sometimes on sloping roofs and vertical surfaces.	It is called 'built-up' because it involves putting layers of felt on top of each other using bitumen to form a waterproof surface.
Bricklayer	Uses bricks and blocks to construct interior and exterior building walls as well as tunnel linings, archways and ornamental brickwork.	Working on new buildings, extensions, maintenance and restoration of older buildings. Building foundations, bringing brickwork and blockwork up to damp-proof course level. Working at height from trestles, hop-ups and scaffolds. Construction of drainage and concrete work.
Construction operative (labourer)	Provides varied support on site in a number of areas.	Skills include concreting, form working, steel fixing, kerb laying and drainage work. Some operatives specialise in areas such as spraying or repairing concrete.
Plasterer	Applies plaster to interior walls so that decoration can be applied.	Renders external walls and sometimes lays floor screeds. Some specialise in fibrous plasterwork. Others work in maintenance, conservation and restoration of existing buildings.
Painter and decorator	Applies paint, wall coverings and other materials to the inside and outside of buildings.	Work can include maintenance, conservation and restoration of existing buildings as well as on new-build, commercial and industrial sites. Wall coverings need to be accurately measured, cut and hung using the correct adhesive. Some decorators help clients with interior design.
Wall and floor tiler	Fixes tiles on exterior and interior walls, as well as on other surfaces such as floors and swimming pools.	On maintenance and restoration work, most surfaces need to be repaired. Tiles are normally laid on adhesive, cut along the edges and around obstacles, and finally grouted up. Setting out the tiles is critical when intricate and complex patterns are being produced. Tiles come in a variety of shapes, sizes, textures and colours and materials (e.g. ceramic, stone, terracotta).
Floor layer	Prepares and levels floors for new and old buildings. The most common materials are carpet, cork, plastic and timber.	Prepares floor measurements and plans laying to avoid waste. Different techniques are used for different materials. These include gripper fixings, glues, adhesives and secret nailing.
Plant operator	Uses cranes and other plant for moving and transporting construction materials.	Cranes can be mobile, track-mounted or tower cranes. Transporting plant includes excavators, specialised earth-moving equipment, fork-lift trucks and power-access equipment.

Figure 3.6 Main tasks of construction trades (*cont'd*)

(*continues overleaf*)

Craft or trade	Main aspects of job	Typical job details
Scaffolder	Erects scaffolding and working platforms for construction workers, fixes edge protection. Scaffolders need to be competent persons who have been properly certificated to do their work.	Scaffold needs a firm foundation and careful planning. It can be made from traditional tubes and fittings or may be a purpose-made system. Working platforms are made using scaffold boards. Access is usually provided with ladders. Hoists, special loading platforms, safety nets and guard rails are all features of scaffolding erection. Most scaffolds need to be lifted and modified as construction work progresses.

Figure 3.6 Main tasks of construction trades (*cont'd*)

Other occupations

There are a number of other occupations that play an important role in the overall construction process.

Architect

Architects plan and design buildings. The range of their work varies widely and can include the design and procurement (buying) of new buildings, alteration and refurbishment of existing buildings and conservation work. An architect's work includes:

- meeting and negotiating with clients
- creating design solutions
- preparing detailed drawings and specifications
- obtaining planning permission and preparing legal documents
- choosing building materials
- planning and sometimes managing the building process
- liaising with the construction team
- inspecting work on site
- advising the client on their choice of a contractor.

Project manager/clerk of works (client)

The project manager or clerk of works takes overall responsibility for the planning, management, coordination and financial control of a construction project. They work for architects, clients such as local authorities, or as a consultant.

A project manager will ensure the client's requirements are met and that the project is completed on time and within budget. Depending on the project, their responsibilities can start at the design stage and continue through to completion and handover to the client.

A project manager's/clerk of work's job includes:

- representing the client's interests
- providing independent advice on the management of projects
- organising the various professional people working on the project
- making sure that all the aims of the project are met
- ensuring quality standards are met
- keeping track of progress
- accounting, costing and billing.

Structural engineer

Structural engineers are involved in the structural design of buildings and structures such as bridges and viaducts. The primary role of the structural engineer is to ensure that these structures function safely. They can also be involved in the assessment of existing structures, perhaps for insurance claims, to advise on repair work or to analyse the viability of alterations and adaptations.

Building surveyor

Building surveyors are involved in the maintenance, alteration, repair, refurbishment and restoration of existing buildings. A building surveyor's work includes:

- organising and carrying out structural surveys
- legal work including negotiating with local authorities
- preparing plans and specifications
- advising people on building matters such as conservation and insulation.

Building-control officer

Working for local authorities, building-control officers ensure that buildings conform to regulations on public health, safety, conservation and access for the disabled. The job involves the inspection of plans and of work-in-progress at various stages relating to Building Regulations. A building-control officer's work includes:

- checking plans and keeping records of how each project is progressing
- carrying out inspections of foundations, drainage and other major building elements
- issuing a completion certificate when projects are finished
- carrying out surveys of potentially dangerous buildings
- meeting with architects and engineers at the design stage
- using technical knowledge on site.

Facilities manager

A facilities manager ensures that buildings continue to function once they are occupied. He or she is responsible for maintaining the building and carrying out any changes needed to ensure the building continues to fulfil the needs of the organisations using it. A facilities manager's work includes:

- planning how the inside of the building should be organised for the people occupying it

- managing renovation works

- managing routine maintenance

- managing the installation and maintenance of computer and office equipment

- managing the building's security

- managing the cleaning and general upkeep of the building

- negotiating with contractors and service suppliers.

Building-services engineer

Building services include: water, heating, lighting, electrical, gas, communications and other mechanical services such as lifts and escalators. Building-services engineering involves designing, installing and maintaining these services in domestic, public, commercial and industrial buildings. A building-services engineer's work includes:

- designing the services, mostly using computer-aided design packages

- planning, installing, maintaining and repairing services

- making detailed calculations and drawings.

Most building-services engineers work for manufacturers, large construction companies, engineering consultants, architects' practices or local authorities.

Their role often involves working with other professionals as part of a team on the design of buildings, for example with architects, structural engineers and contractors.

When working for a consultant, this job is mainly office-based at the design stage. Once construction starts, there will be site visits to liaise with the contractors installing the services. When working for a contractor, the building services engineer may oversee the job and even manage the workforce and is therefore likely to be site-based. When working for a services supplier, the role will require being involved in design, manufacture and installation and may involve spending a lot of time travelling between the office and various sites.

How plumbing relates to the construction industry

Plumbing businesses often work for construction companies on a subcontracting basis. This means that the main contractor is a construction company, and that they contract directly with the customer or client. The main contractor then sublets a part of the contract, for example the plumbing work, to a plumbing contractor (subcontractor).

The plumbing subcontract deals directly with the construction company and not the client. The tender price, contract details and payment of work are between the plumber and the construction company.

Plumbing is not the only part of the contract that is subcontracted out. Other areas include:

- electrical
- heating and ventilating – domestic, industrial and commercial
- refrigeration and air-conditioning
- service and maintenance
- ductwork
- gas – domestic, industrial and commercial.

FAQ

Why is communication so important to me? I am only going to be a plumber.

Communication is vital for the passing and receiving of information between one or more people, which achieves understanding and gets the required result. This can be done verbally, in written form or through body language. There will be many examples when you need to communicate clearly in order to get a job done as required by the customer.

Knowledge check

1 The majority of the construction industry is made up of businesses employing how many people?

2 What is the ongoing management of all aspects of a large building or estate called?

3 You've set up your own business and don't want to be liable for its debts should it fail. What category of business would you choose?

4 List three tasks that Construction Managers cover as part of their job.

5 Quantity Surveyors ensure the technical aspects of the construction projects are correct. TRUE or FALSE?

chapter 4

Key plumbing principles

OVERVIEW

Key plumbing principles covers the science of plumbing and acts as an introduction to the basic materials, theories and concepts that you will encounter and work with on a daily basis. This chapter includes:

- **Plumbing materials**
  - Standards and Regulations
  - materials and their properties

- **Plumbing science**
  - units and measurement
  - properties of water and heat
  - capillary action and siphonage
  - principles of electricity
  - electricity generation supply and controls

- **Hand and power tools**

Plumbing materials

At the end of this section you should be able to:

- name and describe the properties of the materials used for plumbing systems
- specify the correct type of materials for particular applications and methods of protecting them from damage.

A plumber's job involves the installation, maintenance and service of the following systems:

- hot and cold water
- central heating
- above-ground drainage
- sheet lead weathering
- rainwater
- below-ground drainage (connection only).

The plumber will work with a range of materials used for system pipework, fittings and components, including:

- copper
- steel
- lead
- brass
- plastics
- ceramics.

It is important that you have a good understanding of the **properties** of these materials and their suitability for the type of work for which they will be used.

Standards and Regulations in the use of materials

Materials used in everyday plumbing work are required to meet minimum standards of performance. It is also important that there is standardisation for the sizes and dimensions of fittings and components. Imagine what it would be like if you could buy a range of 15 mm fittings that all had a slightly different internal diameter to the external diameter of a 15 mm pipe!

British Standards

You may be familiar with the BSI kitemark. This symbol can be used only on materials and equipment that meet the standards of the British Standards Institution (BSI), the organisation for standards in the UK.

The BSI ensures standards of quality and also sets standard dimensions for such items as pipes and fittings.

Figure 4.1 The BSI kitemark

British Standards all start with the letters BS followed by the number of the standard. For example BS6700 is one of the main standards for the plumbing industry. It's a specification for design, installation, testing and maintenance of services supplying water for domestic use within buildings and their **curtilages**.

We will refer to BS6700 often throughout this book.

Definition

Properties – physical and working attributes of materials

Did you know?

The BSI is an independent organisation and was set up in 1901 under a Royal Charter

Definition

Curtilage – an area attached to a dwelling house and forming one enclosure within it

Codes of Practice

Codes of Practice (CoPs) make recommendations related to good practice. While they are not legal documents, CoPs are widely used by specifiers such as clients and architects.

The Water Supply (Water Fittings) Regulations 1999

These Regulations apply only to England and Wales, and have replaced Water Bye-laws. The Regulations are national Regulations made by the Government's Department of the Environment, Food and Rural Affairs (DEFRA), while the Bye-laws that they replace were made locally and applied in that area, for example, Yorkshire Water, Thames Water, Northwest Water etc.

We will discuss these Regulations in more detail later. We mention them here because they cover materials and substances in contact with water.

European Standards

These start with the letters EN followed by the standard number, in the same way as British Standards. Where a product is certified to an EN standard it means the manufacturer will have taken the product through a series of tests that are regularly checked under EC Quality Control Schemes.

International Standards

International Standards start with ISO followed by the number, for example ISO9000, which refers to a standard for quality.

Water Regulations Advisory Scheme (WRAS)

Formally known as the Water Bye-laws Scheme, the WRAS has been carrying out fitting testing for many years and will continue to advise on Water Regulations in the future. As part of their work they produce a *Fittings and Material Directory*, which lists all approved fittings and is an important guide for all who aim to comply with or enforce The Water Regulations. Products approved by WRAS carry the symbols shown in the illustrations.

Figure 4.2 WRAS approved product symbol

Materials and their properties

The properties of materials relate to things such as how strong they are, how well they conduct heat or electricity or how flexible they are. We will examine the different properties materials have before looking at common plumbing materials in more detail.

Materials are classified according to a variety of properties and characteristics. Properties can be measured by the way materials react to a variety of influences.

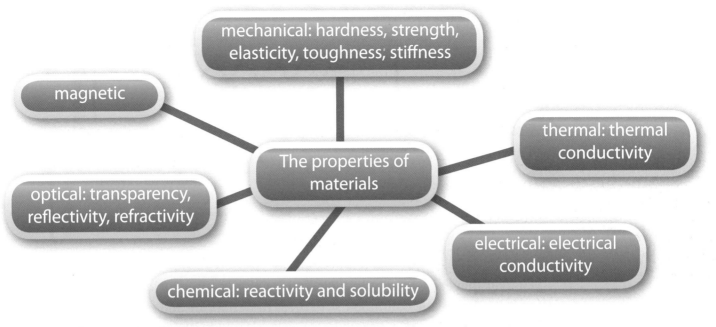

Figure 4.3 The properties of materials

Hardness

Hardness is a measure of a material's resistance to permanent or plastic deformation by scratching or indentation. It is an important property in materials that have to resist wear or abrasion – moving parts in machinery, for example – and frequently needs to be considered along with the strength of materials. Hardness is measured on a scale of 1 to 10 based on the hardness of 10 naturally occurring minerals.

1	Talc
2	Gypsum
3	Calcite
4	Fluorite
5	Apatite
6	Feldspar
7	Quartz
8	Topaz
9	Corundum
10	Diamond

Figure 4.4 Mohs' scale

Strength

The strength of a material is the extent to which it can withstand an applied force or load (stress) without breaking. The load is expressed in terms of force per unit area (newtons per square metre, N/m^2), and can be in the form of a:

- compression force, as applied to the piers of a bridge or a roof support
- tensile or stretching force, as applied to a guitar string, tow rope or crane cable
- shear force as applied by scissors or when materials are torn (see figure 4.5).

Materials are therefore described as having compressive, tensile or shear strength.

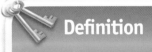

Definition

Thermal conductivity – how well or poorly a material will conduct heat

Electrical conductivity – how well or poorly a material will conduct electricity

Did you know?

The hardness scale is called Mohs' scale. It is named after the German mineralogist Frederich Mohs (1773–1839), who developed the scale in 1812

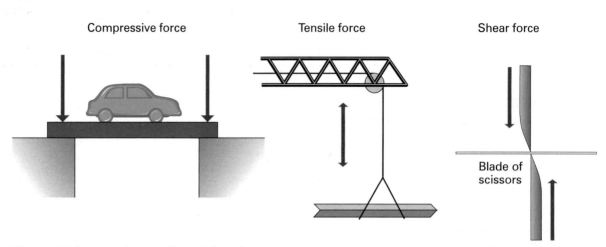

Figure 4.5 Compressive, tensile and shear forces

Materials that can withstand a high compression loading include cast iron, stone and brick, hence the common use of these materials for building purposes. However, they are brittle and will break if subjected to high tension. If a building is to be designed to resist tensile strain – in an earthquake-prone area, for example – steel, which has high tensile strength, would be a more suitable building material.

Elasticity and deformation

Almost all materials will stretch to some extent when a tensile force is applied to them. The increase in length on loading, compared to the original length of the material, is known as strain. As loading continues, a point is reached when the material will no longer return to its original shape and size on removal of the load, and permanent **deformation** occurs; the material is said to have exceeded its elastic limit or yield stress, and is suffering plastic deformation – it has been stretched irreversibly. Eventually, at maximum stress, the material reaches its breaking point – its ultimate tensile strength – and failure or fracture rapidly follows. This sequence is illustrated for a variety of materials below.

Find out

What can we learn about the materials from the stress/strain diagram?

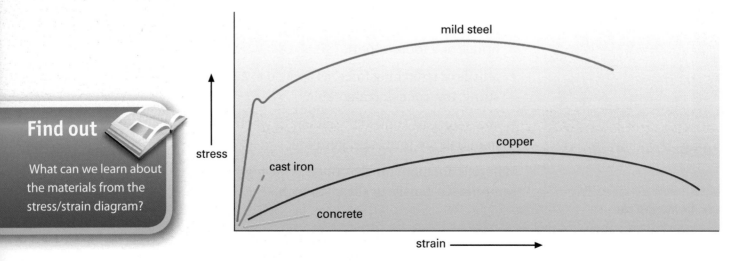

Figure 4.6 The effects of stress and strain

- Mild steel has little elasticity but has the highest yield stress of all the samples; it is fairly ductile, i.e. it has a large range over which it can sustain plastic deformation, and it has the highest ultimate tensile strength.

- Cast iron is brittle – it has the least elasticity of the four samples, and no ability to sustain plastic deformation, although its tensile strength is higher than that of concrete.

- Copper has little elasticity but is the most ductile of the four samples. It has an ultimate tensile strength less than half that of mild steel.

- Concrete has little elasticity, and the lowest tensile strength of the four samples.

Elasticity	The ability of a material to resume its normal shape after being stretched or squeezed
Plasticity	The exact opposite of elasticity – a material which does not return to its original shape when deformed
Ductility	The ability of a material to withstand distortion without fracture, such as metal that can be drawn out into a fine wire (plastic deformation under tension)
Durability	A material's ability to resist wear and tear
Fusibility	The melting point of a material, i.e. when a solid changes to a liquid
Malleability	The ability of a metal to be worked without fracture
Temper	The degree of hardness in a metal
Tenacity	A material's ability to resist being pulled apart
Thermal Expansion	The amount a material expands when heated

Pipework materials

There is no perfect pipework material that is suitable for all applications; different materials perform better in relation to different factors and conditions such as pressure, type of water, cost, bending and jointing method, corrosion resistance, expansion and appearance. There are two basic types of pipework material:

- metal

- plastic.

Metal

Metals rarely occur in their pure form. More often they occur as ores, which are compounds of the metal and have unwanted impurities. To produce the required metal, a process of smelting is necessary. Metals commonly used in the manufacturing industry include iron, copper, lead, tin, zinc and aluminium.

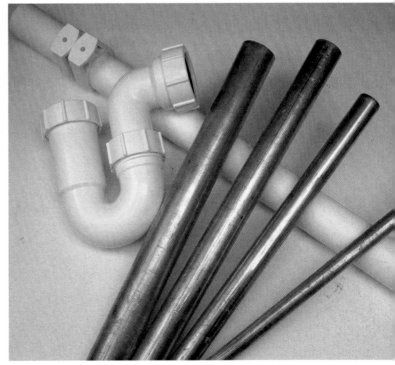

Copper and plastic pipes

The most common method of producing metals is by removing the oxygen from the ore by a process known as reduction.

The industrial production of iron and steel

Iron ore (haematite – iron oxide) is loaded into a blast furnace along with coke and limestone.

Hot air is blasted into the base of the furnace and carbon from the coke reacts with oxygen from the air to form carbon monoxide.

Carbon monoxide reacts with oxygen from the haematite (iron oxide) to form carbon dioxide and iron.

Limestone combines with impurities in the ore (mainly silicates) to form slag.

The molten iron is tapped from the base of the furnace and solidifies into billets known as 'pigs' – hence the term 'pig iron'.

At this point the iron is impure. To form steel, which is an alloy of iron and carbon, it is necessary to reheat the iron to drive off the impurities, and then to add up to 1.5 per cent of carbon. Other metals can give the steel particular properties – the addition of chromium will produce stainless steel, for example. Alloys can be produced either by mixing different metals or by mixing metals with non-metallic elements, such as carbon.

Metals commonly used in the plumbing industry

Alloys

An alloy is a type of metal made from two or more other metals.

Some commonly used alloys		
Brass	Copper and zinc	Used for electrical contacts and corrosion-resistant fixings (screws, bolts etc.) and pipe fittings
Bronze	Copper and tin	Used for decorative or artistic purposes and corrosion-resistant pumps
Solder	Lead and tin, tin and copper	Used for electrical connections Used as a jointing material
Duralumin	Aluminium, magnesium, copper and manganese	Used in aircraft production
Gunmetal	Copper, tin and zinc	Used for underground corrosion-resistant fittings

Figure 4.7 Alloys

Copper

Copper tube has been used as a material for pipework for over 100 years. It is a malleable and ductile material which you will use frequently throughout your plumbing career. There are four main types of copper tube used in the plumbing industry:

- R250 half hard (Table X) is the copper tube most commonly used above ground for most plumbing and heating installations. It is fairly rigid and will usually need to be bent using a bending machine

- R290 hard (Table Z) is a more rigid copper tube. Its increased hardness means that the walls of the pipe don't need to be as thick, so the internal diameter of the tube or bore can be wider than that of the R250 type of tube; the tube cannot easily be bent

- R220 soft coils (Table W) is a copper tube used for micro-bore pipework, typically on central heating systems

- R250/220 soft and half-hard straights (Table Y) is a softer copper tube which is most commonly used underground for the supply of water.

The outside diameter is the same for each type of pipe, but there are differences in the internal bore due to the variations in the pipe wall thickness.

Remember

Table X and Table Z types of copper tube are not suitable for underground use

Steel

1 Low carbon steel

Low carbon steel (LCS) or mild steel is an alloy made from iron and carbon. It is frequently used in the plumbing and heating industry and is manufactured to BS1387. The tube comes in three grades of weight: light, medium and heavy. As with copper tube, the outer diameter is similar, but the internal bore and wall thickness varies.

Grade	Wall thickness	Bore	Colour code
Light LCS tube	Thin walls	Larger bore	brown
Medium LCS tube	Medium walls	Medium bore	blue
Heavy LCS tube	Thick walls	Smaller bore	red

Did you know?

Galvanised tubes have an outer and inner layer of zinc, which prevents oxidation or rusting

Light LCS is usually used for conduit. As a plumber, you will come across it on occasion, but you will work far more frequently with medium and heavy LCS. Medium and heavy LCS tubes are used for water-supply services, as they are capable of sustaining the pressures involved. When LCS tube is used for domestic water supplies, it must be galvanised.

2 Stainless steel

Stainless steel is the most recently developed pipe material used for water services. It is a complex alloy made up of a number of elements, as shown in the table.

Element	%
Chromium	18
Nickel	10
Manganese	1.25
Silicon	0.6
Carbon	0.08
Iron	70 (approx)
Sulphur	trace
Phosphorus	trace

Figure 4.8 Composition of stainless steel

Did you know?

The main use of lead in plumbing today is the weathering of buildings

The tube has a shiny appearance due to the chromium and nickel content and is protected from corrosion by a microscopic layer of chromium oxide, which quickly forms around the metal and prevents further oxidisation. This tube is produced with bores of 6mm to 35mm and has an average wall thickness of 0.7 mm. The outside diameters are similar to those of R250 copper tubes.

Stainless steel is commonly used where exposed pipework and sanitary appliances are needed, as it is a very strong metal (much stronger than copper) and is easy to clean. Stainless steel is also commonly used for:

- sink units
- urinal units and supply pipework
- commercial kitchen discharge pipework.

Lead

Lead is a very heavy, valuable metal which requires specialist handling. It is one of the oldest known metals and is highly ductile, malleable and corrosion-resistant. During your career, you will come into contact with lead in sheet form, which is used for weatherings on buildings. It was used in the past for mains, sanitary and rainwater pipework, but this practice stopped in 1986 with the Model Water Bye-laws, due to the possibility of lead poisoning. It has now been superseded by the use of such materials as plastics.

Cast iron

Cast iron is an alloy of iron and is approximately 3 per cent carbon. It is very heavy but quite brittle, although it can stand up to years of wear and tear. It has been used in the plumbing industry for many years for above- and below-ground sanitary pipework. You will probably come into contact with it on older properties and new industrial/commercial properties.

Plastics commonly used in the plumbing industry

Plastics (polymers) are products of the oil industry.

Ethene, a product of crude oil, is a building block of plastics. It is made up of carbon, hydrogen and oxygen atoms. Molecules of ethene (monomers) can link together into long chains (polymers) to make polythene (poly + ethene) when they are heated under pressure with a catalyst. If the ethene monomer is modified by the replacement of one of the hydrogen atoms with another atom or molecule, further monomers result, producing other plastics. This process is called polymerisation.

There are two main categories of plastics used in the plumbing industry:

- thermosetting plastics
- thermoplastics.

Fractionating crude oil

Petroleum and other hydrocarbons are made from crude oil through a process called **fractionation**. Crude oil is heated up and the vapour produced is fed into a large tower called a fractionating column; different products condense out at different temperatures, at different levels in the tower:

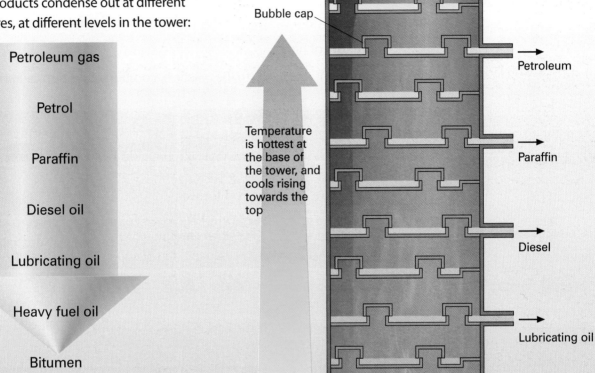

Petroleum gas

Petrol

Paraffin

Diesel oil

Lubricating oil

Heavy fuel oil

Bitumen

Fractionation

The chemical differences between the fractions result from the number of carbon atoms present in the molecules (groups of atoms). Heavy fuel oils have 20–30 carbon atoms per molecule, while petrol has 5–10 carbon atoms.

In a separate process, those molecules with 8–12 carbon atoms per molecule (i.e. between paraffin and petrol) are chemically 'cracked', or split, into smaller units, one of which is ethene.

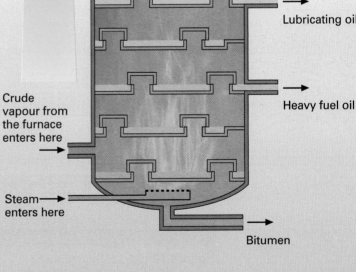

Bubble cap

Petroleum Gas

Petroleum

Paraffin

Diesel

Lubricating oil

Heavy fuel oil

Bitumen

Temperature is hottest at the base of the tower, and cools rising towards the top

Crude vapour from the furnace enters here

Steam enters here

Figure 4.9 Fractionating column

Thermosetting plastics are generally used for mouldings. They soften when first heated which enables them to be moulded, then they set hard and their shape is fixed; it cannot be altered by further heating. WC cisterns can be made of thermosetting plastic.

Thermoplastics can be resoftened by heating them. Most of the pipework materials you will come into contact with fall into this category. The different types of thermoplastics share many of the same characteristics:

- strong resistance to acids and alkalis
- low specific heat (that is, they do not absorb as much heat as metallic materials)
- poor conductors of heat
- affected by sunlight (this leads to the plastic becoming brittle, also called degradation).

Material	Max usage Temp °C	Main plumbing industry purpose
Polythene – low density	80	Flexible pipe material used to channel chemical waste
Polythene – high density	104	More rigid, again used for chemical or laboratory waste
Polypropylene	120	Tough plastic with a relatively high melting temperature, can be used to channel boiling water for short periods of time
Polyvinyl Chloride (PVC)	40–65	One of the most common pipework materials, used for discharge and drainage pipework
Unplasticised Polyvinyl Chloride (UPVC)	65	More rigid than PVC, used for cold water supply pipework
Acrylonitrile Butadiene Styrene	90	Able to withstand higher temperatures than PVC, used for small-diameter waste, discharge and overflow pipework

Figure 4.10 Types of thermoplastic

Other materials relevant to the plumbing industry

Ceramics include those products that are made by baking or firing mixtures of clay, sand and other minerals: bricks, tiles, earthenware, pottery and china. The kiln firing process fuses together the individual ingredients of the product into a tough matrix. The main constituent of all these products is the element silicon (Si) – clay is aluminium silicate; sand is silicon dioxide (silica). Ceramics also includes those products made by 'curing' mixtures of sand, gravel and water with a setting agent (usually cement) to form concrete, or mortar, using a sand, water and cement mixture.

Glass is also produced by the melting together of minerals. The basic ingredients are sand (silicon dioxide), calcium carbonate ($CaCO_3$) and sodium carbonate ($NaCO_3$). The resulting mixture of calcium and sodium silicates cools to form glass. Again, additives can change the character of the product: boron will produce heat-resistant 'Pyrex'– type glass, and lead will produce hard 'crystal' glass.

Types of glass – pyrex, lead crystal and normal

Plumbing science

At the end of this section you should be able to:

- state the standard units of measurement
- describe the principles of mass, weight and density in relation to solids, liquids and gases
- describe pressure, temperature and heat transfer
- explain pH values and the effects of corrosion on plumbing systems and components
- describe capillary action, siphonage and the effects of water flow
- demonstrate understanding of electricity generation, basic circuits and electrical units
- state types of domestic circuits and types of fuses
- state the importance of earth continuity and earthing systems.

Units of measurement

In the UK there are two principle systems of measurement: metric and imperial. The units of measurement you will come across in the plumbing trade will usually be metric (metres, kilograms etc.), although you may hear reference to imperial measurements such as feet, inches, pounds and ounces.

The standard international measurement system, commonly known as SI units, predominantly uses metric measurements. The table overleaf gives an indication of the basic SI units and appropriate metric and imperial equivalents:

Did you know?

SI stands for 'Système Internationale d'Unités'

Description

Attribute	SI unit	Abbreviation		Imperial unit(s)	Imperial abbreviation	Conversion
Length	metre	m		inches, feet, miles	ins, ft.	1 in = 2.54 cm 1 ft = 0.3048 m
Mass	kilogram	kg		ounces, pound	ozs., lb.	1 oz = 28.35 g 1 lb = 0.4536 kg
Time	second	s	60 s = 1 min 60 mins = 1 hour			
Electric current	ampere	A				
Temperature	kelvin	K	degrees Centigrade (°C) K = °C + 273.15	degrees, Fahrenheit	°F	°C = ⁵⁄₉ (°F − 32)
Angle	radian	rad	1 rad = 57 degrees (°) 1 degree = 60 mins (°) 1 min = 60 seconds (°)			
Area	square metres	m²	1 hectare = 10,000 m²	square inches, acre		
Volume	cubic metres	m³	cc = cm³	cubic inches		
Capacity	litre	l	1 ml = 1 cc or cm³	pints, gallons	1 pint = 0.5663 l	
Speed	metres per second	m/s		miles per hour feet per second	mph fps	
Acceleration	metres per second per second	m/s² or ms²				
Force	newtons	N		pounds per square inch	lb/in²	

Figure 4.11 SI Units

FAQ

Why do I need to know all this stuff about science?

This all affects the way in which plumbing systems are chosen, installed and work, so it is essential to have a good awareness of the key science principles.

Mass and weight

In its simplest terms, mass is the amount of matter in an object and is measured in grams (or for larger weights, kilograms). Under normal circumstances and as long as it remains intact, an object should always maintain the same mass. For example, a nail will maintain the same mass whether it is:

- on a workbench
- on the moon.

But will it weigh the same?

The weight of an object is the force exerted by its mass as a result of acceleration due to **gravity**. On earth all objects are being accelerated towards the centre of the planet due to the earth's gravitational pull. The 'pull' exerted by gravity on the mass of an object is known as its weight, which is measured in newtons. A newton is a unit of measurement equivalent to 1 metre per second (m/s) per 1 kg of mass.

Therefore: weight (in newtons) = mass × acceleration due to gravity

To give an example of the difference between mass and weight we need to go to the moon:

On earth

The gravitational pull of the earth is 9.8 m/s^2

Therefore an object with a mass of 1 kg on earth would weigh 9.8 newtons.

On the moon

The gravitational pull of the moon is approximately 1.633 m/s^2

Therefore an object with a mass of 1 kg on the moon would weigh 1.633 newtons.

The mass of the object does not change whether it is on the earth or the moon, but the weight of an object changes considerably because of the reduced gravitational pull of the moon.

Density

Density of solids

Solid materials, which have the same size and shape, can frequently have a completely different mass. This relative lightness or heaviness is referred to as **density**. In practical terms the density of an object or material is a measure of its mass (grams) compared to its volume (cm^3) and can be worked out using the following formula:

$$density = \frac{mass}{volume}$$

The densities of all common materials you'll come into contact with during your plumbing career are known. Lists of these comparative densities can be found in a variety of reference sources.

Find out

Find a list of comparative densities from a reference in your learning centre or local library

Did you know?

Water is at its maximum density at 4°C

Remember

The effect of density changes in relation to temperature changes can be put to good use in making plumbing systems work

Did you know?

Any substance with a relative density of less than 1.0 will float on water; substances with a higher relative density will sink

Density of liquids

Like solids, liquids and gases also have differing densities depending on the number of molecules that are present within a particular volume of the substance. As a plumber it will be important to understand the density of water and that this changes with the water's temperature. Water is less dense when it is heated:

- 1 m³ of water at 4°C has a mass of 1000 kg
- 1 m³ of water at 82°C has a mass of 967 kg

This is because heat energy excites the molecules so that they move further apart and the water becomes less dense.

Relative density

Relative density (also occasionally known as specific gravity) is an effective way of measuring the density of a substance or object by comparing its weight per volume to an equal volume of water. Water has a relative density (specific gravity) of 1.0. For example:

- 1 m³ of water has a mass of 1000 kg
- 1 m³ of mild steel has a mass of 7,700 kg.

The mild steel is 7.7 times heavier than water and therefore has a relative density of 7.7.

Density of gases

In the same way that water is said to have a relative density of 1.0 to enable comparisons of relative density for solids and liquids, air has a relative density of 1.0 to enable comparisons between gases which are classified by whether they are lighter or heavier than air. Think about helium-filled balloons for instance – do you think helium is lighter or heavier than air?

Pressure

Pressure is defined as force applied per unit area and is measured in newtons per square metre (N/m²), a unit also known as a Pascal (Pa).

You will probably come across other terms used to identify pressure such as the 'bar' or 'pounds per square inch' (lbs/in²). These can be expressed as:

- 1 bar = 100,000 N/m²
- 1 lbs/in² = 6894 N/m²

Pressure is therefore a measurement of a concentration of force. The effect of a concentration of pressure can be seen if water flowing through a pipe is forced through a smaller gap by reducing the diameter of the pipe. (Think about a hosepipe, and how to maximise the force of the jet of water.)

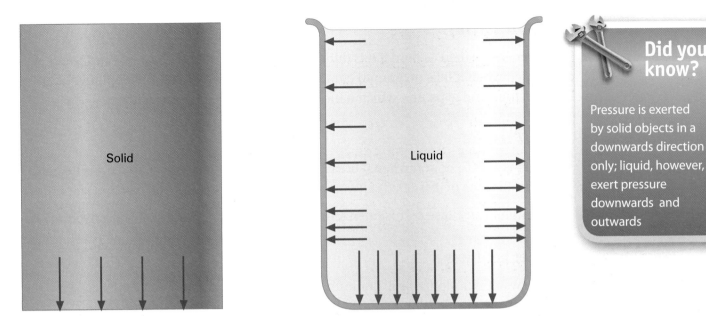

Figure 4.12 The effects of pressure in solids and liquids

Pressure can also be lowered by 'spreading' the applied force over a wider area. For example, rescue teams will often spread themselves over fragile roofs or on thin ice to minimise the chance of the surface giving way.

As a plumber you will need to have a basic understanding of the effects pressure has on the pipes and fittings you will install. The internal pressure in a pipe or vessel will be affected by what is being transported (water or gas), and must be considered when deciding which material and which size of pipe or vessel should be used.

Atmospheric pressure

The pressure exerted by the weight of the earth's atmosphere pressing down on the ground varies depending on height above sea level. The pressure at the top of Mount Everest is not as high as the pressure in the bottom of a valley below sea level (such as the Great Rift Valley in Africa). The pressure at sea level is 101,325 N/m2 (approximately 1 bar).

Plumbers must be aware of the effects of atmospheric pressure to ensure that they avoid creating 'negative' pressure or vacuums within pipework systems. Negative pressure can damage components.

Properties of water

Water is a chemical compound of two gases: hydrogen and oxygen (H_2O). It is formed when hydrogen gas is burned.

One of the most important properties of water is its solvent power. It can dissolve numerous gases and solids to form **solutions**. The purest natural water is rainwater collected in the open countryside. It contains dissolved gases such as nitrogen, oxygen and carbon dioxide, but this does not affect its potability (suitability for drinking).

Water may be classified as having varying degrees of **hardness** or softness.

Hard water

Water is classified as hard if it is difficult to obtain a lather with soap. Hard water is created when it falls on ground containing calcium carbonates or sulphates (chalk, limestone and gypsum), which it has dissolved and taken into solution.

Soft water

Water is said to be soft when it is easy to produce a lather with soap. This is because of the absence of dissolved salts such as calcium carbonates and calcium sulphates.

Soft water can cause corrosion in plumbing components because it is relatively acidic

Water hardness

Hard water is undesirable in domestic installations as it can produce limescale in pipework, heating equipment and sanitary appliances. This can lead to high maintenance costs. Hard water also requires the use of much more soap and detergent for washing purposes, as the 'hardness' makes it far more difficult to produce a lather.

Water hardness can be described as temporary or permanent.

- Permanent hardness is a result of ions of nitrates and sulphates. It makes it difficult to form a lather and *cannot* be removed by boiling.

- Temporary hardness is a result of the amount of carbonate ions in the water. Temporary hardness *can* be removed from the water by boiling, which results in the carbonate being precipitated out as limescale. This hard scale accumulates inside boilers and circulating pipes, restricting the flow of water, reducing the efficiency of appliances and components and ultimately causing damage and system failure.

pH value and corrosion

The term **pH value** refers to the level of acidity or alkalinity of a substance. As a plumber, you will need to be particularly aware of the potential effects the acidity or alkalinity of water can have upon materials, appliances and components.

Both acids and alkalis can cause **corrosion** and thereby damage plumbing materials; metals are at particular risk.

Did you know?

Rainwater collected in towns contains higher percentages of dissolved substances such as soot and other pollutants from the atmosphere

Did you know?

Pure water has a neutral pH of around 7

FAQ

Why is it more difficult to obtain a good lather in different parts of the country?

This is because the hardness of water varies in different parts of the country. The harder the water, the more difficult it is to obtain a lather. Is the water where you live soft or hard?

All water has a 'pH value'. Rainwater is naturally slightly acidic, due to small amounts of carbon dioxide and sulphur dioxide in the atmosphere being dissolved into it, forming very weak carbonic and sulphuric acids. The pH value of groundwater is affected by the different rock types it passes though. For instance, water with dissolved carbonate from chalk or limestone is alkali.

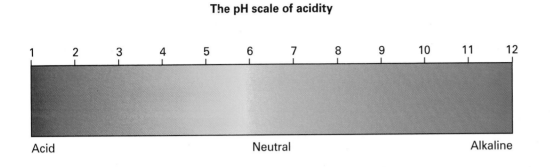

The pH scale of acidity

The main causes of **corrosion** are:

- the effects of air
- the effects of water
- the direct effects of acids, alkalis and chemicals, e.g. from environmental sources
- electrolytic action.

Atmospheric corrosion

Pure air and pure water have little corrosive effect, but together in the form of moist air (oxygen + water vapour) they can attack **ferrous** metals such as steel and iron very quickly to form iron oxide or rust. The corrosive effects of rusting can completely destroy metal.

Various other gases (carbon dioxide, sulphur dioxide, sulphur trioxide), which are present in our atmosphere, also increase the corrosive effect air can have on particular metals, especially iron, steel and zinc. These gases tend to be more abundant in industrial areas as they are often waste products from various industrial processes.

Coastal areas also suffer from increased atmospheric corrosion due to the amount of sodium chloride (salt) from the sea which becomes dissolved into the local atmosphere.

Non-ferrous metals, such as copper, aluminium and lead, have significant protection against atmospheric corrosion. Protective barriers (usually sulphates) form on these metals to prevent further corrosion. This protection is also known as **patina**.

Corrosion by water

Ferrous metals are again particularly vulnerable to the effects of corrosion caused by water. These are commonly seen in central heating systems as black ferrous oxide and red rust build-up in radiators. A by-product of this process is hydrogen gas, which accumulates in the radiator, leading to the need to be 'bled'. See Chapter 9.

Definition

Ferrous – metals that contain iron

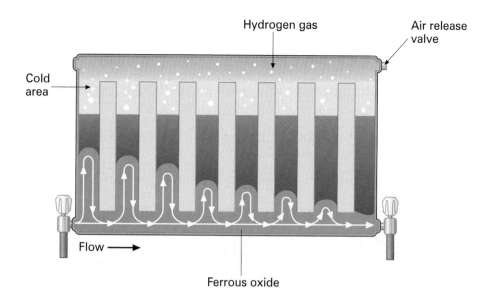

Figure 4.13 Corrosion effects in steel panel radiators

In certain areas of highly acidic water, copper may become slightly discoloured. This won't affect the quality or safety of the drinking water. However, in areas where lead pipework is still in use, if the water is very soft (acidic) there is a risk of it dissolving minute quantities of lead, thus contaminating the water – with potentially toxic effects, especially for children.

Corrosive effects of building materials and underground conditions

Some types of wood (such as oak) have a corrosive effect on lead, and latex cement and foamed concrete will adversely affect copper. Certain types of soil can damage underground pipework. Heavy clay soils may contain sulphates which can corrode lead, steel and copper. Ground containing ash and cinders is also very corrosive as they are strongly alkaline; if pipes are to be laid in such ground they should be wrapped in protective material.

Electrolytic action and corrosion

Electrolytic action describes a flow of electrically charged ions from an **anode** to a **cathode** through a medium known as the **electrolyte** (usually water) as shown in figure 4.14 on page 101.

Electrolytic corrosion takes place when the process of electrolysis leads to the destruction of the anode. The length of time it takes for the anode to be destroyed will depend on:

- the properties of the water that acts as the electrolyte: if the water is hot or acidic the rate of corrosion will be increased

- the position of the metals that make up the anode and the cathode in the electromotive series.

Did you know?

A simple battery uses electrolysis to produce electricity

The electromotive series

The list below shows the common elements used in the plumbing industry; the order in which they appear indicates their electromotive properties.

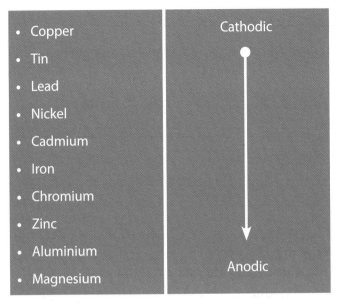

- Copper
- Tin
- Lead
- Nickel
- Cadmium
- Iron
- Chromium
- Zinc
- Aluminium
- Magnesium

Cathodic

Anodic

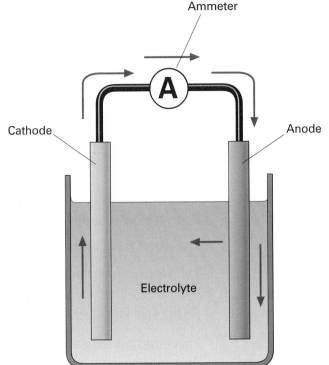

Figure 4.14 The process of electrolysis

The elements higher up in the list will destroy those lower down through the process of electrolytic corrosion. The further apart in the list the materials appear, the faster the corrosion will take place. For example, copper will destroy magnesium at a faster rate than lead will destroy chromium.

As a plumber you will need to be aware of the potential for electrolytic corrosion when two very dissimilar metals such as a galvanised tube and a copper fitting are in direct metallic contact. If these metallic elements are then surrounded by water (of a certain type) or damp ground, a basic electrical cell is effectively created and electrolytic corrosion can take place. See also dezincification in Chapter 8.

Find out

Try to think of some practical examples of electrolytic corrosion

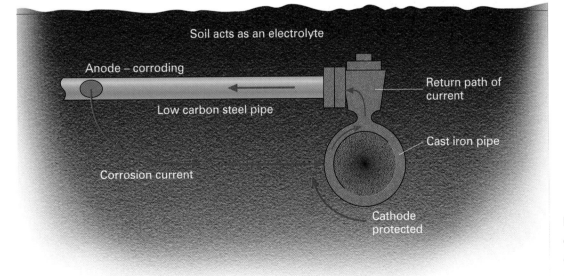

Figure 4.15 Example of possible electrolytic corrosion in pipework systems

Properties of heat

Difference between heat and temperature

The main difference between heat and temperature is that heat is recognised as a unit of energy, measured in joules (J).

- temperature is the degree of hotness of a substance
- heat is the amount of heat energy (J) that is contained within a substance.

For example, imagine an intensely heated short length of wire and a bucket of hot water:

The wire has a temperature of 350°C

The water has a temperature of 70°C

The wire is far hotter, but actually contains less heat energy.

Measuring temperature

The SI unit of temperature measurement is the degree kelvin, but the unit you'll deal with most frequently is degrees Celsius (or centigrade), written as °C.

Temperature is measured using thermometers. There are many types, but the most common depend upon the expansion of either a liquid or a **bi-metallic strip**.

The two most common liquids used in thermometers are alcohol and mercury. These liquids are used because they expand at a uniform rate when exposed to heat.

Bi-metallic strips work on the principle of thermal expansion and contraction (covered in more detail later) and the fact that some metals expand and contract at a faster rate than others. Bi-metallic strips are frequently used in thermostats, where they will bend (see illustration) when a particular temperature is reached. This process will break an electrical circuit and turn off the heating.

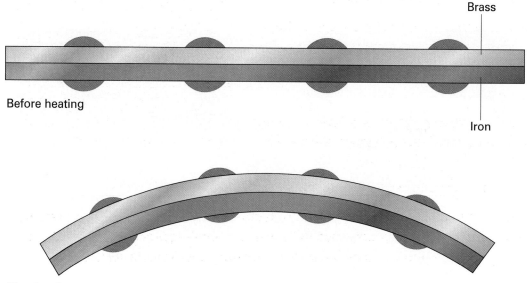

Brass

Iron

Before heating

After heating

Figure 4.16 A bi-metallic strip. The brass expands more than the iron on heating, thus bending the strip.

Specific heat capacity

The **specific heat capacity** is the amount of heat required to raise 1 kg of a material by 1°C. The heat required differs from material to material. For example, while it would require 4.186 kJ to raise the temperature of water by 1°C, only 0.385 kJ would be needed to raise the temperature of copper by 1°C.

Specific heat values vary as the temperature changes.

Thermal expansion and contraction

Most materials will expand when heated. This is because all substances are made up of molecules (groups of atoms) which move about more vigorously when heated. This in turn leads to the molecules moving further apart from each other – resulting in the materials taking up more volume.

As the material cools, the molecules slow down and move closer together; thus the material gets smaller or contracts. The amount that the material expands in length when heated can be calculated using the following formula:

length (m) × temperature rise (°C) × coefficient of linear expansion

This table shows the coefficient values for some of the most common materials used in the plumbing industry.

Example:

Find the amount a 6 m-long plastic discharge stack will expand due to a temperature rise of 19°C.

6 × 19 × 0.00018 = 0.02052 m or 20.52 mm

We need to take this expansion and contraction into account in plumbing systems that are constantly being subjected to heating and cooling processes – if we don't, the system or component may break down, causing, for example, leakage.

Material	kJ/kg °C
Water	4.186
Aluminium	0.887
Cast iron	0.554
Zinc	0.397
Lead	0.125
Copper	0.385
Mercury	0.125

Figure 4.17 Specific heat values

Material	Coefficient °C
Plastic	0.00018
Zinc	0.000029
Lead	0.000029
Aluminium	0.000026
Tin	0.000021
Copper	0.000016
Cast iron	0.000011
Mild steel	0.000011
Invar	0.0000009

Figure 4.18 Coefficient values

Find out

What do you think happens when materials are cooled down?

Safety tip

Plastics tend to expand the most and we quite often need to leave an expansion gap in the pipework system when working with plastics to prevent the material from failing

Heat transfer

As a plumber you will need a good understanding of the methods of heat transfer because you will be dealing with the effects of this process on a daily basis. There are three methods of heat transfer.

- conduction

- convection

- radiation.

Conduction

Conduction is the transfer of heat energy through a material. It takes place as a result of the increased vibration of molecules, which occurs when materials are heated. The vibrations from the heated material are passed on to adjoining material, which then heats up in turn. Some materials are better at conducting heat than others. For example:

- metals tend to be good conductors of heat

- wood is a poor conductor of heat.

Gases and liquids also conduct heat, but poorly. In the plumbing industry, conduction is principally between solids. Of the metals commonly used by plumbers, copper has a higher conductivity than steel, iron or lead. Wooden, ceramic and plastic materials, which are poorer conductors of heat, are known as **thermal insulators**.

Convection

Convection is the transfer of heat by means of the movement of a locally heated **fluid** substance (usually air or water). As a fluid is heated, the process causes expansion, which in turn causes a lowering of density. The less dense warm fluid begins to rise, and is replaced by cooler, denser fluid from below. Eventually, convection currents are set up which allow for a continuous flow of heat upwards from the source. Examples of systems that use convection currents for heat transfer are:

- convector heaters, which warm the air at one place in a room; the resulting convection currents transport the heat around the room

- domestic hot water systems, which depend on convection currents to transfer heat from an immersion heater (similar to the 'element' in an electric kettle) to the rest of the water in the hot tank (cylinder).

It is easy to demonstrate the 'updraught' part of a convection current by hanging a piece of light material above a convector heater. The movement of the hanging material will clearly show the presence of rising currents of warm air. The illustration overleaf shows a practical example of the use of convection currents.

Find out

How good are the following materials as heat conductors: plastic, rubber, ceramic, wool, carbon?

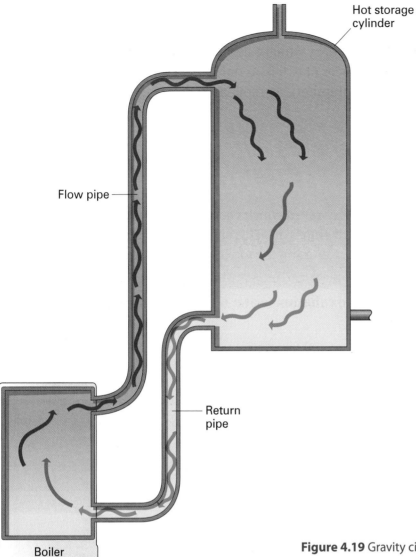

Figure 4.19 Gravity circulation taking place owing to the effects of convection

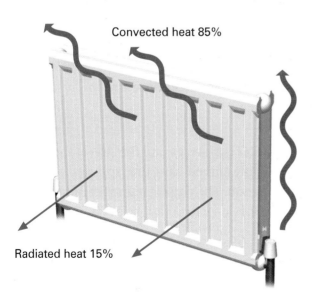

Figure 4.20 Heat production from panel radiator

Radiation

Radiation is the transfer of heat from a hot body to a cooler one without the presence of a material medium (other than air), by means of 'heat' waves. Heat radiation can be felt as the 'glow' from a fire or as the heat from the sun.

Heat radiation is better absorbed by some materials than others; in these instances colour can be an important factor. Dull, matt surfaces will absorb radiated heat more efficiently than shiny, polished surfaces.

Remember

Most domestic 'radiators', such as those found in central heating systems, in addition to radiating heat energy will also warm a room by convection

Capillary action

Capillary action is the process by which a liquid is drawn or hauled up through a small gap between the surfaces of two materials. The relevance of this phenomenon is especially important for plumbers, as capillary action can affect the way in which water can get into buildings.

Forces of attraction

Surface tension

Surface tension describes the way in which water molecules 'cling' together to form what is effectively a very thin 'skin'. This can be demonstrated by filling a glass beaker right to the top and examining the top of the glass. The water will appear above the upper limit of the glass – why doesn't it spill down the side?

The answer is because of the **cohesion** (water molecules 'sticking together') as a result of surface tension.

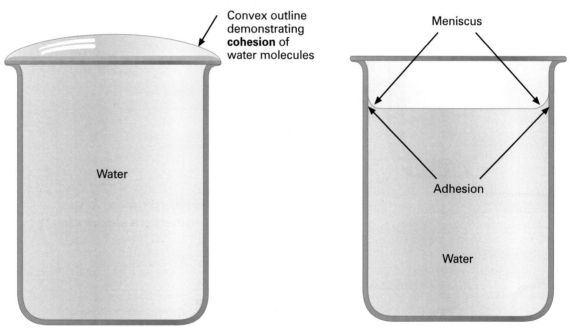

Figure 4.21 Surface tension and cohesion

Figure 4.22 Adhesion

Adhesion

Adhesion is the force of attraction between water molecules and the side of the vessel the water is contained in. This leads to the slightly curved 'skin' that appears when water is held in a vessel as shown in the illustration. The correct name for this skin is a meniscus. These processes of adhesion and cohesion cause capillary action.

Practical examples of capillary action

Plumbers must consider the possible affect of capillary action when planning lead roofing work. Water can find its way into a building's lapped roof if materials are close together. The diagram shows how this occurs and how it can be remedied.

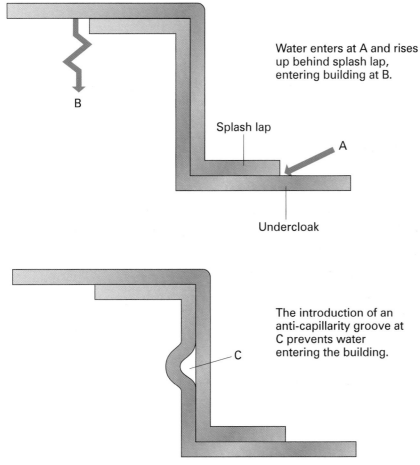

Water enters at A and rises up behind splash lap, entering building at B.

B

Splash lap

A

Undercloak

The introduction of an anti-capillarity groove at C prevents water entering the building.

C

Figure 4.23 Avoiding capillary action in fixing sheet leadwork

Capillary action can occur in 'S bend' drainage traps, which are found under sinks. If a length of waste material, such as a piece of dishcloth, becomes lodged in the 'S bend', capillary action could take place. This can lead to the loss of seal at the bottom of the trap, allowing bad smells from the drainage pipe to filter back into the home.

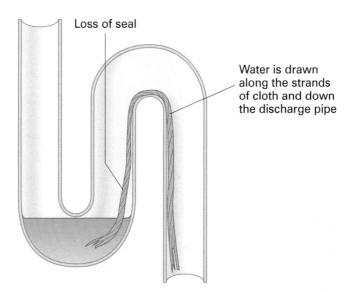

Loss of seal

Water is drawn along the strands of cloth and down the discharge pipe

Figure 4.24 Trap seal lost due to capillary action

Siphonage

Principles of siphonage

Atmospheric pressure is the key to siphonage, which is used in a number of plumbing applications. The illustration will help you to understand how siphonage occurs.

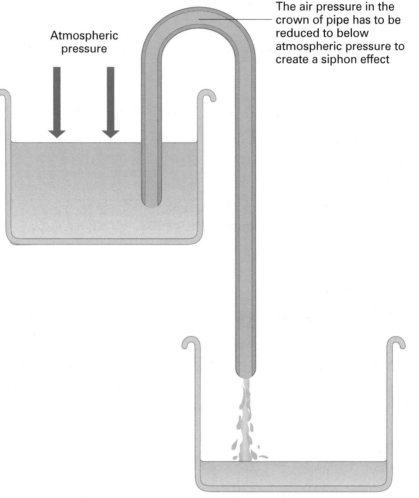

Figure 4.25 The principle of the siphon

Basically, siphonage works when atmospheric pressure is able to force water through a channel (such as a length of hose or a run of pipe), often seemingly against the pull of gravity. For siphonage to work as illustrated, the air pressure in the channel must be reduced to below that of atmospheric pressure.

Practical uses of siphonage

The principle of siphonage is used in many areas of plumbing, especially sanitation.

Siphonic WCs, such as the single trap and the double trap, use siphonic action to create a negative pressure below the trap seal of the WC to help clear the contents of the pan. We will discuss this further in a later section.

Did you know?

Siphonage has been used in WC flushing for many years

Water flow

The rate of water flow through pipework is affected by friction. This is an important consideration when designing pipework installations. To get an idea of how friction works, compare the speed of a car driving freely through a tunnel with the speed of a car that is in contact with the tunnel walls. As you can imagine, the result of rubbing on the wall will cause a considerable reduction in speed.

Pipe walls have the same effect on water flow, i.e. the water is 'rubbing' on the pipe. This friction is further increased if the inside surface of the pipe wall is rough. This effect is known as frictional resistance or frictional loss.

Comparison of flow in smooth and rough pipework

- Water flowing through a 10 m length of 25 mm diameter pipe would flow at approximately 22 litres/minute for copper and plastic pipes, which have a smooth internal surface.

- Water flowing through a 10 m length of 25 mm diameter pipe would flow at approximately 18 litres/minute for galvanised LCS pipe, which has a rougher inner surface.

Principles of electricity

Try to imagine a world without electricity. Everything from computers to stereos, washing machines to lighting and heating relies on it.

So what precisely is it? You can't see it with the naked eye; you cannot smell it or hear it. But if you misuse or touch it, it can hurt or even kill. Because of its possible dangers, a deeper understanding of the subject is required.

What are electrons?

Molecules and atoms

Every known substance is made up of molecules. A molecule is a very tiny part of matter which can only be seen using special microscopes. Molecules are always in a state of rapid motion, and the ease with which they move around determines the form of the substance they make up:

- When molecules are densely packed together, their movement is restricted and the substance formed by them is called a **solid.**

- When the molecules are less densely bound together, and where they can move more freely, the substance is called a **liquid.**

- A substance that allows the molecules almost unrestricted movement is known as a **gas**.

These three conditions, solid, liquid and gas, are sometimes known as the three states of matter.

Remember

Frictional resistance is also increased by the number of fittings in a pipework run

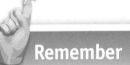

Remember

Gas, oil, solid fuel and LPG central heating systems contain electrically operated components, such as an electric pump to circulate the water around the system

Definition

Electricity – the flow of electrons through a conductor

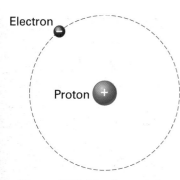

Electron

Proton

Figure 4.26 A hydrogen atom

Molecules are made up of **atoms**, and we need to take an imaginary look inside the atom to begin to understand what electricity is. Atoms are the smallest parts of matter that can be subdivided, and they are not solid. They have a **nucleus** at their centre, made up of very tiny particles, known as **protons** and **neutrons**. Protons are said to contain a positive charge (+) and neutrons are electrically neutral (that is, they have no charge). We can think of neutrons as the 'glue' that holds the nucleus together. Around the nucleus orbits a third type of particle – **electrons** – which contain a negative charge (–).

You may remember from your school science studies that like charges repel each other (+ and + or – and –) and unlike charges attract each other (+ and –)

All atoms contain equal numbers of protons and electrons, and, in this unaltered state, the matter is said to be electrically neutral (in other words, no electricity is flowing). In some cases, it is possible to add or remove electrons, leaving the atom with a positive (+) or negative (–) charge.

Now to the part that assists the definition of electricity. Consider our own solar system, where the planets (including earth) rotate around, or '**orbit**', the sun.

In each tiny atom, there is the nucleus containing protons and neutrons (sun) and electrons (planets), which orbit the nucleus. Sometimes these electrons 'break free' and flow to a neighbouring atom. It is these wandering, or free, electrons, moving through the material structure, that give rise to the electricity.

What are conductors?

In some materials, it is very easy to get the electrons to move; these materials are called **conductors**. In other materials, it is very hard to get the electrons to move; these materials are called **insulators**. Examples of good conductors of electrons are copper and aluminium (gold is probably the best conductor, but it is too expensive to use in everyday installations). Typical insulators are wood, plastic and rubber.

Measuring electricity

For practical purposes, three things need to be present to create an electrical circuit:

- current
- voltage
- resistance.

Current

If electricity is the movement of electrons through a conductor, we need to be able to measure *how much* electricity will flow in any given circuit in a given time and to control how much electricity is flowing to create a practical electric circuit. A single electron, because it is only a tiny part of an atom, is much too small a quantity to have any practical use or to usefully measure. Instead, we 'group' millions and millions of electrons together into useful amounts and then measure the groups. These 'groups of electrons' are known as coulombs, and a coulomb is an extremely large number of electrons.

Did you know?

A coulomb consists of approximately 6,240,000,000,000,000,000 electrons

As plumbers, we can use water analogies to help us to understand more about electricity. If we consider the electron as a single drop of water, ask yourself whether it is practical to measure water flow in terms of drops of water flowing in a second. It is more practical to measure *water current flow* in litres per second.

Similarly, an electrician would consider the flow of electricity through a conductor in terms of coulombs (millions of drops) per second, which brings us nicely to one of the main properties of an electric circuit: the electric **current** flow, measured in **amperes** (usually abbreviated to amps and given the symbol I).

Voltage

The next essential ingredient is **electro-motive force** measured in terms of the number of joules of work required to push one coulomb of electrons along the circuit. It is measured in joules/coulomb, more commonly refered to as the **volt**. When a coulomb of electrons leaves a battery or generator it has a potential energy, but as it travels round the circuit this energy is used up. The amount of energy used up by one coulomb in its passage between two points in a circuit is known as the **potential difference**, measured in volts.

To continue with our water comparison, voltage can be thought of as the pressure that pushes the water (electricity) around the system. By applying a voltage to the end of a conductor, we provide an electrical pressure that causes a current to flow. The voltage may be supplied from a battery or a mains supply. A current that is produced by connecting a battery to a circuit is called **direct current d.c**. The electricity that we use in our homes and which is produced by power stations is called **alternating current a.c**. There's more on the differences between them later.

Resistance

The third item we need to consider to understand electricity properly is **resistance**, given the symbol R. In our water comparison think about water being supplied to a bath via a cistern using gravity. The head of water in the cistern will provide the pressure (voltage in electrical terms). With the tap open, the water will flow into the bath at a fixed rate per second (current in electrical terms). In order to flow from the cistern to the bath the water passes through a pipe of a certain diameter. If we increase the size of the pipe (and tap), the natural resistance to the flow of water will decrease and the bath will fill more quickly. If we decrease the size of the pipe, the resistance to the flow will increase and the bath will take longer to fill. This resistance to current flow in an electrical circuit is measured in units called ohms (sometimes denoted by the Greek letter omega, Ω) and given the symbol R. In plumbing terms, the bigger the pipe the lower the resistance to water flow; in electrical terms, the bigger the conductor the lower the resistance to current flow.

Definition

Ampere – a flow of one coulomb in one second. In other words, the quantity or amount of electricity that flows every second

Remember

We measure the amount of current flowing in amps using an instrument called an ammeter, which is connected into a circuit in series (more about series circuits later)

Remember

We use an instrument known as a voltmeter to measure the voltage of a circuit. Voltmeters are connected into a circuit in parallel (more later)

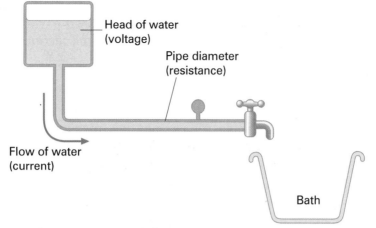

Figure 4.27 Analogy of a water system as an electrical circuit

Gravity drives the water through this systems, but for this to happen there must be a difference in level between the supply (the tank) and the load (the bath). This gives the equivalant of potential difference. The tap performs the same function as an electric switch.

One very important point to note here is that ammeters and voltmeters are connected into live electrical circuits to measure current and voltage, and they pick up the electricity that they need to operate from the live circuit.

The three major components that make up an electric circuit – voltage, current and resistance – are interrelated, and, if we know any two of the quantities, we can calculate the third by using a basic rule known as 'Ohm's Law'.

Ohm's Law states the relationship between the electrical quantities as: voltage equals current multiplied by resistance; can be expressed by the equation:

$$V = I \times R$$

where V = voltage, I = current and R = resistance. It can also be shown in a simple form, often called the 'Ohm's Law Triangle'.

Using the Ohm's Law Triangle

If you know two of the quantities, you can find the third by covering up the quantity that you are looking for with your finger and calculating the other two.

Suppose you want to determine the voltage, V, of a circuit and you know the current, I, and resistance, R.

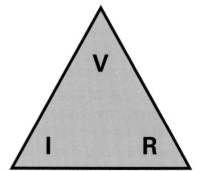

Figure 4.28 Ohm's Law Triangle

Covering the V with your finger shows you that you need to multiply the current I by the resistance, R, to find voltage, V.

To find the current I, cover I and you will see that you now need to divide voltage, V, by resistance, R, as V is shown above R.

For example:

A circuit has a resistance of 120 ohms and a current of 2 amps is flowing. What would be the voltage applied to the circuit?

$V = I \times R = 2 \times 120 = 240$ volts

How an electric circuit works

If we were to bend a conductor, such as a copper wire, into a loop and connect a battery across the two ends, we would have a complete electric circuit and a current would flow. Unfortunately this would not create a practical circuit that we could use. A single length of copper wire has a very low resistance on its own (depending on the length and cross-sectional area it may have only a few millionths of an ohm of resistance). If we were to connect this across a normal domestic voltage supply of 230 volts, using Ohm's Law we could show that, as $I = \frac{V}{R}$ and R is very low, thousands of amps of current will flow, causing serious damage to the circuit. The flow of electric current causes a corresponding heating effect, and the higher the current the higher the heating effect, causing a serious fire risk.

Even if we could have the circuit described above it would still not be practical because there is no load to limit the current, no way to switch the current off and no protection for the circuit from the high current flowing due to the very low cable resistance.

For all practical electric circuits, therefore, we must have:

- conductors through which current can flow (usually copper wire)

- a source of supply (such as a battery) connected across the ends to provide a potential difference (voltage) and to make a complete circuit

- a load such as a lamp or a device that needs electric current to make it work but which will also act to limit the amount of current flowing

- a device such as a fuse or miniature circuit breaker to protect the circuit should too much current flow

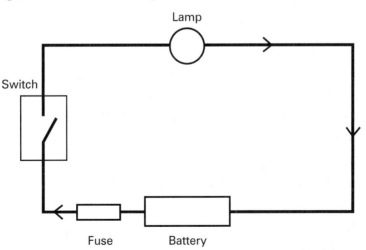

Figure 4.29 A simple electrical circuit

- a switch for opening and closing the circuit to switch the current on and off.

The diagram shows a practical electric circuit containing a battery as a supply source, a lamp as a load, a switch to open and close the circuit and a fuse to protect the circuit from excess current flow.

Fuses and circuit breakers

Why are fuses needed?

Fusing is a safety measure which aims to prevent a high electrical current passing through wires that are not designed to carry such large charges. Why is this important? Because when a current flows, it causes a heating effect and a temperature rise in the wire; if a current that is too high for a wire is passed through it, the overheating that results presents a serious risk of fire.

How do fuses help?

The various different types of fuse that exist all contain fuse wire which will melt or 'blow' if an electric current above a specified level is passed through the wiring. Fuses come in different sizes to protect against different levels of current.

Rewirable fuses   Cartridge fuses

Miniature circuit breaker (mcb)

In modern electrical installations you will also come across miniature circuit breakers (mcbs). An mcb is a device that will trip a switch to break the electrical circuit if excessively high current is detected. When electricity flows, it also produces a magnetic field around the conductors as well as the heating effect previously mentioned. The mcb uses a combination of the heating and magnetic effects to break the circuit under fault conditions. Mcbs are more accurate and expensive than fuses but are re-settable and are now found in all electrical consumer units within newer domestic properties.

You will also come across residual current devices (RCDs). These are very sensitive devices providing a high degree of protection to high-risk parts of electrical systems, such as plug-socket outlets and electric showers. An RCD measures the difference between the current in the electrical conductors in the system, e.g. live and neutral, and measures changes in the electrical current. If a small change occurs, the system is automatically disconnected. A typical operating current for an RCD in the event of a fault would be 30 mA (that is, 30 one thousandth parts of one amp.) This may seem very low, but it has been calculated that the average person can 'feel' electricity at current levels of around 1 mA.

Fuse rating

Ensuring that the appropriate size of fuse is used – called the fuse rating – can be worked out using the simple formula:

amps = watts ÷ volts

Most items of electrical equipment – lamps, televisions, washing machines, hi-fi systems, vacuum cleaners – are rated in watts, which is the amount of electrical power that the item consumes. Frequently in domestic environments fuses are overrated: i.e. a fuse with too a high rating is used for safety reasons. For example, if a lamp contains a 100-watt light bulb, the fuse rating would be calculated as follows:

100 ÷ 230 (the voltage of domestic mains supply) = 0.434 amps

Manufacturers of fuses do not produce fuses rated at 0.434 amps, and the next practical size up would be used, in this case a 3-amp fuse. In this example, the bulb may be used in a stand-alone table lamp fitted with a cord and 13-amp plug top, which should be fitted with a 3 A fuse. Plugs are normally supplied with a 13 A fuse, but this is unnecessarily high in this example and provides insufficient protection for the user. Technically, this set up would be classed as unsafe.

If the lamp were being used as one of a number connected to a fixed lighting circuit, the entire circuit would usually be fitted with a 6A fuse or circuit breaker.

There is potential for confusion here, because 1 amp of current is more than sufficient to kill at 230 volts (the average that a person can 'feel' is 1 mA).

> **Remember**
>
> The purpose of the fuse is mainly to protect the wiring; an earth is provided to protect the user. We will learn more about earthing later

Circuits

There are two basic types of electrical circuits:

- series
- parallel.

It is also possible to connect components into a circuit using a combination of these two ways. These are called 'series/parallel' circuits, and they will be considered later when we start to look at control circuits for central heating and boiler installations etc.

Series circuits

If we take a number of different resistors (items of electrical equipment such as lamps etc.) and connect them together end to end and then connect the free ends to a battery, we will find that the current will have only one route to take around the circuit. This type of connection is known as a series circuit. Most switches and controls such as thermostats are usually connected into circuits in series, for example in those used for central heating systems.

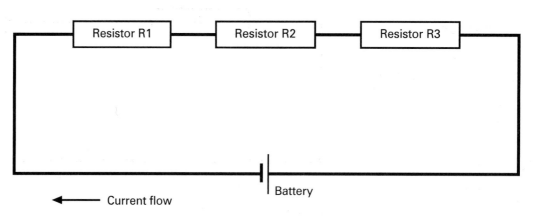

Figure 4.30 The series circuit

The main features of a series circuit are:

- the total resistance of the circuit is found by simply adding together all of the resistances within the circuit:

  R = R1 + R2 + R3

- the total current flowing (I) can be found by dividing the supply voltage by the resistance

  $$\mathbf{I = \frac{V}{R}}$$

(you will recognise this as Ohm's Law)

- the current flowing will have the same value at any point in the circuit

- the potential difference or voltage across each resistor is proportional to its resistance. We know from our study of electrons that we use voltage to push electrons through a resistor. How much voltage we use depends on the size of the resistor. The bigger the resistor the more voltage we use for a given current (revise Ohm's Law).

Therefore the voltage across resistor R1 = I (remember that the current is the same at all points of a series circuit) × value of R1.

Considering the circuit shown in the illustration:

V1 = I × R1   V2 = I × R2   V3 = I × R3

- the supply voltage of the circuit (V) is equal to the amount of the individual voltages across each resistor, as shown by the formula:

  V = V1 + V2 + V3

Calculations with series circuits

Example:

Two resistors of 6.2 ohms and 3.8 ohms are connected in series with a 12V battery, as shown in the diagram. Calculate:

(a) total resistance of the circuit   (b) total current flowing

(c) the potential difference across each resistor.

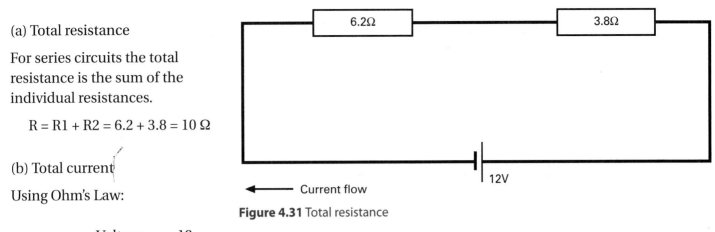

(a) Total resistance

For series circuits the total resistance is the sum of the individual resistances.

$$R = R1 + R2 = 6.2 + 3.8 = 10\ \Omega$$

(b) Total current

Using Ohm's Law:

Figure 4.31 Total resistance

$$\text{Current} = \frac{\text{Voltage}}{\text{Resistance}} = \frac{12}{10} = 1.2\ \text{Amps}$$

(c) Potential difference across each resistor

$$V = I \times R$$

Therefore: Across R1: $V1 = I \times R1 = 1.2 \times 6.2 = 7.44$ volts

Across R2: $V2 = I \times R2 = 1.2 \times 3.8 = 4.56$ volts

Parallel circuit

An alternative way to connect components into a circuit is via a parallel circuit. These are commonly used for lighting and power circuits, as one important feature of this type of connection is that the same voltage appears across each of the resistances in the circuit. This is important in

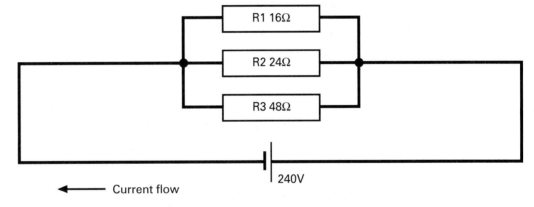

Figure 4.32 The parallel circuit

lighting circuits where we want all the lamps to burn with the same brilliance and not to burn dimly because the available voltage is shared between them.

In a parallel circuit a number of resistors are connected together so that there are two or more routes for the current to flow as shown in the illustration.

In a parallel circuit the current 'splits up' and divides itself among the various branches of the circuit. However, while the current divides within a parallel circuit, the voltage across each branch remains the same. This means that we can take components out of the circuit and the remaining ones will still work. In a series circuit, if we removed a single component the circuit would be broken and would not work.

It is this feature that makes parallel circuits popular, and indeed all lighting and power circuits in domestic premises are connected in parallel; only the switches and certain other control devices are connected in series.

The main features of a parallel circuit are:

- the total current (I) is found by adding together the current flowing through each of the branches

$$I = I1 + I2 + I3$$

- the same voltage will occur across each branch of the circuit

$$V = V1 + V2 + V3$$

the total resistance can be found by using the formula:

$$\frac{1}{R} = \frac{1}{R1} + \frac{1}{R2} + \frac{1}{R3}$$

Some calculations with parallel circuits

Using the example shown above, calculate the total current flowing through the circuit.

$$\frac{1}{R} = \frac{1}{R1} + \frac{1}{R2} + \frac{1}{R3}$$

Therefore:

$$\frac{1}{R} = \frac{1}{16} + \frac{1}{24} + \frac{1}{48}$$

And therefore:

$$\frac{1}{R} = \frac{3 + 2 + 1}{48}$$

Giving us:

$$\frac{1}{R} = \frac{6}{48}$$

Turn this upside down to find R:

$$R = \frac{48}{6} = 8 \text{ ohms}$$

Now, using Ohm's Law:

$$I = \frac{V}{R}$$

We see that:

$$I = \frac{240}{8} = 30 \text{ amps}$$

Direct and alternating current

The illustration shows the flow of electrons in both a.c. and d.c. circuits.

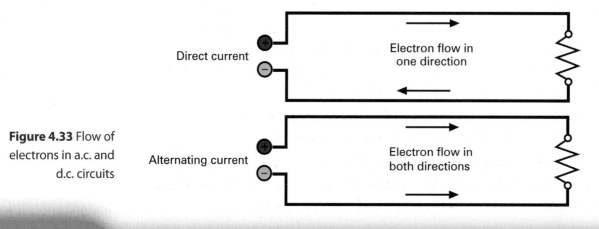

Figure 4.33 Flow of electrons in a.c. and d.c. circuits

Direct current — Electron flow in one direction

Alternating current — Electron flow in both directions

Direct current (d.c.)

In a direct current electrical circuit, the electron flow is in the same direction all the time. One example would be from the anode to the cathode of a battery around a simple circuit.

Direct current from batteries is produced by a chemical reaction where plates containing dissimilar metals are placed in a solution known as an 'electrolyte'. When the battery is charged and a load is connected across it, a current flows. Batteries and direct current supplies are generally only used where very small amounts of electricity are required – for example in torches, light current back-up power supplies and battery-operated power tools.

As plumbers you will probably only use direct current when using small batteries or perhaps in heating or boiler control circuits where the direct current has been produced from an alternating current source using a transformer to reduce the voltage and a rectifier (which is a solid state electronic device) to convert the alternating current to direct current.

Alternating current (a.c.)

Most of the work that plumbers do and the installations that plumbing systems are connected into will be supplied with alternating current. The supply in a domestic building will almost always be single phase and supplied at 230 volt and a frequency of 50 Hertz.

We said earlier that electricity was invisible to the naked eye and we could not see it or smell it. This is true, but there is an instrument known as an oscilloscope with a screen like a small TV which allows us to look at electricity.

If we look at direct current we see a straight line where the electricity flows in one direction only from one pole of the battery to the other, as shown in the diagram.

If we look at alternating current, however, we see a shape that is called a 'sine' wave .

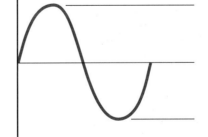

Figure 4.34 The alternating current sine wave

The sine wave represents one complete 'cycle' of alternating e.m.f. which would be 'induced' into a coil of copper wire when it is rotated within a magnetic field. During the top half of the cycle, the current will flow in one direction, and during the bottom half of the cycle, the current will flow the other way (hence 'alternating'). In the UK, the electricity supply has 50 of these single cycles produced every second. This is known as the frequency of the supply, and the unit used for frequency is the Hertz, (abbreviated to Hz). The frequency is therefore stated as 50 Hz.

Whereas direct current is caused as a result of a chemical reaction in a battery, alternating current is produced as a result of **electromagnetism**:

- As well as the heating effect already mentioned, all electrical currents produce a magnetic force. This is the basic principle that underpins the creation of almost all the electricity used today, and is known as electromagnetism.

Definition

Hertz – the number of cycles of a.c. that are produced every second

Did you know?

Sine is a short name for 'sinusoidal', which you may have come across in school during trigonometry

- The application of this fact was in principle demonstrated by Michael Faraday in the 1830s. He discovered that electricity could be generated by moving a magnet in and out of or around a coil of wire wound on to a soft iron core.

- The same effect is present if we move a coil of wire on an iron core within a fixed magnetic field.

Did you know?

The term 'generator' is sometimes used instead of alternator. Generator is an old term and refers to a machine that produces 'direct current'. All modern power stations produce 'alternating current' and use alternators.

Definition

Thermal – heat

Definition

Kinetic – movement

Generation of electricity

Almost all electricity used in domestic and industrial premises is generated at power stations. There are two types of power station, but they all utilise the same principle of a turbine turning a three-phase alternator, which is the actual machine that produces the electricity.

Power stations are either 'thermal' or 'kinetic'; the name refers to the fuel or process used to turn the turbine that turns the alternator and so makes electricity.

Most commercial power stations in use today are of the thermal type, where a fuel such as gas, oil, coal, nuclear fusion etc. is used to heat water, producing high-pressure steam which drives the turbine.

Alternative energy – a wind farm

In recent years, as fossil fuels such as oil and gas have begun to run out, there has been increased research into the use of kinetic power stations. One specific type of kinetic power station which can be seen more and more frequently is the wind turbine, where the wind is harnessed to turn a giant propeller blade – which then turns an alternator to produce electricity. Wind turbines utilise a renewable energy source: the earth's fossil fuels will be used up eventually, but wind is unlikely to stop blowing in the foreseeable future!

A nuclear power station

Another type of kinetic power station is water driven (known as hydroelectric stations). They harness the power of water when it falls from a great height, as down a mountain or from a dam, but some experimental stations are now using wave power from the sea to turn the alternator.

The basic components of a thermal power station are shown in the illustration.

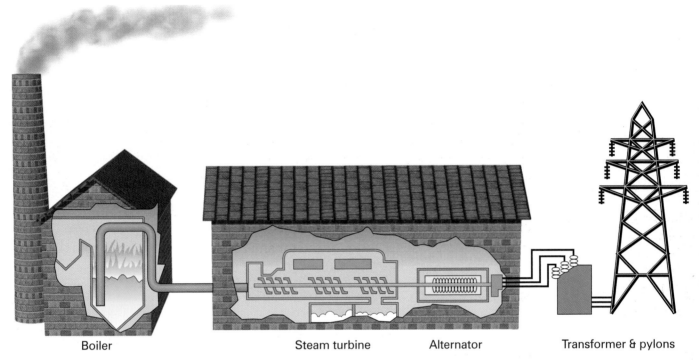

Boiler   Steam turbine   Alternator   Transformer & pylons

Figure 4.35 Simple thermal power-station layout

Supplying electricity

The alternator in a power station produces electricity at high voltage which is then fed through devices called transformers. These increase the voltage to even higher levels before passing the electricity into the National Grid. Electricity is transported through the National Grid at voltages of between 130 and 400 kV. The voltage is raised to these levels so that losses are reduced when the electricity is distributed.

Electricity is 'taken' from the National Grid through a series of appropriately located distribution stations. These transform the grid supply back down to 11 kV and distribute electricity at this level to a series of local sub-stations.

At sub-stations the 11 kV supply is transformed down to 400 V and then distributed through a network of underground radial circuits to customers. In rural areas, this distribution sometimes takes place using overhead lines. The electricity is generated and distributed in the form of a 'three-phase' supply, which is used without modification by some large factories and industrial sites.

The supply that we are familiar with in our homes, however, is a single-phase supply, derived from one of these three phases, and it comes into the main intake point of a home as a 230 V 50 Hz supply with phase and neutral conductors and a main earth point provided.

Did you know?

The National Grid is a network of nearly 8,000 km (5,000 miles) of overhead and underground power lines that link power stations together throughout the country

Sub-stations can be seen dotted throughout our cities. They are normally connected together on a ring-circuit basis

Generally at a domestic property main intake point you will find the following:

- a sealed overcurrent device (fuse or circuit breaker) which protects the supply company's cable.

- a metering system to determine the customer's (consumer's) electricity usage.

The consumer's installation must be controlled by a main switch, which should always be located as close as possible to the supply-company equipment. In the average domestic installation, this device is merged with the means of distributing and protecting the final circuits in what is known as the consumer unit (formerly known as a fuse box).

The diagram illustrates the final part of the electrical supply journey.

A domestic electricity meter

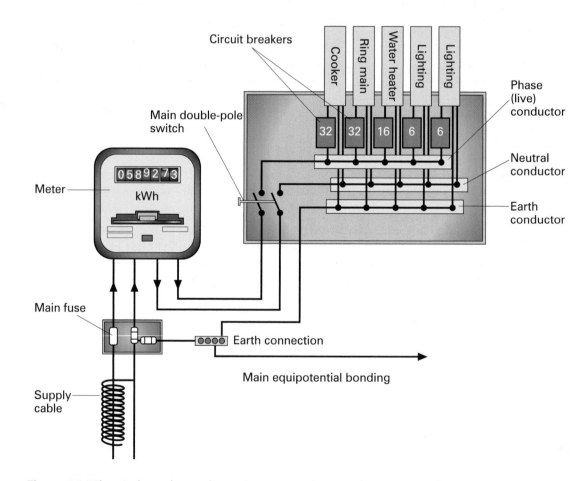

Figure 4.36 Electrical supply to a domestic property showing the meter installation and consumer unit

Basic domestic circuits

As a domestic plumber you will usually come into contact with three types of wiring circuits.

Lighting circuit

This is a radial circuit which feeds each overhead light or wall light in turn. To allow the lights to be turned off and on, the live or phase wire is passed through a wall-mounted switch. Two-way switches are also used (at the top and bottom of stairs or in long passageways), and these require special switch controls.

The lighting circuit is usually fed by a 1.5 mm twin and earth PVC-insulated cable and is protected by a 6 amp mcb or a 5 amp fuse at the consumer unit. You will often find that the lighting in domestic houses is split into two separate circuits, one for upstairs and one for downstairs.

Ring main circuit – 13 amp socket outlets

The sockets used in domestic properties feeding televisions and stereos etc. will normally be 13 amp sockets fed from a continuous ring circuit. As with the lighting circuit, cables circulate from the consumer unit round each socket, but the difference is that they return to the consumer unit forming a continuous loop – hence the term ring main. The ring main permits the cables to be kept to an optimum size as electricity is permitted to flow in two directions to reach the socket.

The ring-main circuit is fed using a 2.5 mm twin and earthed PVC cable and is protected by a 32 amp mcb or a 30 amp fuse.

Spur outlets

Spur outlets are usually used to connect into a ring-main circuit on an existing system (you would not usually encounter spurs on new installations) where it is inconvenient to place a socket from the ring main using the conventional two cables. The spur is connected to the ring main through a junction box, or is wired directly from the back of an existing socket. Spurs can be either fused or non-fused.

The diagram overleaf shows examples of all the features described above.

Definition

Radial circuit – starts at the consumer unit and connects to every point on the circuit, terminating at the last one

Definition

Ring circuit – connects the consumer unit and every point on the circuit, and returns to the consumer unit, creating a ring

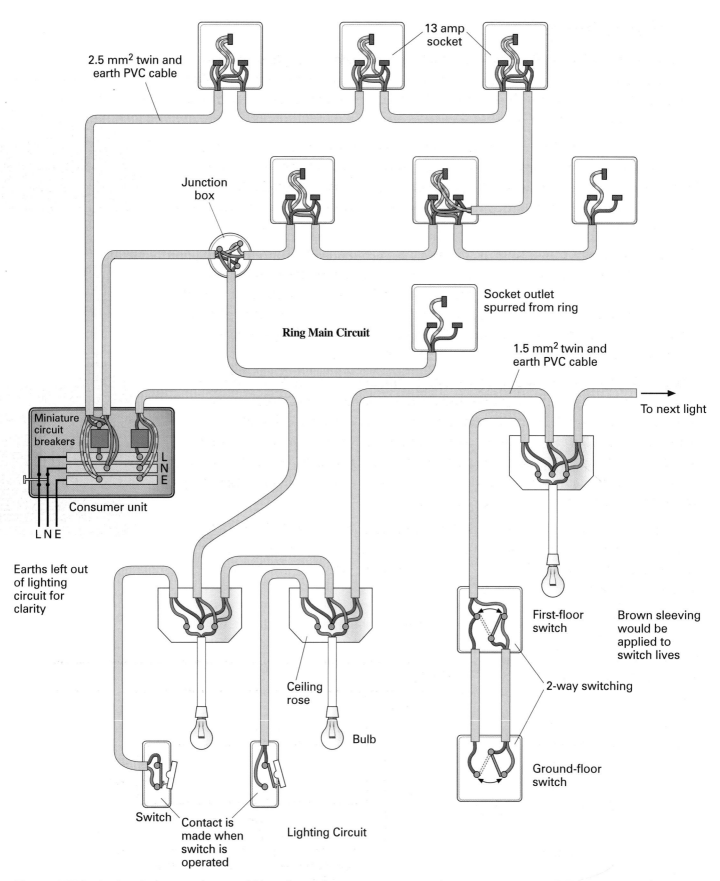

13 amp socket

2.5 mm² twin and earth PVC cable

Junction box

Ring Main Circuit

Socket outlet spurred from ring

1.5 mm² twin and earth PVC cable

To next light

Miniature circuit breakers

L
N
E

Consumer unit

L N E

Earths left out of lighting circuit for clarity

First-floor switch

Brown sleeving would be applied to switch lives

2-way switching

Ground-floor switch

Ceiling rose

Bulb

Switch

Contact is made when switch is operated

Lighting Circuit

Figure 4.37 Basic electrical system layout within a domestic property

Earth continuity

In order for electricity to flow, we know that we must have a complete circuit of conductors, a voltage source to create a potential difference across the ends of the circuit and, for the circuit to be of practical use, a known value of circuit resistance to limit the amount of current flowing.

In the case of a short circuit, when the known circuit resistance is by-passed or breaks down, very high currents can flow, causing serious fire risks due to the heating effect of the current and danger to life from the risk of electric shock. It is for this reason – to protect people from danger – that we **earth** electrical equipment and installations.

The earth can be thought of as a very large conductor at zero potential (voltage). By connecting together all of the metalwork of an installation (other than the metalwork such as copper circuit conductors which are designed to carry current) we ensure that dangerous potential differences cannot exist between metal parts of the installation and earth (ground). If they did, persons touching the 'live' metalwork could get a serious electric shock or even be killed.

One simple way to remember what earthing is all about is to think of electricity as being lazy. It is always flowing around a circuit through a load with a known resistance to control the current, and it flows from a high potential to a low potential. As it is lazy, it will always try to find the path of least resistance to get to the zero potential of earth. If it can, it will do that through you (you have a low resistance) when you touch live un-earthed metalwork, instead of fighting its way around the circuit through the high resistance.

Earthing is designed to provide a very low-resistance path to earth through the exposed installation metalwork so that, in the event of a fault, a very high fault current will flow which will almost instantaneously operate the circuit protective devices such as fuses and circuit breakers.

The definition of earthing from the IEE Regulations (BS 7671 – the electricians, 'rule book') is:

> *the act of connecting the exposed conductive parts of an installation to the main earthing terminal of that installation*

To more fully understand the concept of earthing, study the illustration.

Safety tip

Electrical installations must be earthed to prevent electric shocks

Remember

Electricity is lazy – it will always find the path of least resistance to get to earth

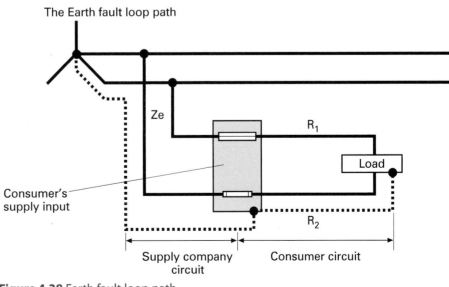

Figure 4.38 Earth fault loop path

Figure 4.39 shows a three-phase supply (the star-shaped bit at the top left of the diagram) which is supplied, by the local electricity authority, to a local sub-station. One 'phase' conductor (single 230 volt) supply is taken into the premises to the consumer's supply input, and a second 'neutral' conductor is also provided. This is connected at the sub-station to the centre of the three-phase supply, which is called the 'star point'. This star point is connected to earth at the sub-station.

You will also notice a dotted line (the earth conductor) which is taken, again from the star point, into the consumer's supply input. At the consumer's end, the earth conductor is connected to a separate terminal called the **main earthing terminal**. It is important to notice at this point that the neutral and earth conductors are connected together at the same point at the electricity authority's supply but are connected to separate terminals when they arrive inside the premises. It is the supply authority's job to maintain the value of the star point as near to zero (earth) potential as possible so that a good low-resistance earth path is always available.

The consumer's responsibility starts inside the premises, where care must be taken that all of the earthing conductors connect to the main earthing terminal to maintain the low-resistance path. The total earth path, all around the installation and back to the star point of the transformer, is known as the '**earth fault loop**' and is given the symbol Ze.

During normal operation, the current will flow through the fuse from the consumer's supply along the phase conductor R1, through the load which has a separate earth connection, and return via the neutral. We know that the neutral is connected to earth back at the sub-station, so the only high potential difference that exists is between the phase conductor R1 and earth.

Imagine now that the load (which we will assume to be an electrical appliance, such as a kettle or toaster) has not been earthed and an internal fault develops which will allow electricity to flow in the casing of the appliance. If the consumer were to come along and touch it, the result could be death!

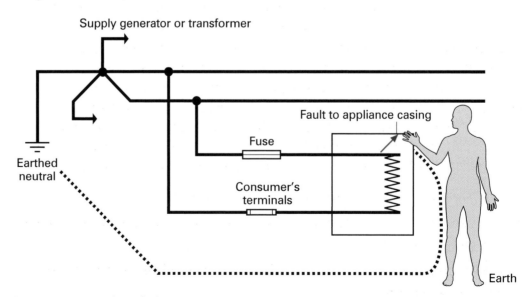

Figure 4.39 An earthing fault

Because the appliance was not earthed, the fault current flowed through the consumer to the low-resistance earth return path. (Remember: why would the lazy electricity struggle to flow through the high-resistance load of the appliance when it can get to the cosy earth through you?)

The type of installation discussed in this section is known as a **TN-S system** and it is very common in this country. As a plumber, you do not need an in-depth knowledge of different types of installation earthing arrangements, but a brief explanation may be useful to you for reference purposes.

The most common types of installation system are listed below, with a brief explanation of each.

TT system

The first T in each of the systems stands for TERRA from the French word *terre*, which means 'earth'. The second T means that the exposed metalwork is connected directly to earth by a separate earth electrode (usually a copper rod driven into the earth). This system is most common in rural areas, where the electricity is often fed by local overhead lines.

TN-S system

Again the T is for TERRA. N means that the exposed metalwork is connected to the main earthing terminal. S means that a separate earth conductor is used throughout the installation. This is a very common system, but it is gradually being overtaken by the TN-C-S system.

TN-C-S system

This is sometimes called 'Protective Multiple Earthing' or PME for short. The T is for TERRA. N means that exposed metalwork is connected to the main earthing terminal. The third letter, C, means that for some part of the system (usually in the supply section) the function of the neutral conductor and earth conductor are combined in a single common conductor. The final letter, S, means that for some parts of the system (usually in the consumer's part of it) the functions of neutral and earth are performed by separate conductors. This type of system is provided in most new housing developments and is the most commonly installed system for new-build environments. While it is very effective, there is a small number of potentially very serious risks with this type of system if it is not installed or maintained properly. For this reason, certain rules are laid down for TN-C-S systems. The main ones are:

1 The neutral conductor must be earthed at a number of points.

2 Neutral and phase conductors must be made of the same material and be of the same cross-sectional area.

3 The neutral conductor must not be fitted with any neutral link or device that can break the neutral path.

You almost certainly will not remember all of the above, and the information is provided for reference purposes only in case you need it.

Bonding and temporary bonding

We have studied earthing and why it is essential to provide a good earth on an installation. Another related term is '**bonding**' and this is something that you will almost certainly come across in your work as a plumber.

When we talk about earthing we generally refer to the designed earthing arrangement within an electrical system where earth conductors are connected to earth terminals in equipment and appliances specifically provided for the purpose.

However, remember that earthing is designed to provide a very low-resistance path to earth through the exposed installation metalwork so that, in the event of a fault, an extremely high fault current will flow, which will almost instantaneously operate the circuit protective devices such as fuses and circuit breakers.

When considering this low-resistance path to earth, we must also think about all of the metalwork inside a modern house that could conduct electricity in the event of a fault occurring. The spidergram lists a number of conductors of electricity that are not specifically designed to carry current.

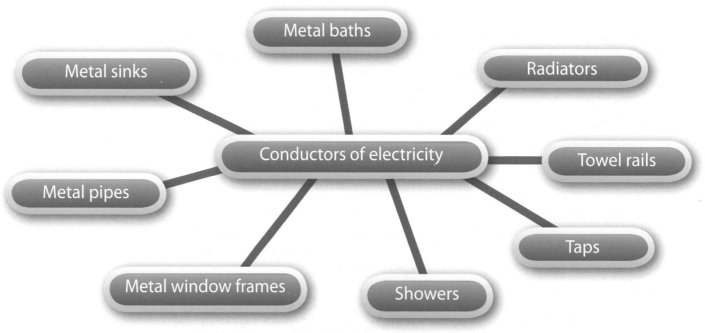

Figure 4.40 Conductors of electricity in a modern house

The use of plastic and non-metallic plumbing fittings and PTFE tape makes the need to connect all of these devices together with low-resistance copper wire even greater, as we cannot safely assume that the total metallic pipework system provides a low-resistance return path for fault currents. The process of connecting these devices together and into the earthing system is called bonding.

Consider, for example, a hot-water radiator central heating system in a house. We could be forgiven for assuming that as the whole system is metal and a good conductor of electricity, then the total resistance must be low.

Now think about it more deeply. Each of the valves connected into the system is likely to have PTFE tape around the thread to improve the water seal. PTFE tape is an excellent insulator. Also it is possible that the system may contain one or more plastic fittings on the cold-water feed side – and again, plastic is a good insulator.

To prevent any build-up of potentially dangerous voltages should a fault occur on the electrical system, all the metalwork is connected together using conductors and 'earth clamps', or clips, in a process known as bonding. It is sometimes called '**equipotential bonding**' because all the metalwork is kept at equal potential so that dangerous potentials cannot exist.

It is a requirement of the electrical wiring regulations BS7671 that all exposed metalwork in a building must be bonded together and connected to the earthing block within the consumer unit. The following bullet points briefly describe the steps that must be taken to ensure the safe earthing of all metallic materials within domestic properties.

- Gas, oil and water pipes can provide a path for stray electrical current. This could lead to corrosion of the pipework and also the potential of electric shock for anyone touching or removing a section of the pipework.

- The bonding of all exposed metal components in a dwelling that are not part of the electrical installation is known as equipotential bonding. The equipotential bonding conductor should be found close to the consumer unit.

- In certain areas of domestic property supplementary bonding may be required. Supplementary bonding is required to link sections of central heating or cold- or hot-water pipework where the metal pipework has been separated by a plastic fitting or length of pipe. This will ensure earth continuity throughout the property.

- When maintenance processes are being undertaken and it is necessary to remove a length of metal pipework, it is essential that the earth continuity be maintained. This is achieved by 'bridging' the gap exposed by the removed section of pipe with a temporary bonding wire. It is vital that the temporary bond is securely fixed in place before the length of pipe is removed.

- Earth clips should be used when connecting bonding wire to pipework. These are designed to indicate clearly the importance of the connection and to show that it ensures a safe electrical connection.

The following illustration shows a typical bonding arrangement for domestic premises. All of the incoming services to the property, such as gas, water and electricity, are bonded together close to the point of entry to the building and are connected to the main earthing terminal.

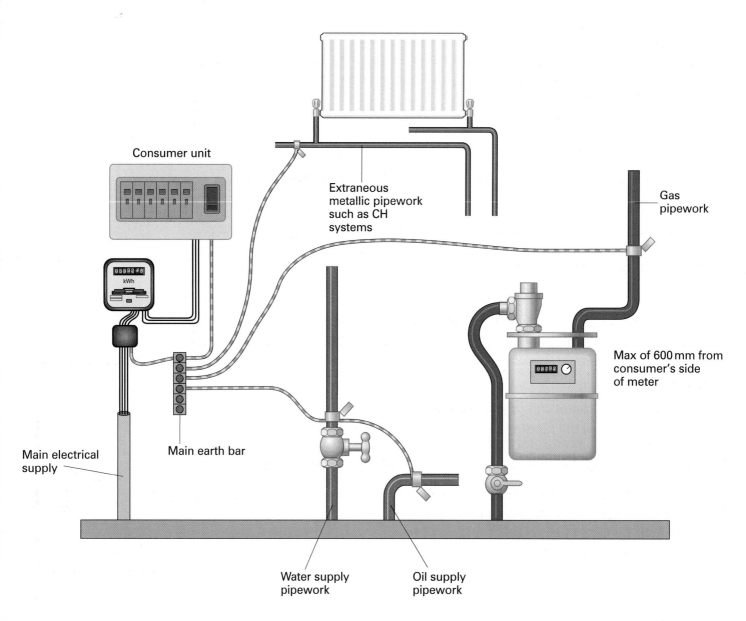

Figure 4.41 Example of the main bonding system in a domestic property

Hand and power tools

At the end of this section you should be able to:

- select and describe tools commonly used in the plumbing industry
- explain safety and maintenance requirements of hand and power tools.

As a plumber, you are required to measure, mark out, cut, fabricate, joint and fix a range of materials. In most cases, this involves the use of tools. Good-quality tools of a well-known brand are expensive but are a good investment.

Once you have put together a toolkit, keep the tools clean and well maintained. This should ensure a long life for them so that you do not have to keep buying replacements.

Hand tools

Tool safety and maintenance checklist

✓ Make sure your tools are cleaned regularly

✓ Lubricate the working part of tools

✓ Once cleaned, lightly coat the tools with an oil spray to prevent rusting

✓ Don't overdo the oiling

✓ Always use the right tool for the job. Screwdrivers are not chisels!

✓ Keep file or rasp teeth clean using a wire brush

✓ Never use tools if they have split handles

✓ Always replace used, worn or defective hand tools.

Hacksaws

- Make sure the teeth are pointing away from the forward cut

- Don't use hacksaws with defective or worn teeth

- When using a large hacksaw, make sure the blade is tightened correctly

- Use the right blade type for the job – 32 teeth per inch for light-gauge pipe such as copper and plastics, and 24 teeth per inch for heavier-gauge LCS.

Hacksaw

Pipe cutters

- Replace damaged or blunt cutter wheels

- Use the correct blade for the material being cut

- Make sure that the wheel and the rollers are lubricated and move freely

- Use a pipe cutter or de-burring tool to de-burr the inside of pipework.

Plastic pipe cutters   Pipe cutter

Wood chisels

- Never use chisels with split handles

- Keep chisels sharp using a grinder or whetstone

- Always keep the plastic guard on the chisel

- Make sure handles aren't loose.

Bevelled wood chisel

Cold chisels

- Keep the cutting edge sharp using a grinder
- Keep the striking end of the chisel free from the 'mushrooming effect', again using a grinder
- Don't use a chisel that has been ground down so much that it has become too small to handle safely.

Cold chisels

Hammers

- Make sure the head is fitted correctly to the shaft
- Don't use a hammer with a defective shaft.

lump hammer

Ball pein hammer

Cross pein hammer

Pipe grips and wrenches

- Keep the teeth free from jointing compounds. If they're clogged up, it could cause the tool to slip
- Once the teeth become worn, the tool should be replaced
- Check for wear on the ratchet mechanism when using pump pliers. These often slip when under pressure

- Be careful when loosening a joint or pipe that is difficult to move. It might give suddenly, and you could damage your hands, or even pull a muscle.

Adjustable spanner

Screwdrivers

- Keep flat-ended screwdrivers for slotted screws to a uniform thickness
- Never use a screwdriver with a defective handle.

Normal slotted and Philips screwdrivers

Stubby PZD/SDV, Philips and normal screwdrivers

Power tools

The main power tools used in plumbing are:

- power drills
  110 V

- cordless powered screwdrivers

- combined cordless drills and screwdrivers

- power saws.

Power drills – 110 V

There is a wide range of power drills on the market. Their power varies with the size of the motor, and this will have a bearing on what each drill can do. Some have a 'hammer action' which, when engaged, makes it easier to drill through masonry. Not only does the drill rotate, but it also moves fractionally backwards and forwards at high speed, giving the effect of hammering the drill into the material.

Typical power ratings range from 620 W to 1400 W. Most drills are variable speed and some have reversible action.

Standard rotary hammer drill 110V

Cordless drills and screwdrivers

The photograph shows a typical example from a very wide range. Most drills are combined so they can be used as a drill and screwdriver. They are powered by batteries and are usually supplied with two batteries and a charger, so that one is working and one is charging. These tools are popular because they are much safer than ones with cords, and you don't need to carry a transformer around. Also, if there is no electricity on site you don't have to revert to hand tools.

Cordless drill/screwdrivers are available in a range of voltages: for example 12 V, 14 V, 14.4 V and 24 V.

Battery-operated hammer drill

Power saws

Most power saws run off 110 V. They are used by plumbers for taking up floorboards or sheets to install pipework under floors. Make sure that nails or screws are avoided when sawing. If possible, take up a board by hand to check what is beneath the area where you are going to use the saw. Always make the cutting depth the same depth as the floor thickness.

Battery-operated circular saws are also available, usually with an 18 V motor.

Another type of saw that plumbers often use for cutting wood is the jig-saw, as shown in the photograph. Similar care should be taken with this type of saw.

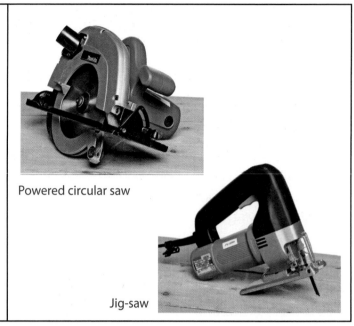

Powered circular saw

Jig-saw

Safety tip

Never use a circular saw without a guard

Safety tip

Remember to wear safety goggles when using drills and saws. These will protect your eyes from dust and any splinters of material that might fly off while working

Find out

What age do you have to be to use cartridge-operated tools?

Checklist of power-tool use

- You should have been properly instructed on the use of the power tool before using it on site.

- Each power tool should be visually inspected for signs of damage prior to its use – if it is damaged it should not be used.

- All electric tools should be double insulated or incorporate an earth cable.

- Battery-powered tools are preferable to mains-operated as they are safer. If you use the mains or a temporary power supply, always use 110 V tools as opposed to 240 V.

- Check that electrical cables are not damaged or worn.

- Check that plugs are not damaged.

- Check for test labels to show the equipment is safe to use. All electrical equipment should be P.A.T. tested in accordance with your employer's procedures. P.A.T. tests are maintenance records of all portable electrical equipment to ensure it is in safe working order – the maximum interval between tests for equipment used for construction is usually 3 months.

- Cartridge-operated tools are covered under the section on Health and Safety. You *must* receive proper instructions on their use before using them.

FAQ

How can it be that a large 1200 litre storage cistern in a bungalow gives less pressure than a small 200-litre storage cistern in a house?

It is all to do with the height of water, *not* the volume. The storage cistern in the house will provide twice the pressure of that in a bungalow due to twice the 'head' but of course in the example given it would empty much quicker as it contains only 200 litres as opposed to 1200 litres.

Knowledge check

1   What does the abbreviation WRAS stand for?

2   Lead is a very malleable metal. What does 'malleable' mean?

3   Which metal is used to galvanise low carbon steel?

4   Which type of plastic is used most frequently in the manufacture of plumbing materials?

5   What are the units of mass, weight and pressure?

6   If you know that the water in a domestic property contains a high percentage of calcium carbonate, would you expect the water to be hard or soft?

7   What is the pH scale used to measure?

8   What is the name for the protective sulphate coating that forms around non-ferrous metals and protects them from further corrosion?

9   What is the material that is destroyed through electrolytic corrosion?

10   What are the principles behind capillary action?

11   What is the unit of measurement for electrical current?

12   State Ohm's Law.

13   What is the voltage of electricity found in domestic property?

14   What is a P.A.T. test used for?

chapter 5

Common plumbing processes

OVERVIEW

This chapter is one of the most important in the book. It provides you with the knowledge to support the basic skills you will need as a competent plumber:

- **Tube bending and marking out**
  - bending methods – copper pipe
  - bending methods – low carbon steel pipe

- **Pipework materials and jointing processes (Part 1)**
  - copper tube and fittings
  - methods of jointing copper tube

- **Pipework materials and jointing processes (Part 2)**
  - LCS and fittings
  - methods of jointing LCS pipe
  - plastic tube and fittings
  - methods of jointing plastic pipes
  - jointing different materials

- **Fixing devices, pipe supports and brackets**
  - fixing devices
  - clips and brackets

- **Associated trade skills**
  - lifting floor surfaces
  - cutting holes in the building fabric

- **Generic systems knowledge**
  - pipework and plumbing symbols
  - measuring jobs and estimating materials
  - documentation
  - work programmes

Tube bending and marking out

At the end of this section you should be able to:

- describe the methods of bending copper and low carbon steel tubes
- select the correct tool or equipment for a given pipe material
- describe common faults and defects in tube bend and their causes.

Bending, measuring and marking out are essential basic skills for any plumber. We will concentrate here on bending copper pipe – which can be done by hand or machine – and low carbon steel pipe, which is usually bent by hydraulic machines.

There are also steel pre-formed 90° brackets for tighter bends, into which the pipe can be clipped.

Bending pipe, rather than using fittings, has the following advantages:

- it produces larger-radius bends than elbow fittings (larger-radius bends have less frictional resistance)
- using bends costs less than using fittings
- long sections of pipework can be prefabricated before installation, saving time.

Did you know?

Plastic pipe can be bent, but its use in plumbing is usually restricted to small-bore polythene pipes which can be positioned into large-radius 90° bends, or offset by hand and then clipped into position

Bending methods – copper tube

The pipe grade suitable for bending is R250 grade X. R290 grade Z is tempered, and has thin walls and cannot be bent. R250/220 grade Y is softer, with thick walls, and is supplied in coils; it is not usually used internally in dwellings. There is also R220 grade W, which is used for micro-bore installations.

Copper pipe can be bent by:

- hand
- machine.

Bending by hand

This is a method of bending pipe when carrying out maintenance and repair work. You may be working in a loft space, for example, where it would be much quicker to fabricate the pipe in situ. The best way of doing this is by using a **bending spring**.

Bending springs can be used externally or internally. In either case you pull the bend against your knee to get the desired angle. When using an internal spring there are a few things to remember.

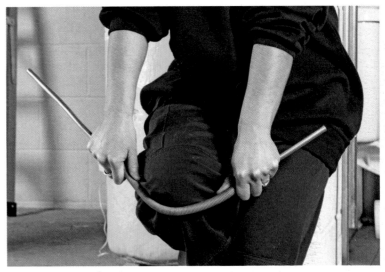

Using an external spring

- Don't try and bend a piece of pipe that is too short. You would have to apply a good deal of physical pressure, which could lead to injury, and kink the pipe, or both.

- Don't make the radius of the bend too tight, or you could find that you can't remove the spring.

- Always pull the bend slightly further than the required angle, and then pull it back to the required angle. This will help to release the spring.

- The spring has a 'hook' at the end. This can be twisted clockwise and will tighten the spring coil inwardly, again making it easier to remove the spring.

Using an internal spring

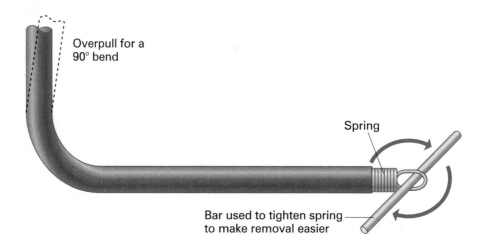

Overpull for a 90° bend

Spring

Bar used to tighten spring to make removal easier

Figure 5.1 Pipe bending with internal spring

Spring pipe-bending can be set out with the same accuracy as machine bending. All bends appear to gain length when bent. This is because, before the pipe is bent, the length A to B to C is longer than the actual bend A to C. It is possible to work out accurately the length of pipe to be bent in relation to fixed points by finding out the length of the pipe actually occupied by the bend.

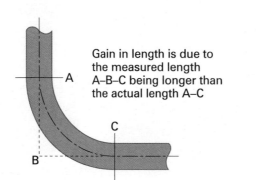

Gain in length is due to the measured length A–B–C being longer than the actual length A–C

Figure 5.2 Apparent gain in length when tubing is bent

This is done by deciding the centre-line radius of the bend. This is usually 4 times the diameter of the pipe (4D).

Next, you apply the following formula:

$$\frac{Radius \times 2 \times 3.14}{4}$$

For example, for a 15 mm diameter pipe:

the radius of the bend = 4D or 4 ×15 = 60 mm

Now, to find the length of bend, use the formula:

$$\frac{Radius \times 2 \times 3.14}{4}$$

Substituting for the radius:

$$\frac{60 \times 2 \times 3.14}{4}$$

this gives 94.26, rounded to 95 mm. This is the length of pipe required to form the bend. The illustration shows how this is applied.

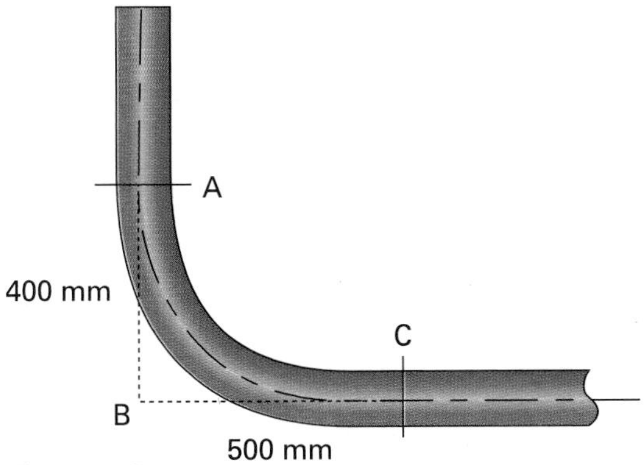

The dimensions for a 15 mm pipework installation are given above. The total length of pipe required, allowing for the bend, would be 400 mm + 500 mm less 95 mm.

Figure 5.3

400 mm
500 mm
A
B
C

Bending by machine

This is the most common method used for bending copper tube. Bending machines can be either hand held (**hand bender**) or free-standing (**stand bender**), and they work on the principle of leverage.

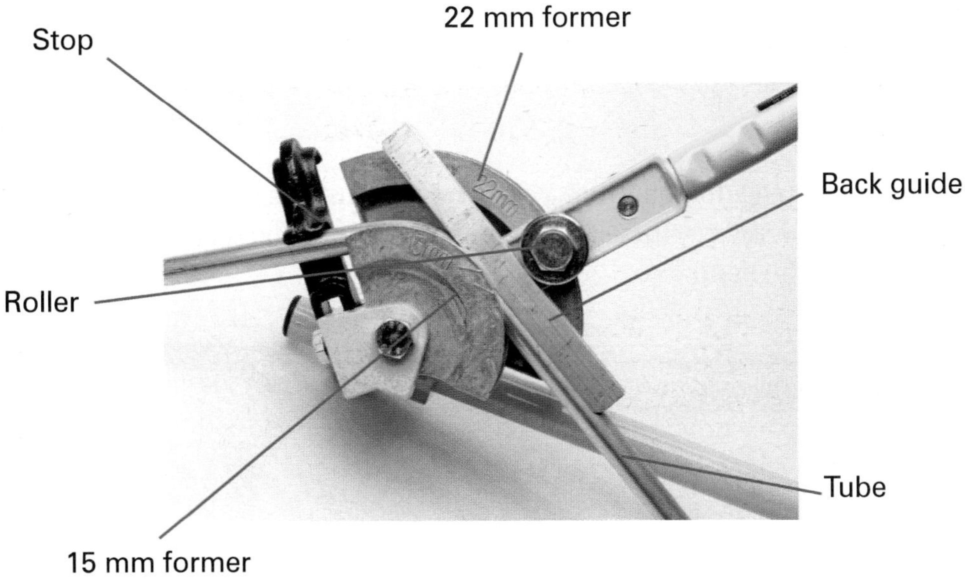

Small mini-bore (8/10mm) bender

Stop
22 mm former
Back guide
Roller
Tube
15 mm former
Hand bender

Stand bender

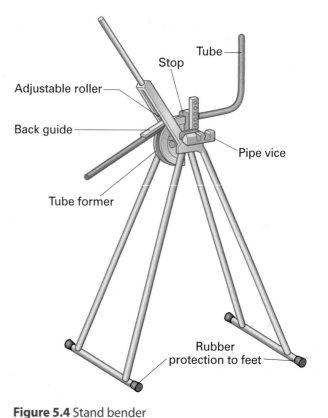

Figure 5.4 Stand bender

The small hand-held bender is used for pipe sizes of 15 mm and 22 mm, and is light and portable. The free-standing bender can handle pipes up to 42 mm and uses a range of sizes for the back guide and former. It is important that the machines are set up properly. If the roller is adjusted so that it is too loose, this will cause rippling on the inside of the radius of the pipe. If it is too tight, it will reduce the pipe diameter at the bend. This is called **throating**.

Types of machine bend

There are three main types of bend:

- 90° or square
- offset
- passover.

90° or square bend

You may be given a measurement from a site drawing, but it is more likely that you will take a site measurement yourself. Measurements should be taken from a fixed point to the back of the bend.

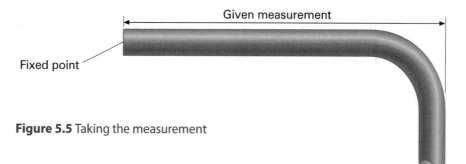

Figure 5.5 Taking the measurement

A pencil mark is made from the fixed point (end of tube) to the required length.

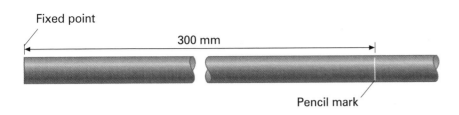

Figure 5.6 Marking and tubing

The tube is set up in the machine with the mark squared off from the outside of the former.

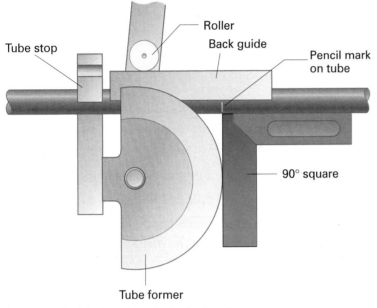

Figure 5.7 Positioning the tube in the bender

Then, the back guide is positioned, the roller correctly adjusted and the bend pulled. Return bends can be made using the same technique, only now the back of the first bend becomes the fixed point.

Offsets

The first set is made to the desired angle. The first angle is not critical, but will usually depend on the profile of the obstacle you want to offset the pipe around.

Now you have made the first set, the tube is reversed and returned to the machine.

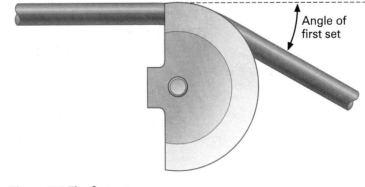

Figure 5.8 The first set

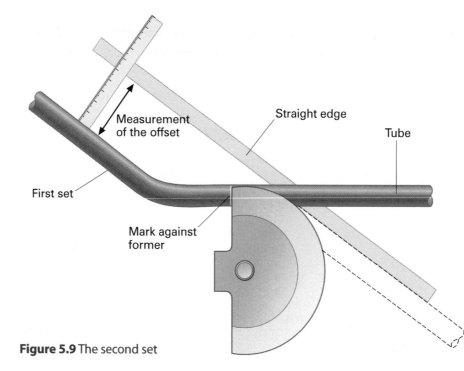

Figure 5.9 The second set

A straight edge is placed against the former, parallel to the tube, and the measurement for the offset is taken from the inside of the tube to the inside edge of the straight edge.

Once the tube has been adjusted against the stop of the machine, it is a good idea to mark the edge against the former, just in case of any movement.

Note the dotted line on the illustration showing the finished position of the second set. This shows that the measurements are taken from the inside to the back of the tube, so in effect it is a centre-to-centre line measurement.

Passover bends

These are used to clear other obstacles, such as another pipe, and could be a passover offset or a crank passover bend.

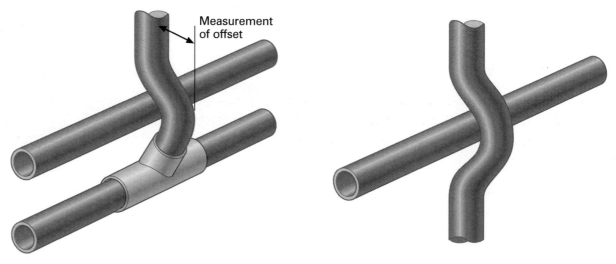

Figure 5.10 Passover offset

Figure 5.11 Crank passover bend

The measurements for a passover bend are taken in the same way as for an ordinary offset. The angle of the first pull will be governed by the size of the obstacle it has to 'pass over'.

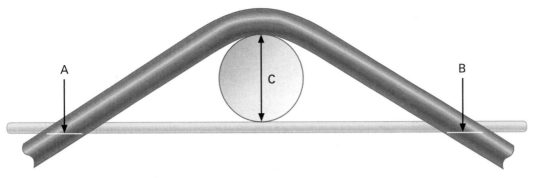

Figure 5.12 Measurement of crank passover

A straight edge is placed across the bend at the distance of the obstacle and the pipe marked. This will be the back of the finished offset.

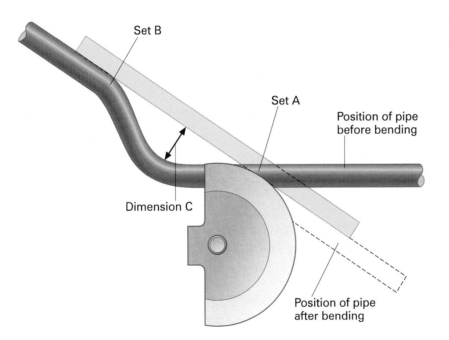

Figure 5.13 The final set in a crank passover bend

The pipe is then returned to the machine, and when the first mark lines up with the former, it can be pulled to the position shown; the pipe is then turned around in the machine, the second mark lined up in the former and the pipe pulled to complete the passover.

Remember

Make sure the first bend is not too sharp or it will be difficult to pull the offset bends

Did you know?

Another technique is sand-loaded heat bending, but this method is no longer commonly used in plumbing in the UK

Bending methods – low carbon steel pipe

Steel pipes are usually bent using a hydraulic pipe bender.

Only medium- and heavy-grade tubes are used in plumbing systems; they are manufactured to BS1387, and are supplied in black-painted or galvanised coatings. Galvanised pipe should not be bent, as it this will cause it to lose its coating. It is unlikely, however, that you will come across this material other than on maintenance work. You are more likely to work on medium-grade tube.

Hydraulic bending machine

Bending with a hydraulic machine

Hydraulic machines are needed to bend low carbon steel tubes, owing to the strength of the material and the thickness of the pipe. For this reason, the pipe does not need to be fully supported with a back guard, unlike copper pipe. Hydraulic bending machines are used to form most bends, including 90° and offsets.

The hydraulic mechanism is usually oil-based, and as liquids are incompressible, once under pressure they can exert a considerable force on the pipework.

Forming a 90° bend

Six steps to forming a 90° bend:

1 Mark a line on the pipe at a distance from the fixed point, where the centre line of the finished bend is required.

2 From this measurement you deduct the nominal bore of the pipe. This is because there is a gain in length of one pipe diameter when bends are made.

3 Make sure the correct size former is in the machine.

4 Put the pipe in the bending machine and line up the mark with the centre of the former.

5 The machine can then be worked to apply pressure and bend the pipe to 90°.

6 Due to the elasticity of the metal, you need to take it to approximately another 5° over 90° to allow it to 'spring back'.

Five steps to forming an offset

It is a good idea to make a template from steel wire to help you achieve the required offset profile. The method of marking out the offset is similar to that of copper:

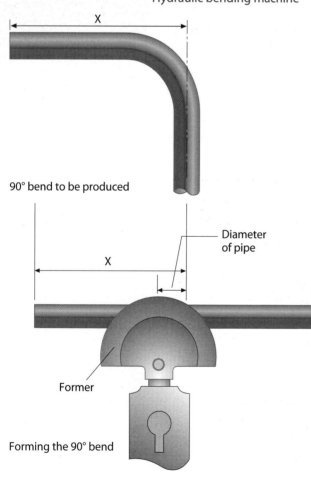

Figure 5.14 Forming the 90° bend

1 Mark off the required measurement for the first set onto the pipe.

2 Place the pipe in the machine, but do not make any deduction. The measurement X mm is from the fixed end of the pipe to the centre of the set.

3 Pull the first set to the required angle.

4 Take the pipe from the machine and place a straight edge against the back of the tube. The measurement of the offset is marked at point A.

5 Replace the tube in the machine and line the mark up with the centre of the former. The second set is pulled and checked against the wire template. Again, allow a 5° overpull for the spring back.

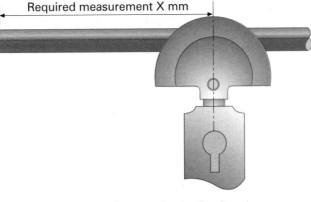

Required measurement X mm

Figure 5.15 Locating the pipe for the first bend

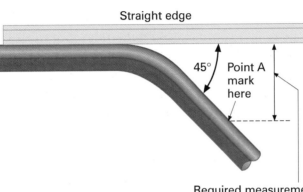

Straight edge

45° Point A mark here

Required measurement of offset

Figure 5.16 Setting out for the second set

Remember

During the bending operation, the pipe can become wedged into the former. Do not hammer the former or the pipe to remove it. You can place a timber block on the pipe and give it a sharp tap with a hammer. Alternatively, remove the pipe and former. Place a wooden block on the floor and strike the end of the pipe on the block. Try to hold the end of the former so that it doesn't fall to the ground

Faults and defects

The main problems encountered with low carbon steel pipe bending are usually down to incorrect marking out, for example not making the deduction for the nominal bore, so the bend does not fit when it is pulled. In the case of offsets, if the pipework is not set up level in the machine, then the resulting offset will be screwed or twisted.

FAQ

How can I make copper tube easier to bend by hand?

By annealing the copper tube first. This is a process of heating the copper tube cherry red and then cooling in water/allowing to cool. The tube will be in a much softer state for bending with a spring.

Pipework materials and jointing processes (Part 1)

At the end of this section you should be able to:

- describe the pipe jointing process for copper tube
- select the correct fitting for a given jointing task
- select the correct pipework material for an installation.

Pipes, fittings and jointing materials acceptable for Water Regulation purposes are listed in the Water Fittings and Materials Directory. BS6700 also states the minimum requirements for pipe joints and fittings. You will cover this in much more detail at Level 3, but you should remember that the following factors must be taken into account when selecting materials for use in plumbing systems.

- Affect on water quality.
- Vibration, stress or settlement.
- Internal water pressure.
- Internal and external corrosion.
- Compatibility of different materials.
- Ageing, fatigue, durability and other mechanical factors.
- **Permeation**.

In domestic plumbing installations the main materials that you will work with are copper, low carbon steel and plastic. To a lesser extent, and usually only on maintenance work, you will also come across cast iron and stainless steel.

Copper tube and fittings

Copper tube is available in four grades (under a new BS Standard the terminology used for copper grades has changed slightly, although the grades shown below are still widely used):

Grade X

Grade X (R250 half hard straight lengths) is widely used for domestic installations. It is classified as half hard, and has pipe diameters ranging from 12 mm to 54 mm. Its pipe diameter is always specified as the external measurement, and the tubes are normally available in 6 m lengths, although most merchants will also supply in 1 m, 2 m and 3 m lengths. Grade X tube should not be used underground. Grade X is also available in chromium plate. It is used where pipework is exposed to the eye and an attractive finish is required.

Grade Z

Grade Z (R290 hard straight lengths) is not as popular as grade Y, one of the main reasons being that it is unbendable. This is because its wall is 0.2 mm thinner than grade X, so the pipe is hardened during the manufacturing process to make

Definition

Permeation – the possibility of microscopic particles of oxygen entering the water supply through the external wall of the pipe

Remember

You may find it useful to refer back to the section on plumbing materials in Chapter 4 when reading this

it stronger. Grade Z is also available in 1 m, 2 m, 3 m and 6 m lengths, and in pipe diameters of 12 mm to 54 mm. Like Grade X, grade Z is available in chromium plate.

Probably the most likely place where you would find chrome-plated pipe is in exposed pipe runs to instantaneous showers in a bath or shower room, particularly where the appliance is a new addition to an existing suite.

Find out

Can you think where you might see chrome plated pipework in a domestic dwelling?

Grade Y

Grade Y (R250/220 soft coils) is used for external underground installations. Its wall thickness is 1 mm, making it the thickest grade, and it is classified as fully annealed, which means it is soft. The size range is 15 mm and 22 mm, and it is supplied in 25 m coils. It is easy to form into wide-radius bends by hand, without affecting the bore of the tube, so the use of bending equipment is not required. Grade Y can be supplied in either a blue (water service) or yellow (gas) plastic coating.

Grade W

Grade W (R220 soft coils) is used for micro-bore heating and, like grade Y, is fully annealed. The sizes used for micro-bore are usually 4 mm, 5 mm, 6 mm, 8 mm, 10 mm and 12 mm. The pipe is supplied in coils ranging from 10 m to 30 m. Again, like grade Y, the pipe lengths can be formed into wide-radius bends or offsets without the use of bending equipment. For short-radius bends, a number of purpose-made hand formers are on the market. Grade W can also be supplied with yellow (gas) or white (heating) plastic coating or with air channels in the plastic coating to improve thermal insulation. This is used where pipework is installed internally in solid floors.

Compression joints

Methods of jointing copper tube

There are three main methods of jointing:

- compression joints
- soldered capillary joints
- push-fit joints.

Soldered capillary joints

Compression joints

Compression joints are of two types:

- manipulative – type B
- non-manipulative – type A.

Plastic push-fit joints

Manipulative joint

The illustration shows a typical manipulative fitting detail; these are sometimes referred to as type B.

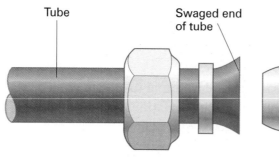

Figure 5.17 Manipulative compression fitting

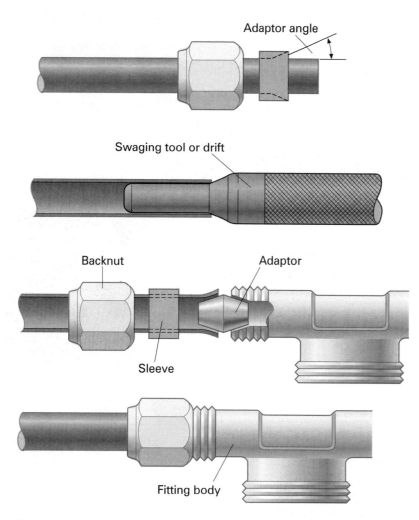

Figure 5.18 Stages of assembly of a manipulative compression fitting

The jointing process is as follows:

1 Cut the tube to the required length, using either tube cutters or a hacksaw.

2 De-burr the inside of the pipe (if cut with tube cutters) or the inside and outside of the pipe (if cut with a hacksaw).

3 Place the nut and compression ring onto the pipe.

4 Using a **swaging tool** or drift, flare out the end of the tube so that it fits the angle of the adaptor.

5 This process is applied to each pipe end being used in the fitting (2 for a straight coupling or elbow, 3 for a tee).

6 The fitting can now be assembled. The angled side of the adaptor fits in to the flared end of the tube, the other side in to the body of the fitting. The joint should be finger tight, with all the fitting components fully engaged. The fitting can then be completely tightened using adjustable grips or spanners.

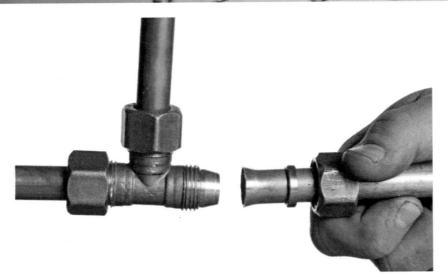

Manipulative joint being coupled

Compression joints, when used underground, must always be of the manipulative type.

Non-manipulative joint

Backnut   Compression ring   Fitting body   Tube

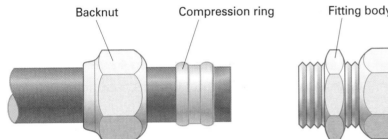
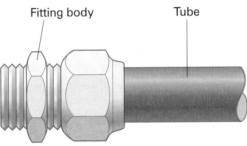

Figure 5.19 Detail of non-manipulative (type A) joint

Sometimes referred to as type A, this is similar to the manipulative fitting. The main difference is that the pipe end is not flared to receive an adaptor; instead, a **compression ring** or **olive** is used to provide a watertight seal on the pipe end. The cutting and deburring process is as before. Again, the joint is hand-tightened first, making sure the pipe is fully pushed into the body of the fitting, before completely tightening with adjustable grips or spanners.

Non-manipulative joint being coupled

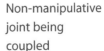

Remember

This type of fitting is suitable for use on grade X, Z and W pipes. Their use is not allowed on underground services

Brass fittings

A range of brass fitting designs, compatible to pipe sizes, is available. For example:

- straight adaptor
  - with parallel male thread
  - with parallel female thread
  - with taper-made thread
- straight couplers
  - equal compression and reducers
  - imperial to metric
  - with drain valve
  - with air release valve
  - compression to solder-ring capillary
- tap connectors
- stop ends

- tees
  - equal
  - reducing
- elbows
  - equal
  - with drain tap
  - with air release valve
  - compression to parallel female thread
  - compression to parallel male thread
- wall-plate elbows
- bent and straight tank connectors.

Soldered capillary joints

Soldered joints can be classified as soft-soldered and hard-soldered. Hard-soldered joints, for example using silver and silver alloys including copper, require heating to a much higher temperature than soft-soldered joints. They are rarely used on the majority of domestic plumbing installations and will not be discussed here.

Soft-soldered joints are made using two types of fittings:

- integral solder ring
- end feed.

Safety tip

Solders containing lead are not allowed on pipework used for either hot or cold water supplies

Find out

How are soldered joints different from other methods?

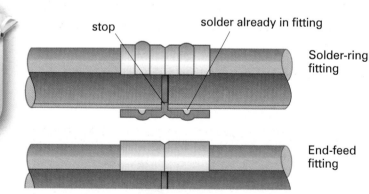

Figure 5.20 Soldered capillary joints

The difference between the two jointing methods is that the solder ring has solder contained in a raised ring within the body of the fitting (integral). When the fitting is heated, the solder melts (at between 180 and 230°C) and the solder is drawn into the fitting by capillary action. The same principle is used for the end feed, except that the solder is end-fed separately from a spool of wire solder.

The jointing process is fairly straightforward:

1   The tube should be cut square and de-burred.

2   Because the inside of the fitting and the outside of the tube are oxidised, the solder will not adhere to the joint correctly; these surfaces should be cleaned to a 'bright shiny light orange' colour (Step 1).

3   Flux is then applied to the surface of the tube and fitting using a brush – not your fingers! (Step 2).

4   The joint is assembled and heat is applied to the body of the fitting (Step 3).

5   If using solder ring, wait until you see a complete film of solder around the end of each fitting. If using an end feed, apply solder from a wire spool and feed it in until the joint is complete.

Remember

Do not use too much solder: it will leave an unsightly appearance when it cools

Only use lead-free solder on hot- and cold-water supplies – a mixture of tin and copper. The spool should be clearly marked *lead free*

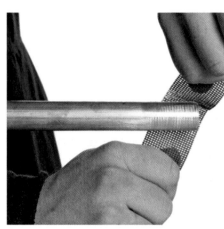

Step 1 Clean and de-burr tube

Step 2 Apply flux using brush

Step 3 Apply heat to body of fitting

6   When the fitting is cool, clean off any excess flux with a cloth. The system should also be flushed to remove flux residue from inside the pipework, as this could corrode the inside of the pipes (Step 4).

The table below provides a summary of what type of fitting can be used on the various grades of copper tube.

Did you know?

There are a number of fluxes on the market that are heat activated. A cleaning action takes place during the heating process, so you don't have to pre-clean the pipe end and fitting. When using these fluxes you should make sure they are non-acidic, non-toxic and WRAS-approved

Jointing method	Grade X	Grade Y	Grade Z	Grade W
Capillary fitting	✓	✓	✓	✓
Type A Compression fitting	✓	✗	✓	✓
Type B Compression fitting	✓*	✓	✗	✗

*Not underground

Figure 5.21 Summary of copper-tube fittings

Capillary fittings

As with compression fittings, there is a vast range of soldered capillary fittings (the same design patterns apply to solder ring and end feed):

- adaptors
  - capillary to male and female thread
- slip couplings
- straight couplings
- tank connectors
- reducing fittings
- metric to imperial converters
- stop ends
- tap connectors
  - straight
  - bent
- return bends
- cylinder unions
- elbows
  - male to female
  - female to female
  - capillary to **BSP**, male or female
- wall connectors (or back-plate elbows)
- tees
  - equal
  - reducing
- offsets

Push-fit joints

There are a number of types of push-fit joints available for use on hot- and cold-water supplies.

Push-fit joints are made from plastic or metal. In the illustration, a grab ring is used to lock the pipe in place, and a neoprene 'O' ring makes it watertight. The fittings are bulky, so they do not look attractive where they are exposed.

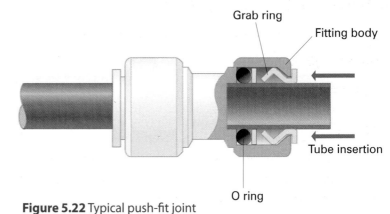

Figure 5.22 Typical push-fit joint

(labels: Grab ring, Fitting body, Tube insertion, O ring)

Pipework materials and jointing processes (Part 2)

At the end of this section you should be able to:

- describe the pipe-jointing process for:
  - low carbon steel pipe
  - plastic tube
  - cast-iron pipe
  - stainless-steel tube.
- select the correct fitting for a given jointing task
- select the correct pipework material for an installation.

Low carbon steel pipe (LCS) and fittings

Often referred to as mild steel, low carbon steel pipe is supplied in three grades:

- light – colour-coded brown
- medium – colour-coded blue
- heavy – colour-coded red.

It is supplied painted in black or with a galvanized coating. Black-painted LCS must only be used on wet heating systems, oil or gas supply pipework, and not on hot and cold supplies.

Generally speaking, light-grade tube is not used for plumbing pipework. You are most likely to work with medium-grade pipes, and occasionally heavy-grade.

Medium and heavy grades are available in approximately 6 m lengths, ranging from 6 mm to 150 mm in diameter, specified as nominal bore. Nominal bore means that the figure is not the actual bore of the pipe, as this will vary depending on the thickness of the pipe wall, which in turn will be determined by the grade.

Methods of jointing LCS pipe

For smaller installations, there are two main jointing methods:

- threaded joints
- compression joints.

Threaded joints

Jointing LCS pipe can be done by cutting threads into the end of the LCS pipe to give a British Standard Pipe Thread (BSPT), then jointing them together with a range of female threaded fittings made from steel or malleable iron. The threads are cut using **stocks and dies** or a threading machine.

The jointing process is as follows:

1   The LCS pipe is cut to length using heavy-duty pipe cutters while the pipe is held securely in a vice. The LCS pipe can also be cut using a large frame hacksaw.

> **Definition**
>
> **Stocks and dies** – the stocks are the body and handle of the tool, the dies are the actual cutter

Pipe vice

Pipe cutter

2 Remove any burrs from the inside and outside of the pipe

3 You need to **chamfer** the end of the pipe that is being threaded to provide a leading edge for the dies to catch the pipe wall.

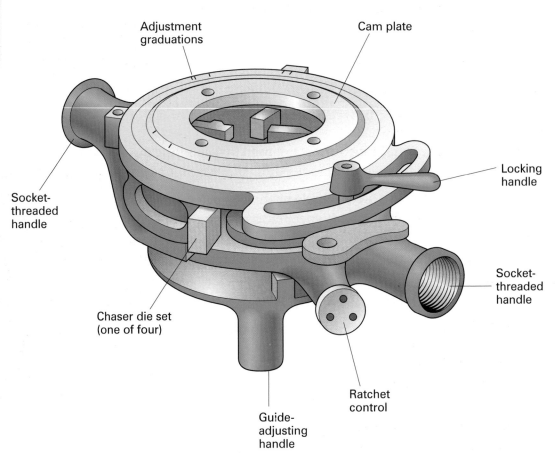

Figure 5.23 Chaser die stock

4 Cutting compound should be applied to the end of the pipe before starting the thread-cutting operation. A thread should be cut that is approximately $1\frac{1}{2}$ to 2 threads longer than the length of the inside of the fitting.

5 The excess cutting compound is wiped off.

6 Threaded pipe sealant, which includes hemp and paste, PTFE sealing tape or gas thread sealing tape, can then be applied to the thread. The fitting is then screwed in place.

Assembling steel pipework using a vice

7 The pipe and fitting are joined together using adjustable pipe grips or alternatively using a short length of pipe as a lever.

Because of the way in which screwed joints are made and installed (rotated on the pipe), it is some times difficult to remove or assemble lengths of pipework. Where this is the case, a union connector, which allows the pipework joint to be 'broken', should be used.

Using a pipe-threading machine

Pipe-threading machines provide a quicker and easier method of forming threads for LCS pipes. 'All in one' machines have a pipe cutter, de-burring reamer, and stock head and dies.

Threaded pipe fittings for LCS

These fittings can be made of steel or malleable iron. Steel fittings can withstand higher pressure but are more expensive than malleable iron. They are manufactured to BS1740 for steel and BS1256 for malleable iron.

Malleable cast-iron fittings are adequate for smaller installations, and as with copper tube fittings there is a wide range available, including:

- bends
    - male to female
    - female to female
- elbows
    - male to female
    - female to female
    - female to female union
    - female to female reducers
- tees
    - female to female to female equal
    - female to female to female reducing and increasing
    - female to female to female pitcher equal
- crosses
    - female to female to female to female equal
- sockets
    - female to female equal
- unions
    - female to female equal
- bushes
- nipples.

Compression joints

There are a number of manufacturers' designs for compression joints; a typical one is shown in figure 5.25.

The fitting is designed to allow steel pipes to be formed without threading. Made of malleable iron, compression couplings use locking rings and seals which are tightened onto the pipe. They can be used on gas supplies and water, and although they are more expensive than threaded joints, they do save time on installation.

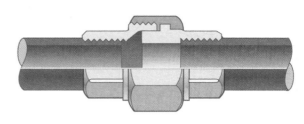

Figure 5.24 LCS union connector

Electric pipe-threading machine

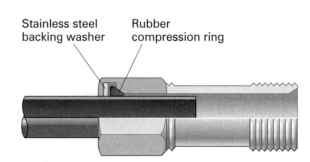

Stainless steel backing washer
Rubber compression ring

Figure 5.25 LCS compression coupling

Plastic tube and fittings

Plastics fall into two main categories:

1 products of the polymerisation of ethers:
   - polythene
   - polyethylene
   - polypropylene.

2 UPVC and ABS.

The first group are used mainly for service-supply pipework (coded blue for underground services), hot water and hot-water heating.

The second group are used mainly for waste distribution pipework and cold-water installations, although the latter will not be discussed here.

The following table summarises the typical jointing methods for plastic pipes.

Types of plastic	Mechanical joints	Solvent welding	Fusion welding	Push-fit 'O' ring*
Polythene	✓	✗	✓	✗
Polyethylene	✓	✗	✗	✗
Polypropylene	✓	✗	✓	✓
UPVC	✓	✓	✗	✓
ABS	✓	✓	✗	✓

*Discharge pipes and overflows only

Figure 5.26 Typical jointing methods for plastic pipes

Methods of jointing plastic pipe

Fusion welding: polythene and polypropylene

This process requires specialist equipment and is used mainly for gas and water mains. It will not be discussed here.

Mechanical jointing

This applies to the jointing of:

- polythene pipework
- UPVC and ABS.

Polythene pipe – compression joint

Used for underground services, this pipe is identified by blue colour coding. It is also available, coded black, for internal use on cold-water services, although for domestic situations it is mainly used underground. The joints are made using gunmetal or brass or plastic fittings.

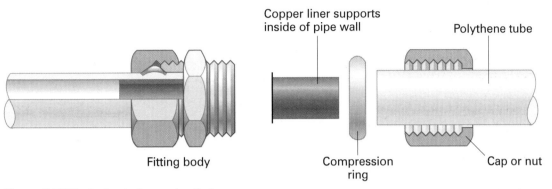

Figure 5.27 Typical polythene pipe fitting

The pipe is cut to length and de-burred. The nut and compression ring are slid onto the pipe, and the copper liner inserted into it. This stops the plastic pipe from being squashed when the nut is tightened onto the body of the fitting and the compression ring applies pressure. Make sure the pipe is fully inserted in the fitting, and hand tighten. Complete the tightening process using adjustable grips or spanners.

UPVC and ABS – compression joint

These joints are restricted to use on waste pipes. Traps are a typical example, using a rubber 'O' ring or rubber plastic washer to make the seal.

UPVC and ABS – solvent-welded jointing

This method is used to joint both UPVC or ABS, using solvent cement. The cement temporarily dissolves the surface of the pipe and fitting, causing the two surfaces to fuse together. The joint sets initially within 5–10 minutes but will take 12–24 hours before it is fully set.

Its use is primarily for joints on soil or waste pipes.

Applying solvent to joint

UPVC and ABS – push-fit joints

These are used mostly on UPVC or ABS soil and waste applications. The pipe is cut to length, making sure it is square and de-burred. The outside edge of the pipe is chamfered to give it a leading edge to make it easier to push into the fitting. This also prevents the 'O' ring being dislodged.

Lubricant is applied to the pipe and fitting, and the pipe is pushed home. Make sure the pipe is withdrawn from the fitting to a length of about 10-15 mm to allow for expansion.

Push-fit connectors

These are also known as **flexible push-fit plumbing systems** for hot and cold water and central heating supplies. A number of manufacturers produce pipe and fittings for these systems, and the fittings can be used on either plastic or copper.

Plastic pipe is supplied in diameters ranging from 10 mm to 22 mm and in 3 m and 6 m lengths, or in coils of 25 m, 50 m and 100 m.

Barrier pipes (which prevent **permeation**) must be used for vented and sealed central heating systems.

They are manufactured from a cross-linked polythene pipe to BS7291.

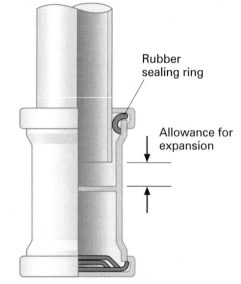

Push-fit joints used for low-pressure plastic pipework

Figure 5.28 Push-fit joint

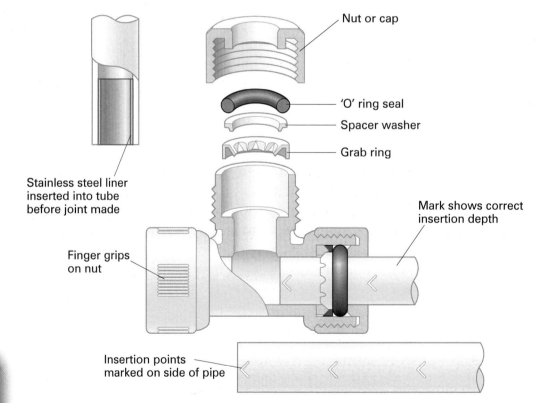

Figure 5.29 Typical de-mountable fitting for polybutylene pipe

A wide range of fittings is available that are similar to their copper equivalents. In a typical system a mix of copper and **polybutylene** will be used: copper for exposed pipework and polybutylene for hidden pipework. The jointing and installation processes are fundamentally different to copper, but generally take less time to complete.

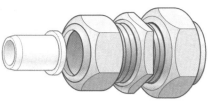
Lead to MDPE

Jointing different materials

Fittings are available that are designed to join copper to plastic; usually they are the plastic push-fit variety. Fittings for jointing copper to LCS are also available.

Although lead pipework is no longer allowed for new installations, it may be necessary to join into an existing supply, either to extend a system where a long run of lead pipework cannot be replaced, or where the joint is at the end of an underground service pipe.

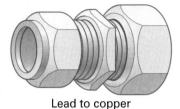
Lead to copper

There are several types of fitting on the market for jointing plastic to lead for below-ground use, and for jointing lead to copper. Only use WRAS-approved products – joint wiping lead to copper is no longer permitted under the Water Regulations.

Stainless steel

Stainless steel is not used on domestic pipework installations. You may, however, come across it on systems installed during the copper shortage of the late 1960s, when stainless steel was used as a replacement.

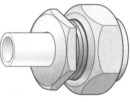
Lead to end feed

Figure 5.30 Fittings to joint different materials

Stainless steel was jointed using compression fittings, and as the outside diameter was that of copper pipe, it would be logical to connect to copper using a compression fitting if carrying out maintenance work.

Cast iron

You are unlikely to install this on new domestic installations, except on historic or listed buildings, but you could come across it during maintenance work, as it was once used extensively on gutters, rainwater pipes and soil pipes. It is still widely used on industrial/commercial buildings due to its resistance to mechanical damage.

Fixing devices, pipe supports and brackets

At the end of this section you should be able to:

- state what the various types of fixing devices are
- select a specific type of fixing device for a particular job
- explain the requirements for supporting pipework
- state what the various types of bracket and clips are used for
- select a specific type of bracket or clip for a particular job.

Find out

The range of sizes for fixing devices is vast, and too detailed to explain here. Have a look at a hardware catalogue so you can see the range for yourself

In this section we will look at the various types of fixing devices used to secure pipework so that it looks neat and is kept in its proper position. Fixings should provide sufficient support to withstand possible accidental damage from people treading on pipework, children pulling at it and so on.

As a plumber you will be required to fix pipework, sanitary ware and appliances to various surfaces. You will also need to know how to re-fit boards and access traps in timber floors.

Fixing devices

These include:

- screws
  - brass wood screws
  - self-tapping screws
  - turn-threaded wood screws
  - steel countersunk screws
  - chipboard screws
  - mirror screws

- plastic plugs
- plasterboard fixings
- cavity fixings
- nails.

Screws

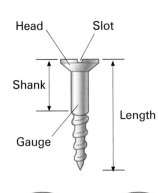

Head   Slot

Shank

Length

Gauge

Slotted

Pozidrive

Philips

Countersunk

Pan head

Mirror screw head

Roundhead

Raised head

Screws are specified by their length in inches, and gauge. The usual lengths used in plumbing range from $\frac{5}{8}$" × 8 for fixing 15 mm saddle clips to $2\frac{1}{2}$ × 12 for fixing radiator brackets.

This is first a rule of thumb, and as a plumber you often have to make a decision about the length and gauge that is right for a particular situation. For example, if the wall you are fixing to is not in good condition, you might have to use a longer screw and thicker gauge. Trial and error and experience are factors here!

Brass or alloy screws are used internally where they could be affected by moisture; this would include sanitary appliances, for example. They are also used externally (mostly in alloy form due to costs) for soil and rainwater fixings and wall-plate elbows for hose union bib taps.

Self-tapping screws are used when fixing into metal sheet. This is not typical in domestic installations, but could apply if you needed to clip to a metal stud partition. This involves drilling a pilot hole in a smaller gauge than the screw, using a steel drill bit, and then screwing into it.

Steel countersink screws are used for general purposes, such as fixing clips and radiator brackets.

Chipboard screws, as the name suggests are used for chipboard fastenings, for example if you had to fix an access trap in a chipboard floor following some installation work.

Figure 5.31 Some commonly used screws

Mirror screws are used where appearance is important, and are often used for fixing timber or plastic bath panels.

Plastic plugs

Plugs come in a range of gauges that are appropriate to the gauge of the screw. They are colour coded for ease of selection.

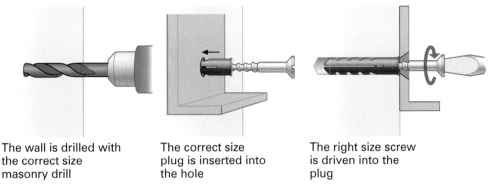

| The wall is drilled with the correct size masonry drill | The correct size plug is inserted into the hole | The right size screw is driven into the plug |

Figure 5.32 Fixing operation using plug and screw

This chart will help you to select the correct drill, plug and screw:

Screw size (gauge)	Drill size (mm)	Plug colour code
6–8	5	Yellow
8–10	6	Red
10–14	7–8	Brown
14–18	10	Blue

Plasterboard fittings

These are used when fixing to plasterboard stud partitions, where there is nothing solid to fix into at the back of the plasterboard. There are several types available, as shown in the illustrations.

When using a spring toggle, drill a hole in the plasterboard big enough to take the toggle when folded. This is inserted through the hole, and when it is in the space behind the board, the toggle is pushed open by the spring.

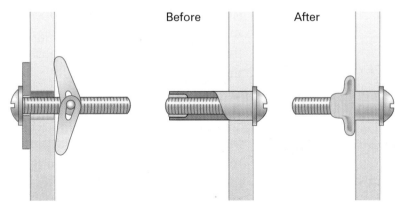

Before   After

Figure 5.33 Spring toggle and rubber-nut fixing

Rubber-nut fixings work on the principle of drawing the nut mounted in the rubber towards the screw head. As it tightens, the rubber is squashed to form a flange at the back of the board.

Cavity fixings work on the same principle, but this time an aluminium body is 'squashed' to form the flange.

When using an all-metal plasterboard fixing, a small pilot hole is drilled into the plasterboard and the complete fixing is screwed into the board. The screw is then removed, leaving a fixing point similar to a plastic plug.

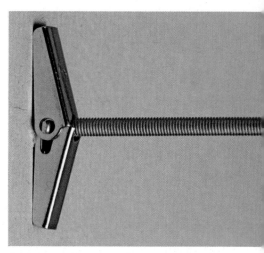

Spring toggle as seen from reverse side

Other fixings

The fixings described so far should be adequate for day-to-day work. There will be occasions, however, when you need larger fixings, perhaps for a heavy appliance, or where the fixing surface is in poor condition.

Coach screws are normally supplied with their own purpose-made plastic plug, and are fixed like a normal plug and screw. The difference is that they are tightened using adjustable grips or preferably a ring spanner.

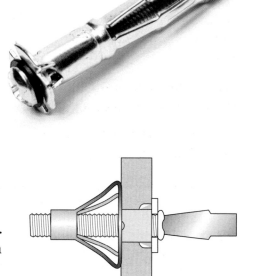

Figure 5.34 Cavity fixing

A wall bolt is an extremely strong masonry fixing. A hole is drilled in the masonry just large enough for the gauge of the bolt. The bolt is inserted and the nut tightened. This pulls the tapered end forward, expanding the segments against the masonry (similar to a cavity fixing). The nut is then removed, the item put in position, and the nut and washer reattached.

Nails

Plumbers use a range of nails, particularly on maintenance, repair and refurbishment work. This includes procedures such as refixing floorboards (although you should use screws where access to pipework is required), pipe boxing, skirtings and boards for clipping pipe runs. If you are working for a small business, you may have to build your own support platforms for cold-water storage cisterns.

You will probably carry the following in your tool bag:

- panel pins
- oval brad/lost head
- masonry nails
- round head, plain and galvanised.

All are available in a range of lengths and diameters.

Clips and brackets

It is likely that the majority of your work will take place in domestic dwellings where the use of copper or plastic clips is adequate for supporting copper and plastic pipework. Once again, there is a range of sizes of clips available.

However, there may be times when you will work on other buildings: schools, hospitals or small industrial units, for example, where the clips or brackets need to be strong and robust.

On large jobs, clips or brackets will be specified, but frequently the plumber has to decide what type of fixing to use.

Plumbers also include low carbon steel pipe (LCS) fixings, which are similar to the brass fixings shown for copper.

The table below lists the recommended spacings for internal pipework fixings:

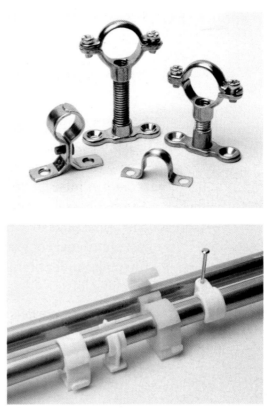

Selection of metal and plastic pipe fittings

Pipe size		Copper		LCS		Plastic pipe	
mm	in	horizontal (m)	vertical (m)	horizontal (m)	vertical (m)	horizontal (m)	vertical (m)
15		1.2	1.8	1.8	2.4	0.6	1.2
22	$\frac{1}{2}$	1.8	2.4	2.4	3.0	0.7	1.4
28	$1\frac{3}{4}$	1.8	2.4	2.4	3.0	0.8	1.5
35	1	2.4	3.0	2.7	3.0	0.8	1.7
42	$1\frac{1}{4}$	2.4	3.0	3.0	3.6	0.9	1.8
54	$2\frac{1}{2}$	2.7	3.0	3.0	3.6	1.0	2.1

Remember

You should always aim to fix pipework to meet these spacing requirements

Associated trade skills

At the end of this section you should be able to:

- describe the procedures for lifting floor surfaces

- state the requirements for cutting holes and notching timber joists

- describe the procedures for cutting holes through a range of building materials

- list and describe the tools used in associated trade work.

I think you will appreciate by now that a plumber's job is both interesting and varied. As a result, you will need to learn some additional skills to carry out work other than just plumbing. For example, you will have to run pipes under timber floors, which involves notching joists.

Crowbar

In existing properties, gaining access to work under the floor will mean lifting floor surfaces, so you will need to know how to do that. Often, both on new work and maintenance work, you will be required to cut or drill holes in brickwork, blockwork, concrete and timber, so that will also be covered.

Lifting floor surfaces

Lifting floorboards using hand tools

The hand tools shown are used for lifting floorboards.

Wood chisel

Pad-saw

Bolster chisel

Lifting a length of floorboard to run pipework through joists

This usually involves lifting a single length of floorboard. If it is a full length it is easier, because you won't have to cut across the board.

Using a hammer and sharp bolster, carefully cut the tongue-and-groove joint down either side of the floorboard. Alternatively, you could use a pad-saw.

Good practice requires punching down the nails to enable the board to be removed. Alternatively, a wrecking bar, or draw bar, is sometimes used to prize up the floorboard and nails. Once the board is partially lifted, if it is pushed down slightly the nail heads will be revealed, allowing their removal with a claw hammer.

If you need to lift only part of a board, you will have to make one, possibly two cuts across a joist, so that when the board goes back it has a firm fixing point. If you cannot locate a joist, you will have to insert timber cleats. You can locate a joist by finding the floorboard nails.

Cross-cuts on a floorboard can be made using an extremely sharp wood chisel or a purpose-made floorboard saw.

Lifting floorboards using power tools

This is done using a circular saw.

Here are a few things to remember about using a circular saw:

- use 110 V supply only!
- the depth of the cut must not exceed the depth of the floorboard
- if possible, remove a trial board using hand tools to check for electrical cables or hidden pipework
- a guard must always be fitted to the saw blade.

Circular saw

The saw can be used to cut down the full length of the tongue and groove on each side of the board. A cross-cut is made, again over a joist, making sure that the blade does not hit the nails.

Cutting traps in floorboards

Cutting a trap uses the same method as removing a single floorboard, either using hand or power tools. A trap requires you to cut more boards, but in shorter lengths.

Replacing floorboards and traps

The floorboard length or trap should be screwed back into position to make future inspection or maintenance easier. Use countersunk wood screws. When refixed over pipework, the board surface should be marked accordingly, e.g. 'hot and cold pipework'.

If it is not possible to find a joist to refit the board or trap, cleats must be used to support the board end.

Cross-cut, making sure the cut is over joist where possible

Figure 5.35 Cutting a trap

Cut through tongue using hand or power tools

The illustration shows a trap or board replacement over joists and using cleats.

Removing chipboard

This is a more difficult job than floorboards, as chipboard is laid in wider sheets.

The best way to remove a chipboard section is by using a circular saw. If a power supply is not available, a section of board can be cut using a floorboard saw. The section of board to be removed should be marked out across the board, creating guidelines to follow for the cut. If a pad-saw is used to make the cut, it is helpful to drill holes in each corner of the area to be lifted in order to start the cutting process.

If this method is used, a new piece of chipboard will be needed to replace the removed section.

Replacing chipboard

Like floorboards, chipboard should be screwed back into position.

Pipe guards

We mentioned earlier that pipe runs under floors should be marked. A more effective way of protecting pipes that pass through joists is to use pipe guards. The illustration overleaf shows a typical example.

Timber joists

It is inevitable that joists will have to be drilled or notched to permit pipe runs under

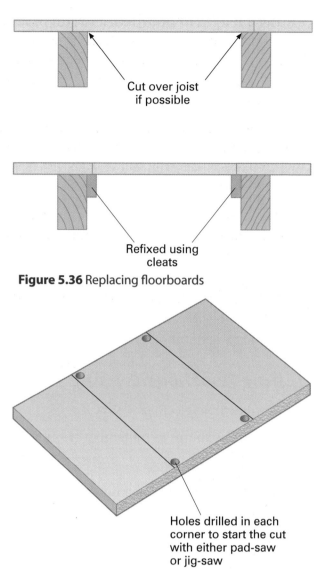

Figure 5.36 Replacing floorboards

Figure 5.37 Preparation for removal of chipboard

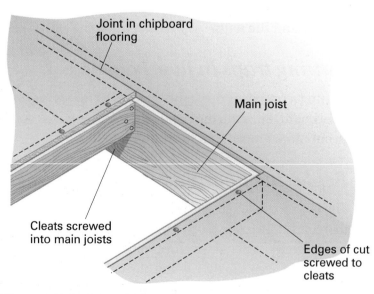

Figure 5.38 Re-fixing chipboard

timber floors. The preferred method would be to drill the joist in the centre of its depth, as this is where there is least stress. In practice, unfortunately, apart from when using plastic hot- and cold-water supply pipe, this tends to be impractical.

The main thing to remember, therefore, for either notches or holes, is not to weaken the joists. This also applies to the distance from the wall where the joist is notched or drilled.

The Building Regulations set out requirements for notching or drilling joists, and these must be followed at all times.

A worked example for notching a joist

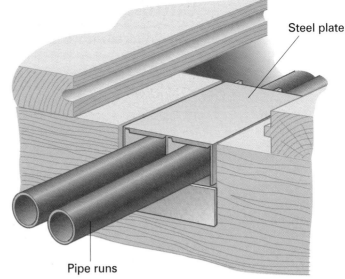

Figure 5.39 Pipe guards

Remember

When drilling a joist, make sure that what you are doing will not weaken the joist

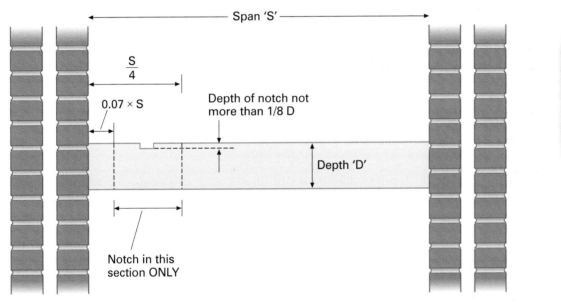

Figure 5.40

Consider a joist 200 mm deep (D) and 2.5 m long (the span, 'S').

Any notch must have a maximum depth of $\frac{H}{8}$ mm. Therefore, for our joist, the depth of the notch is $\frac{200}{8} = 25$ mm.

The minimum distance from wall is $7 \times \frac{S}{100}$ mm. For our joist, this is $\frac{7 \times 2500}{100} = 175$ mm.

The maximum distance from wall is $\frac{S}{4}$ mm from its bearing, giving a maximum span for our joist of $\frac{2500}{4} = 625$ mm.

Joists are normally cut by hand, using a hand or floorboard saw. It is cut to the required depth and width, and the timber notch removed using a hammer and sharp wood chisel.

The width should be enough to give freedom of movement, in order to allow for expansion and contraction, particularly in the hot-water supply.

Cutting holes in the building fabric

By building fabric we mean:

- brickwork
- blockwork
- concrete
- timber.

Hand or power tools can be used to cut holes.

Power tools

The selection of drill will depend on the job in hand.

Rotary hammer drill	500 watt	850 watt	1400 watt Diamond core drill
Brick/block	30 mm	42 mm	152 mm
Concrete	24 mm	30 mm	120 mm

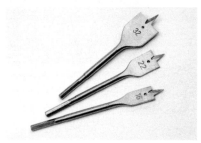

Selection of wood bits

Drill bits are designed for specific tasks and are purpose-made for brick, block, concrete, steel or wood. A diamond-core drill uses either diamond- or tungsten-tipped bits.

Core drills are excellent for drilling through brickwork, blockwork or concrete where you need to pass large-diameter waste or soil pipes, or flue pipework, having the advantage over hand tools in this situation as they provide a much neater finish to the job, and leave less making good.

Wood bits or wood-boring bits

These are useful for drilling holes in joists or other timber constructions for the passage of pipes.

Hole saws

These are handy when drilling through kitchen units in order to pass waste pipes through the side or back of cabinets. They are also used to drill holes in plastic cold-water storage cisterns.

Hole saw set

Hand tools

Tools for brick, block and concrete include cold chisels, brick bolsters (preferably with guard), plugging chisels and club hammers. Tools for timber include claw hammers and wood chisels, and brace and bits for cutting or boring holes in timber – although power tools have generally replaced the brace and bit.

Hand tools, such as club hammers, cold chisels and bolsters are used to cut holes and chases in brickwork, blockwork or concrete. Care must be taken to cut out only the minimum of material required for the passage of the pipe, in order to keep making good to a minimum. Plugging chisels are used to chase out mortar between brickwork, and the plumber does this to let in sheet lead flashings, or for timber-fixing plugs.

A brace and bit, or cordless drill, is used to bore holes through timber in the same way as power tools are used. However, the power drills can use much larger-diameter bits, and do not require as much hard work.

Making good

For most jobs, a mortar mixture of 4 parts sand to 1 part cement will be adequate for pointing any brickwork joints disturbed while doing the job, and for **making good** the gap around the pipe penetration. Mastic sealant, either clear or close to the colour of the pipework, can be used as an alternative.

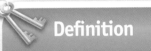

Definition

Making good – repairing and finishing off the brickwork, blockwork and concrete

Generic systems knowledge

At the end of this section you should be able to:

- state the meaning of pipework and system components symbols
- carry out basic site measurements
- carry out basic measurements from drawing details
- estimate the materials required for a job
- describe how to prepare for the installation of systems and components
- state how customers' property can be protected while work is carried out
- state what documentation is used when carrying out a job
- describe the requirements for dealing with customers and co-workers.

Generic systems knowledge refers to knowledge that is 'general' or common to all the systems used in plumbing.

Identification of pipework and use of plumbing symbols

Identification of pipework

British Standards are produced to standardise the colour coding of pipework. This is particularly relevant in larger installations where a number of services are used. In domestic installations, pipework colour coding is not essential, but it is used on service pipes below ground, as follows:

- cold-water service pipes = blue

- domestic natural gas = yellow

- central heating pipework for use in solid floors is covered in white plastic.

Plumbing symbols

In order to provide some standardisation of symbols used on drawings, the BSI has produced a specification for pipework symbols; the main items are shown in the illustration.

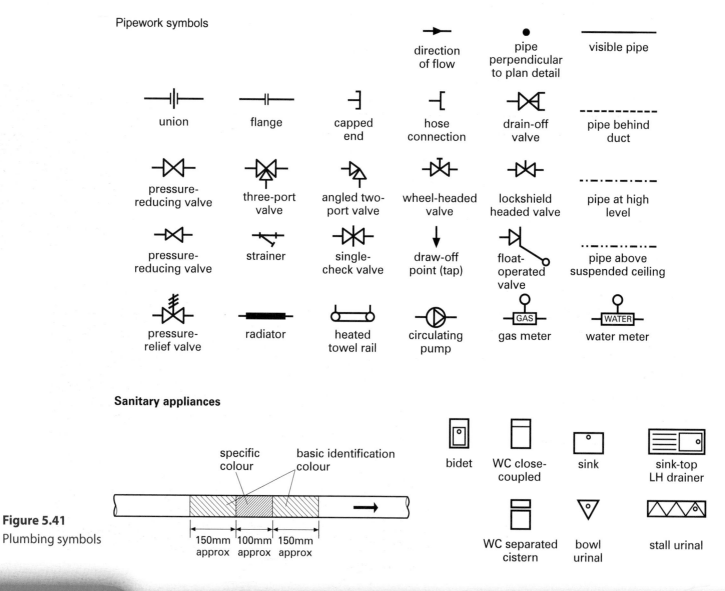

Figure 5.41
Plumbing symbols

Measuring and estimating materials

This topic includes:

- estimating from a drawing
- estimating from site.

Estimating from a drawing

It is unlikely that you will be required to carry out an estimate from a complex services drawing, but you may be required to estimate from a detail such as the one shown here:

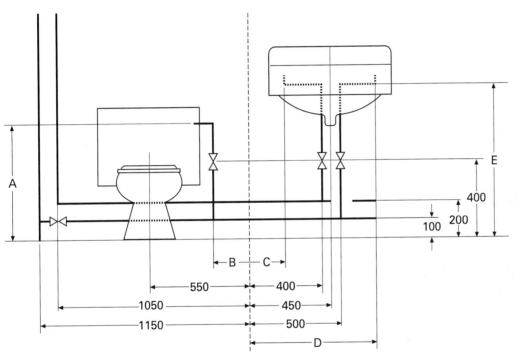

Figure 5.42 Example of line drawing, not to scale

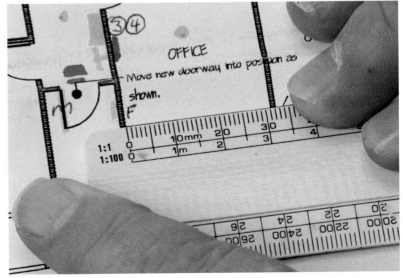

Find out

Using the pipework diagram shown, calculate the total pipe requirement for the installation

On this drawing the dimensions are given. The detail could have been given drawn to scale, in which case you would need to use a scale rule to measure the dimensions.

This photograph shows an example of a scale rule, used to measure dimensions on a drawing.

Using a scale rule

Estimating from site

Some of your estimating will take place on site, without drawings or specifications. On one-off new installations, it will be carried out in two stages:

1 first fix

2 second fix.

The first fix involves installing all the pipework prior to floor covering going down and plasterboard finishing (sometimes referred to as carcassing). The second fix is the completion of the installation – fitting the bathroom suite, other sanitary appliances or radiators.

You will need to have a thorough understanding of the system requirements and where the intended pipe runs are. The next stage is to take measurements, using as a minimum a 3 m, and preferably a 5 m tape, and then recording your measurements in a notebook. An estimate of fittings can also be done at the same time.

Once completed, the dimensions can be totalled to give an estimate of the pipe requirements.

On larger contracts – a large housing project, for example, – the dimensions, pipe and fitting requirements will probably already have been worked out. These are often provided to the plumber in the form of a specific materials pack, booked out from the stores on a house by house basis using a requisition order, so that the stock can be monitored.

Find out

Estimate the material requirement for a typical installation – maybe the installation in your own home

Documentation

You will come across the following general documentation:

- Water Regulations
- requirements of British Standard specifications and in particular BS6700
- Building Regulations, which we mentioned earlier with regard to drilling and notching joists
- Health and Safety regulations.

As well as these documents, which are the 'yardstick' of everything you do, there is other documentation that you need to be aware of. The other documentation that is used in support of plumbing work will vary from job to job, depending on what is being done.

Job specifications

These usually accompany site and services drawings and details, and are mainly used on larger contracts. Job specifications provide details such as:

- type and quality of components, materials and fittings to be installed
- type of clips or brackets to be used
- system test specifications

- any specific installation requirements, e.g. who is to carry out any associated building work, such as cutting holes for pipework, drilling or notching joists and making good.

Remember: specifications form part of the contract documentation; any alteration required to the specification should not be done by you. If, for any reason, a part of the job cannot be done to the specification, you should inform your supervisor or employer immediately.

Quotations

On a smaller job, for example renewing a customer's bathroom suite, it is unlikely that a detailed specification will be produced. Your employer, however, will have issued a detailed quotation that the customer will have agreed. This in itself is a form of contract. Again, any revisions to the original quote must be passed on to your employer, so that they can be agreed with the customer.

Quotations are also used by plumbers' merchants to supply a price for the materials required to carry out the job. This would be followed by an invoice, which is a statement of the materials supplied, requiring payment by the plumber.

An example of the paperwork system

Some of the other important documentation can be illustrated by looking at how the paperwork system works for a bathroom installation.

1 Customer asks plumber for a quote.

2 Plumber visits customer, measures up, estimates materials.

3 Plumber gets a quotation for the materials from the merchant.

4 This enables plumber to finalise quote to customer.

5 In this case, the customer accepts and the quote is given to customer in writing.

6 Plumber orders materials from merchant after the materials quotation is confirmed in writing. This is an order (it may not be confirmed in writing if the plumber has an account with the merchant). Plumber receives delivery of materials, confirmed by a delivery note.

7 Job is carried out. Plumber receives invoices for payment of materials and also invoices the customer for the work carried out. A remittance advice (record of payment) is sometimes issued. A receipt may be given to the customer.

On any job, large or small, if you are advised that the delivery of materials will be delayed, or a particular item is not the one ordered, notify the site manager or your immediate supervisor/employer (so that he or she is aware of potential delays).

Work programmes

Work programmes will be covered in greater depth at Level 3. On a small job, you will have the work programme inside your head, based on agreed start and finish dates. On larger contracts, the approach is more scientific, and a contract programme will have been provided. This could consist of an overall programme for all site trades, as well as a separate programme for each trade. There could be a number of variations on this theme, but the principle of programming is basically 'activity against time'.

Preparing for the installation

Depending on the size of the job, the following checklist will cover most of the preparation items:

* Have you confirmed with the client the start time and date?

* Are all the tools, materials and equipment required to do the job on site?

* Are electrical tools safe to use?

* Is the work area clean and safe to work in?

* If using access equipment, ladders etc., are they safe to use?

* If you are working in a loft, do you have adequate lighting to see what you are doing (torch, inspection lamp etc.)?

* If working in an occupied dwelling, have you checked the work area for any pre-work damage? If you have noticed any damage, have you notified the customer prior to starting the work?

* Again, in an occupied dwelling, have you made sure you have protected the customer's property before you start?
    * use dust sheets (but not where they could be dangerous, e.g. on stairs)
    * request that vulnerable furniture is moved
    * request that carpets are taken up if necessary
    * cover up sanitary appliances if working in the bathroom area.

Safety tip

Never leave tools such as hammers or spanners where they could be accidentally knocked – for example, into sanitary appliances

FAQ

How important is access to pipework and fittings installed in the structure of a building?

It is essential to have reasonable access to plumbing services, otherwise expensive, time-consuming work is needed, much to the annoyance of the customer. A lack of access may be in breach of Regulations.

Knowledge check

1   State three advantages of using pipe bends rather than fittings.

2   What are the main methods of bending copper tube and low carbon steel tube?

3   What is the most suitable grade of copper pipe for use in bending machines?

4   A hydraulic bending machine would be the correct equipment for bending 22 mm copper tube. True or false?

5   State two faults that occur when bending copper tube and two faults when bending low carbon steel tube.

6   What are the three main materials used for new plumbing system installations?

7   What are the two main types of material used for fittings to joint LCS?

8   List four common types of fixings used in plumbing.

9   How are screws specified?

10 Where would you be most likely to use a brass or alloy screw?

11 What type of screw would be used to fix a radiator bracket to a concrete block wall?

12 State three types of fixing suitable for plasterboard stud partitioning.

13 What are the two main types of wooden floor surfaces?

14 A timber joist measures 200 mm deep by 4 m long. What is the maximum depth of notch?

15 Pipework should be sleeved through load-bearing walls. Sketch a sleeve.

16 Give two examples of what a job specification might cover.

17 State what action you should take if the job can't be done to a specification.

18 Tick the two main elements of an installation programme:
    Activity against time
    Activity against speed
    Time against labour
    Time against materials.

19 When carrying out a small job, what is the main thing to remember before leaving a customer's property?

chapter 6

Cold-water supply

OVERVIEW

The supply of fresh wholesome cold water to people's homes is a basic human need. As a plumber, it is your job to get the water to the taps so that it is clean and fit for human consumption, otherwise severe illness can occur. Most people take for granted the supply of cold fresh water to their homes. Few would probably appreciate what goes into providing this service. Although water supply legislation is fully covered at Level 3, it is important that you gain a general understanding relating to all areas of your work at Level 2. In this chapter you will cover:

- **Cold-water supply and treatment**
  - Water Regulations
  - The origin, collection and storage of water
  - Water treatment
  - Water supply and distribution mains connection

- **Cold-water systems – inside the building**
  - Cold-water storage cisterns
  - Frost protection

Cold-water supply and treatment

At the end of this section you should be able to:

- state the Regulations and their purpose
- describe how water is collected, stored and distributed
- describe the cold water service from mains to a dwelling (including meters)

Water Regulations

The purpose of the Water Supply (Water Fittings) Regulations 1999 is to prevent contamination of a water supply, prevent the waste of water, prevent misuse of a water supply, prevent undue consumption of water and prevent erroneous measurement (fiddling the meter).

The Regulations are also designed to permit the introduction of new products and ideas, as well as supporting environmental awareness.

Enforcing the Regulations

The water supplier is responsible for enforcing the Regulations. Under the Water Regulations, water companies are encouraged to set up approved contractors schemes. These require that approved contractors certify to the water company that water fittings installed are in compliance with the Regulations.

Although there is no legal requirement for a person working on water services to be qualified, anyone who carries out such work can be prosecuted for an offence against the Water Regulations and, if convicted of an offence, can be liable to a fine.

The origin, collection and storage of water

Water is clear, tasteless and colourless, and all water sources come from rainfall. Water is defined in chemical terms as a compound of the two gases, hydrogen and oxygen, in the proportion two parts hydrogen and one part oxygen (H_2O).

The classification of water

Probably the most important property of water is its ability to dissolve gases and solids to form solutions; this is referred to as its solvent power and has a bearing on how soft or hard the water eventually becomes. Water is classified according to its hardness. Hard water can result in limescale build-up, as seen in the photograph. This can be tackled by using water treatment methods, which we'll look at later.

Remember

Water supply installations are covered by the Water Supply (Water Fittings) Regulations 1999, which requires water companies to supply 'wholesome' water

Definition

Water undertaker – the legal term for the water companies that supply domestic water

Did you know?

To be an **approved contractor** a company will need to meet a number of key requirements, such as employing competent staff, completing proper records and allowing work to be checked on site

Find out

There's more on water hardness in Chapter 4

Did you know?

Solvent power can be a negative effect for the plumber to deal with as it causes corrosion and 'blocking up' of systems and components

Limescale build-up in pipework

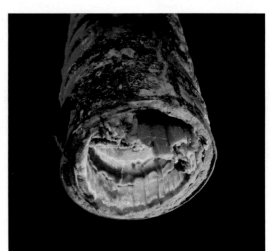

This diagram explains the classification of water.

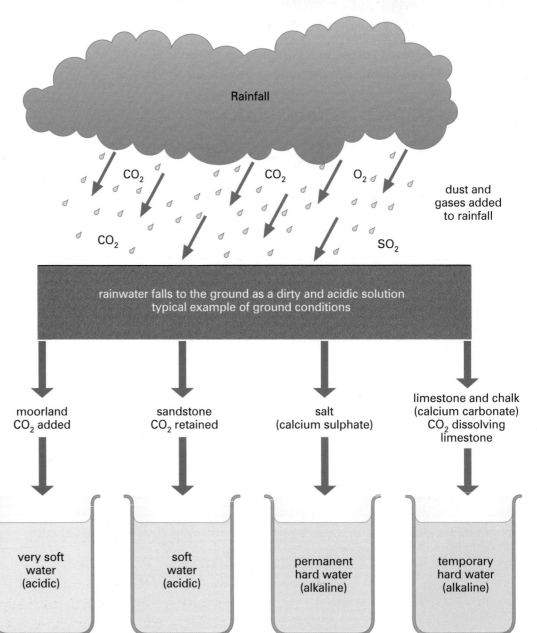

Figure 6.1 Classification of water type

The measurement of the hardness of water is shown as parts per million or alternatively milligrams per litre. This is how hardness is classified:

Water type	Dissolved gases
Soft water	0 to 50 mg/litre
Moderately soft	50 to 100 mg/litre
Slightly hard	100 to 150 mg/litre
Moderately hard	150 to 200 mg/litre
Hard	200 to 300 mg/litre
Very hard	over 300 mg/litre

Figure 6.2 Water hardness

Where does water come from?

Water evaporates from the sea, rivers, lakes and the soil. It forms clouds containing water vapour, which eventually condenses and falls as rain. When it hits the ground, some of the rainwater runs into streams, rivers and lakes, some soaks into the ground, where it will collect temporarily and evaporate, or it will soak away and form natural springs or pockets of water to be accessed by wells (see figure 6.3).

Water companies, for example Severn Trent, Thames Water, etc., obtain their water for public consumption from two main sources:

- surface sources, such as:
  - upland surface water
  - rivers and streams
- underground sources, such as:
  - wells
  - artesian wells
  - springs.

Surface sources

Upland surface water

This category covers **impounding reservoirs**, lakes and natural reservoirs. This source of water is mostly found in the northern part of the UK, as the landscape there is hilly or mountainous, allowing lakes to form naturally, or the damming of streams to form impounding reservoirs. The water quality from this source is good because it is generally free from human or animal contamination. It is usually classified as soft as it runs directly off the ground surface and into the water source, so it is not affected by passing through a particular soil type. Where water comes into contact with peat, it can become acidic.

Remember

All surface and underground sources are dependent upon rainfall

Definition

Impounding reservoir – a man-made reservoir

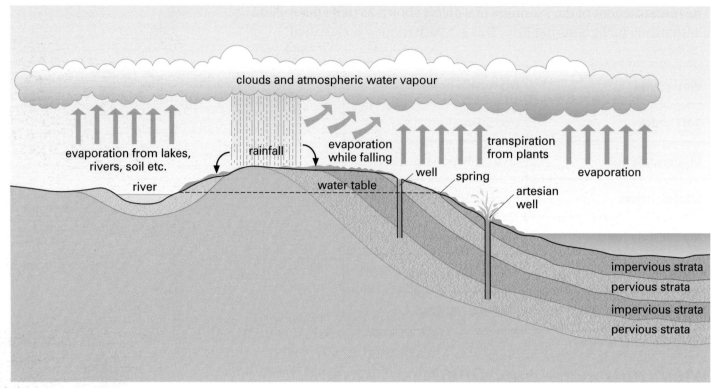

Figure 6.3 The rainwater cycle

Artificial reservoirs

Artificial reservoirs are constructed to meet an ever-increasing demand for water in both domestic and industrial sectors. They are also used in areas where insufficient natural resources exist. Reservoirs are created by flooding low-lying areas of land, normally by damming a water course. The water is classified as soft.

Rivers and streams

The quality of water from rivers varies depending on the location. Water from moorland rivers and streams tends to be relatively wholesome compared to further downstream, where it could become polluted by natural drainage from farmyards, road surfaces and industrial waste. The quality of river water varies in hardness depending on the nature of the ground where it originated. Water from upland river sources is generally soft compared with that of the lower reaches, which is usually hard.

Underground sources

Wells

In the past, before the existence of Water Authorities (water companies) wells were used to provide the water supply to dwellings or small communities. There are two types of wells, shallow and deep. Surprisingly, this classification does not necessarily refer to the actual depth of the well, but whether it penetrates the first impervious stratum of the earth.

Shallow well

Deep well

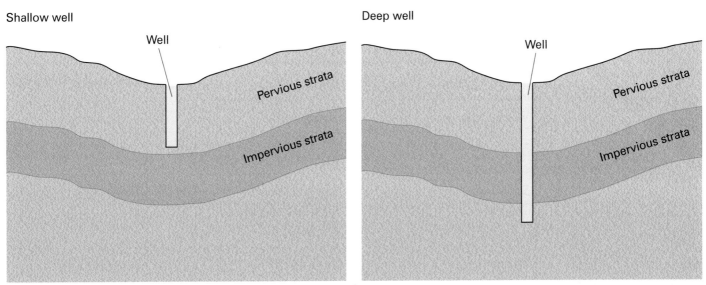

Figure 6.4 Shallow well does not penetrate impervious strata. Deep well penetrates through impervious strata.

Some water companies still retain deep wells as a back up to their supplies from other sources, or as a standby in case of drought. A shallow well has a greater risk of contamination, but water from a deep well should be pure and wholesome.

Definition

Water table – the natural level of water under the earth

Artesian wells

An artesian well penetrates the impervious stratum and enters a lower porous zone containing water. The outlet of the well is situated below the water table, so the water is forced out by gravity through the mouth of the well.

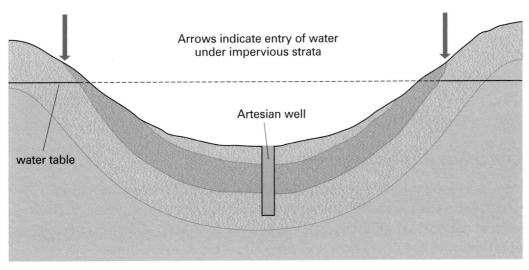

Figure 6.5 Head of water forces supply out through artesian well

Springs

The quality of spring water varies depending upon the route that the water has taken from underground to the surface. If it has travelled for a long distance through rock formations, it will probably be free from contamination, but it is likely to be hard.

The following table summarises the various water sources and their level of contamination before treatment.

Classification	Source	Contamination level
Wholesome	1. Spring water 2. Deep well water 3. Upland surface water	Very palatable
Suspicious	4. Stored rainwater 5. Surface water from cultivated land	Moderately palatable
Dangerous	6. River water 7. Shallow well	Non palatable

Figure 6.6 Summary of water source classification

Water storage

Water is stored by water companies either in its untreated state in impounding reservoirs or lakes, or as wholesome water in service reservoirs. It is worth noting that water companies usually aim to store enough drinking water in their service reservoirs, for emergencies, to maintain supply for about 24 hours. This safeguards against failure of pumps or mains and allows time to repair any faults before supplies run out.

Water treatment

It is the responsibility of the water companies to ensure water is fit to drink. All water must be treated before it is put into the supply system. How it is treated will depend on its source and what impurities it contains. Some impurities are actually essential to our health and will be retained; others are harmful and must be removed during the treatment process.

Deep wells and bore holes

The quality of water from this source is already quite good, due to the natural filtering process as the water passes through the rock strata. In this case, the only treatment needed is sterilisation. Sterilisation serves to keep both the water and supply pipework free from bacteria as the water is piped to our homes.

Rivers, lakes and impounding reservoirs

These are our main sources of supply, but generally speaking this water is dirty and polluted, so treatment needs to be extremely thorough. The treatment process usually involves the water being strained, filtered, chemically treated and sterilised.

Definition

Sterilisation – purification by boiling, or, within the industry, dosing the supply with chlorine or chlorine ammonia mix

Water supply and distribution

Water is supplied to our homes via a network of pipes known as **mains**. These will vary in diameter depending on the purpose of the main and the likely demand of the supply. The diagram shows an example of a typical water-supply system layout.

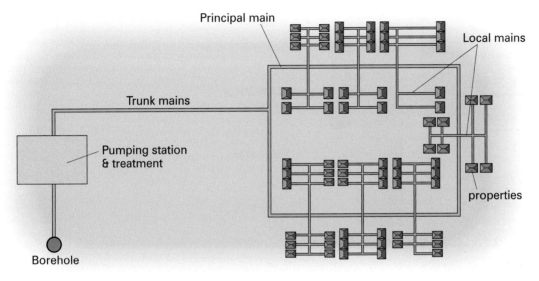

Figure 6.7 A typical mains water distribution system

The local mains shown will provide the 'final leg' of the supply of water to your home.

Mains connection

Water mains are constructed of asbestos cement, steel, PVC or cast iron. PVC is now used extensively on new installations and mains replacements. The connection to the mains is the responsibility of a water company, and new connections to existing mains, operating under live pressure, require the use of specialist equipment. The illustration shows an example of a connection to a cast iron main.

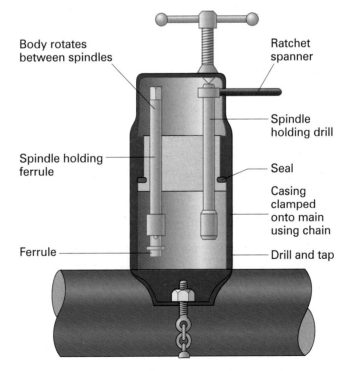

Figure 6.8 Section through a mains tapping machine

The connection to the main is made using a **ferrule**.

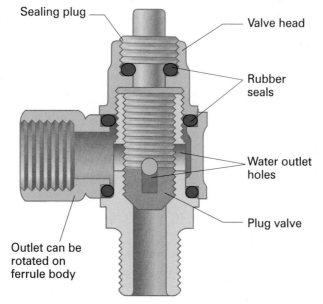

Figure 6.9 Section through a brass ferrule

The supply from the mains to the building

The next diagram shows a full installation from the main to the stop and drain valve wherever it enters the building. A plumber's work usually starts from the external stop valve.

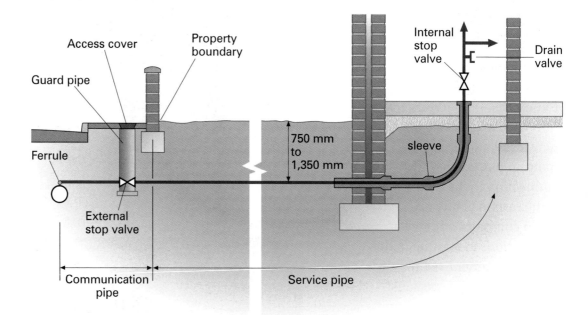

Figure 6.10 Supply pipework to the building

There are some key points to remember about service pipes:

- In order to protect against frost damage, mains and service pipes should be at least 750 mm beneath the surface of the ground.

- The maximum depth of cover should not exceed 1.350 m, as this would prevent ease of access.

- Metal pipes should be protected against possible corrosion from the soil, particularly acidic soils. this can be done by:
  - use of plastic sheathed pipe
  - wrapping the pipe with an anti corrosive tape
  - installing the pipe inside a duct.

- The minimum size allowed for a cold water service pipe to a dwelling is 15 mm.

Installation of external stop valves

As you can see from figure 6.10, the service pipe to the building can be isolated from the main by using the external stop valve. Ease of access to the stop valve is very important, so it should be located in a stop valve chamber, constructed from 150 mm PVC, or earthenware pipe, sited on a firm base, and finished off at the top with a stop valve cover. The stop valve cover is usually made from steel plate, or a combination of steel plate and plastic.

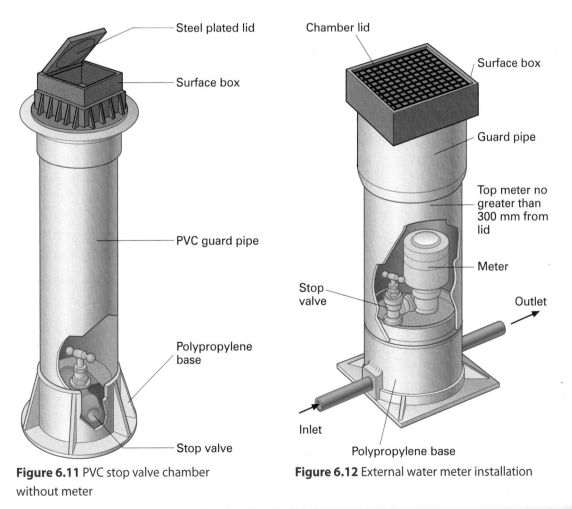

Figure 6.11 PVC stop valve chamber without meter

Figure 6.12 External water meter installation

Figure 6.11 shows a PVC chamber; these are now widely used on new installations. The guard pipe is cut to the desired length to suit the installation as part of the fitting process.

Installation of water meters

The installation of water meters for domestic premises is becoming more widespread, particularly on new housing developments. They measure the amount of water used by the householder, who then only pays for the water used. Water meters can be installed either internally or externally.

External water meters usually incorporate a service stop valve and are found on new domestic installations. The parts come already assembled, with the guard pipe being cut to length as part of the fitting process.

The next two diagrams show the installation of internal water meters. They are normally seen in commercial or larger domestic properties.

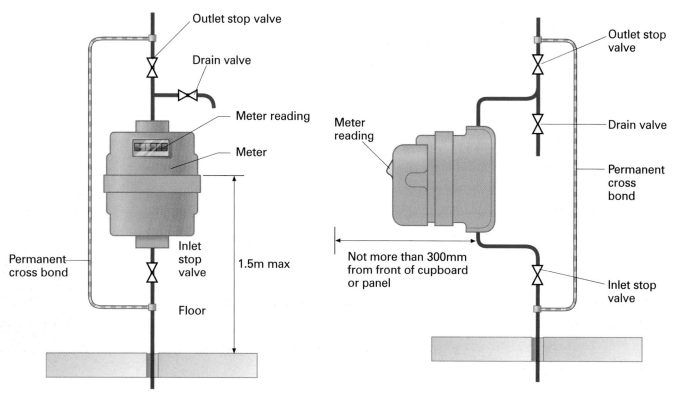

Figure 6.13 Exposed internal water meter   **Figure 6.14** Concealed internal water meter

Cold-water systems – inside the building

At the end of this section you should be able to:

- describe direct and indirect systems
- state the requirements of storage cisterns, pipes and fittings
- explain frost protection measures for cold-water systems

Stop and drain valve

This section deals with the supply from the stop and drain valve inside the building.

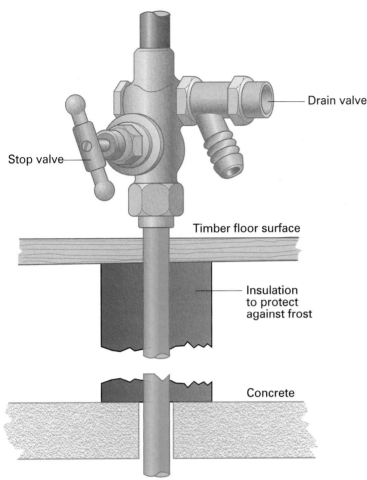

Figure 6.15 Combined stop and drain valve

The purpose of the stop and drain valve is to:

- turn on/off the water supply to pipework fittings and components to enable system maintenance
- drain down all the system pipework fittings and components to enable system pipework repair/replacement.

System pressure and flow rate

The incoming water pressure and flow rate supplied via the main is vital as it has a key bearing on the size of pipework and fittings used in the system. We will look at pipe sizes in more detail at Level 3, however you may be required to assist in taking readings of incoming water pressure and flow rate as part of the installation process.

Water pressure measurement

This is carried out using a pressure gauge as shown on page 188. The gauge usually reads in bar pressure.

Remember

The supply pipe must contain an isolation valve as it enters the property, together with a drain-off facility at the lowest point. These arrangements can be made up using separate fittings as an alternative to the combined valve shown

Did you know?

Other servicing valves are included in the system to isolate components and appliances

Water pressure reading being taken

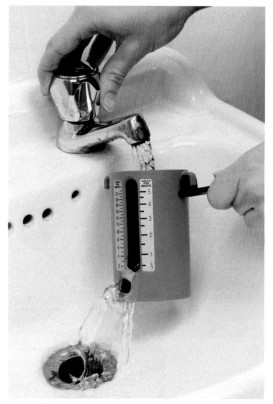

Water flow rate reading being taken with a weir cup

Water flow rate measurement

This is carried out using a water flow measuring device as shown above. The reading is usually quoted in litres per second.

A common fault with domestic systems installation is that water pressure readings are not taken at the right stage in the job – that is at the design or pricing stage. Insufficient supply pressure and flow rate can result in big problems: for example, if the wrong system components, such as combination boilers, are installed, they will not work correctly. If the pressure to an existing dwelling is poor, check first that there is no burst on the service from the external stop tap. You can usually tell by putting your ear against the stop tap key while it is on the stop tap and listening for a hissing noise.

Use of water conditioners

One way to prevent damage to water systems by hard water is to install a water conditioner in the cold water supply. This has the effect of softening the water prior to it being heated, and so reduces scale formation.

There are three main ways to treat water hardness:

- base exchange softeners
- scale reducers
- magnetic water conditioners.

Base exchange softeners

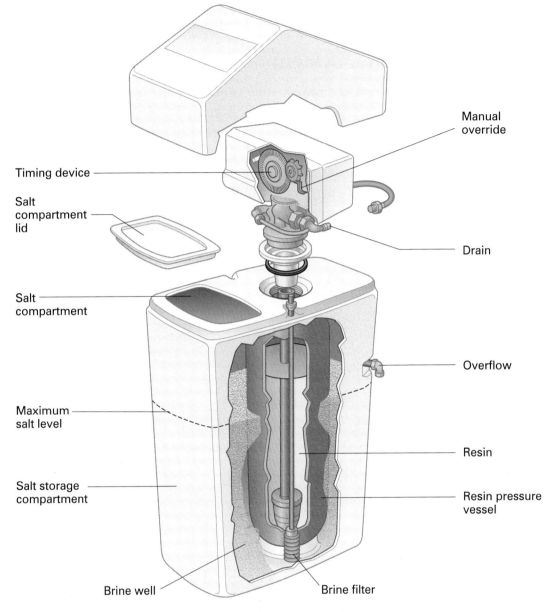

Timing device

Salt compartment lid

Salt compartment

Maximum salt level

Salt storage compartment

Brine well

Manual override

Drain

Overflow

Resin

Resin pressure vessel

Brine filter

Figure 6.16 Section through a typical base exchange water softener

These work by passing the hard water through a tank containing resin particles. The resin attracts and absorbs the hardness salts – mainly calcium and magnesium – from the water. At the same time it replaces them with sodium from the resins. After a while, the resin becomes saturated with hardness salts and needs to be regenerated, using salt solution to put sodium back into the resin. The hardness salts are released from the resin and washed down the drain. The unit requires regular maintenance and checking, an annual service and a check on water hardness needs carried out by a service engineer. The salt, however, can usually be topped up by the end user. BS6700 makes recommendations for the installation of base exchange water softeners. Its main consideration is the prevention of backflow and contamination of the water supply. BS6700 is the key reference source, together with the manufacturer's instructions when installing the water softener.

Remember

BS6700 is the main British Standard detailing the installation requirements for hot and cold water systems – so you'll read about this one a lot!

Further information on scale reduction and magnetic water conditioners can be found in manufacturers' installation instructions. The installation of these units is relatively straightforward, and not considered here in great detail.

Types of cold-water system

There are two types of cold-water system: direct and indirect. In a direct system, all the pipes to the draw-off points (sink, bath, hand basin, WC, etc.) are taken directly from the rising main or service pipe, and operate under mains pressure. In an indirect system, one point – usually the kitchen sink – is fed directly from the rising main, which then supplies the cold water storage cistern. The remaining draw-off points are fed from the cold water storage cistern – hence the term indirect.

The direct cold-water system

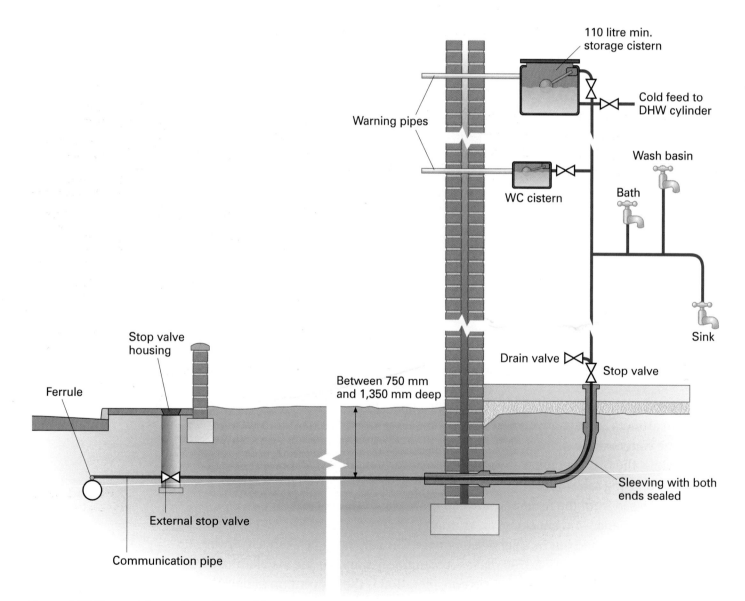

Figure 6.17 Pipework layout for a direct cold-water system

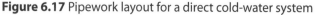

The installation of direct systems is permitted by water companies in domestic properties in medium to high pressure areas where the supply can provide adequate quantities of water at sufficient pressure to meet the building's needs.

The advantages of a direct system are that it is:

- cheaper to install because
  - less pipework is required
  - the storage cistern is smaller (110 litre minimum)
- drinking water is available from all draw-off points
- less risk of frost damage due to a smaller amount of pipework
- less structural support required for smaller cistern.

However, there are disadvantages:

- higher pressure may make the system noisy
- there is no reserve of cold water if the mains or service supply is shut off
- more wear and tear on taps and valves due to high pressure
- higher demand on the main at peak periods.

The direct system is the most commonly installed type of cold water system in domestic properties because its installation is cost effective and there is usually relatively high pressure of supply available.

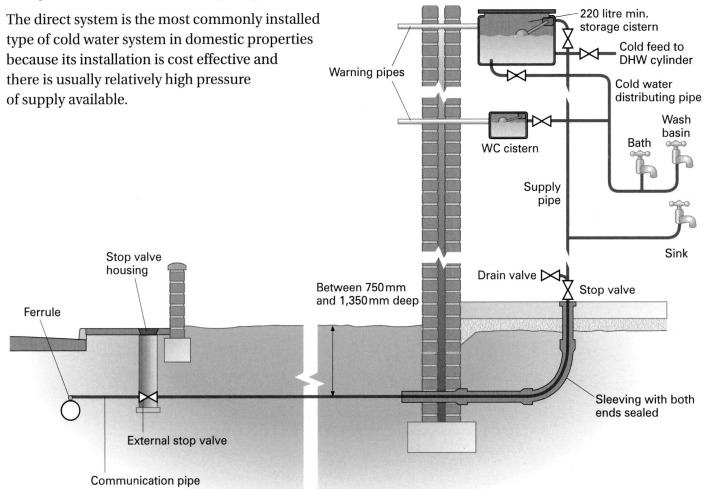

Figure 6.18 Pipework layout for an indirect cold-water system

The indirect cold-water system

The draw-off points in an indirect system are fed indirectly from the cold-water storage cistern, one outlet being fed directly from the supply pipe. The system is designed to be used in low-pressure water areas where the mains supply pipework is not capable of supplying the full requirement of the system. This type of system also has a reserve of stored water in the event of mains failure.

The advantages of indirect systems are that there is:

- a reserve of water should the mains supply be turned off
- reduced risk of system noise due to lower pressures
- reduced risk of wear and tear on taps and valves, again due to lower pressure
- lower demand on the main at peak periods.

The disadvantages are:

- an increased risk of frost damage
- the space occupied by the larger storage cistern (220 litre minimum)
- the additional cost of the storage cistern and pipework
- before the 1999 Water Regulations water might not have been potable.

Cold-water storage cisterns

Under the Water Regulations Schedule 2, paragraph 16, a storage cistern supplying cold-water outlets, or feeding a hot water storage system, should be capable of supplying wholesome water, and therefore various protection measures are included in the design of the cistern to ensure the water supply does not become contaminated.

Cisterns must therefore be:

- fitted with an effective inlet control device to maintain the correct water level
- fitted with service valves on inlet and outlet pipes
- fitted with screened warning/overflow pipes to warn of overflow
- covered to exclude light or insects
- insulated to prevent heat loss and undue warming
- installed so that the risk of contamination is minimised
- arranged so that water can circulate preventing stagnation
- supported to avoid distortion or damage leading to leaks
- readily accessible for maintenance and cleaning.

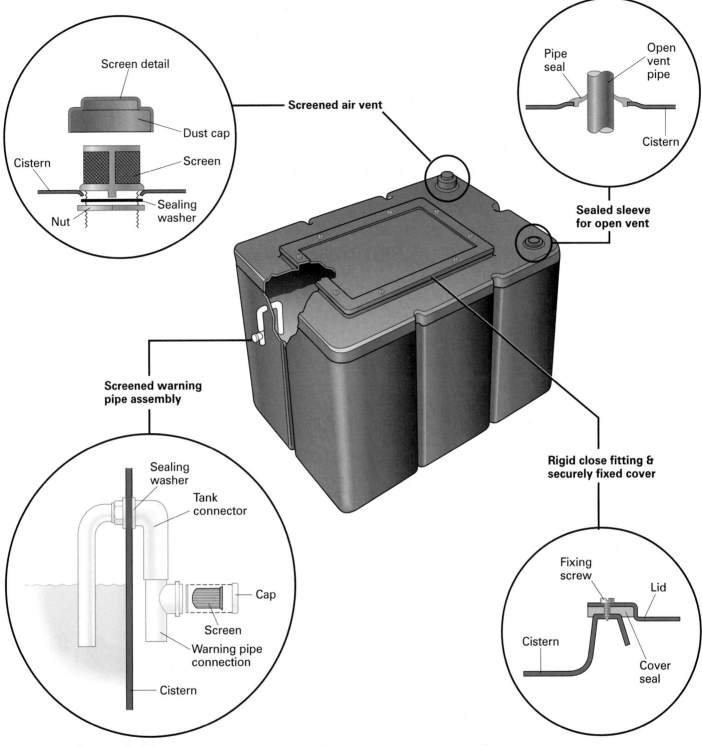

Figures 6.19 Protected cold-water storage

Materials for cisterns

In the past, galvanised low carbon steel was the main material used to make cold water storage cisterns. You might still come across these on maintenance jobs, but most new installations use cisterns made from plastic such as polyethylene, polypropylene and polyvinyl chloride.

The majority of cisterns are polypropylene, because this material allows them to be:

- light
- strong
- hygienic
- resistant to corrosion
- flexible enough to be manoeuvred through small openings.

Cisterns are available in square, rectangular or circular shapes, and are produced in black to prevent the growth of algae. However, because they are flexible, the base of the cistern *must* be fully supported throughout its entire length and width. Holes for pipe connections should be cut out using a hole saw. The joint between the cistern wall and fitting should be made using plastic or rubber washers.

On no account must the hole in the cistern be made by heating a section of pipe and using it to make a hole in the cistern – *this degrades the plastic and will result in cistern failure.*

Connections to cisterns and control valves

Inlet controls

Water Regulations require that a pipe supplying water to a storage cistern be fitted with an effective adjustable shut-off device, which will close when the water reaches its required level. For most domestic applications, a float-operated valve is used. These must comply with BS1212 (parts 1–4), and the following types are available:

- Portsmouth type
- diaphragm valve made of brass
- diaphragm valve made of plastic.

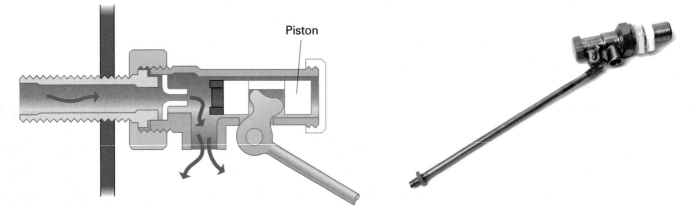

Piston

Figure 6.20 Portsmouth float valve to BS 1212 Part 1

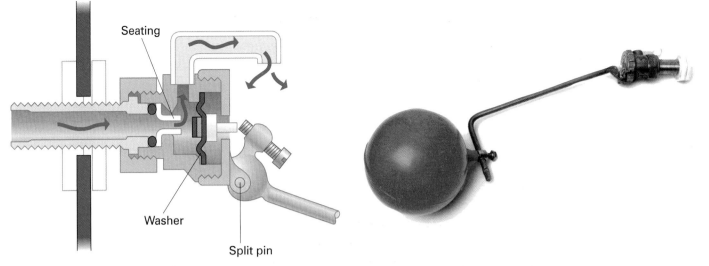

Figure 6.21 Diaphragm float valve to BS1212 Part 2 or 3

The Portsmouth valve is not widely used on new installations as it does not provide an effective air gap between the water level and the point at which the valve discharges, but you may see this valve on existing installations. Brass or plastic diaphragm float valves can be used in any situation. BS1212 (part 4) refers specifically to diaphragm equilibrium float valves (not illustrated) designed primarily for use in WC flushing cisterns; the Portsmouth and diaphragm valves work by the principle of leverage, the equilibrium valve uses the water pressure. However, equilibrium valves are available with a similar external appearance to the Portsmouth and diaphragm valves. Inlet pipes to cisterns must have a servicing valve fitted immediately before connection to the cistern, in order to enable any maintenance to be carried out on the cistern without turning off all the water supply. This also applies to WC cisterns. The valve usually installed is a spherical type plug valve.

Outlet pipes

The outlet pipes should be connected as low in the cistern as possible – there is a move in the new Water Regulations to preferred connections being in the bottom of the cistern rather than the side. This prevents the build up of sludge in the bottom of the cistern. Outlet pipes, such as cold feed and distribution pipes, should be fitted with servicing valves, and these should be located as near to the point of connection to the system as possible. The valve type used in this position is usually a wheelhead gate valve.

Overflow and warning pipes

When water in a cistern rises above a pre-set level, usually due to a faulty float operated valve, the water is allowed to flow through a pipe away from a cistern. An **overflow pipe** is used to discharge water where it will not cause damage to the building. A **warning pipe** is a pipe used to give warning to the occupiers of a building that a cistern is overflowing and needs attention.

Small cisterns of up to 1000 litres (that is most domestic cisterns), must be fitted with a warning pipe and no other overflow pipe.

Larger cisterns (between 1000 and 5000 litres), are fitted with both.

Definition

Overflow pipe – discharges excess water safely

Warning pipe – alerts occupants to a problem

Location of a warning pipe in a small cistern

The warning pipe must be located in the cistern so that a minimum air gap is maintained between its point of discharge and the normal water level in the cistern. The position of the float-operated valve is also crucial, to ensure that a minimum air gap is maintained between its outlet and the spill over level of the cistern at the warning pipe. Note also:

- If the float operated valve becomes defective, the warning pipe should be capable of removing excess water without becoming submerged.

- The warning pipe should fall continuously from the cistern to the point of discharge.

- Warning pipes should discharge where the water will be noticed, usually outside the building.

- Warning pipes should be fitted with a screen or filter to exclude insects.

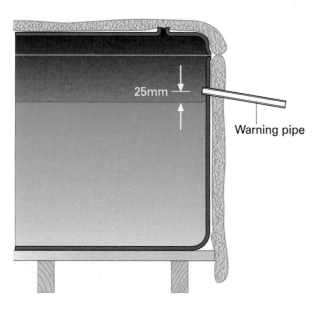

Figure 6.22 Warning pipe positioning

Connecting two or more cisterns together

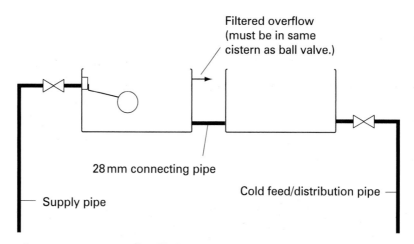

Figure 6.23 An example of linking two or more storage cisterns together

We mentioned earlier the need to avoid contamination and stagnation in cisterns. This can occur when water cannot fully circulate throughout the cistern and is permitted to stagnate. This problem is more likely to occur when two or more cisterns are joined together – perhaps because the access hatch was not big enough. When jointing cisterns we must ensure proper water movement throughout the cisterns to avoid stagnation, as illustrated in figure 6.23.

Particular attention must be given to:

- the provision of inlets to both cisterns fed via float operated valves
- the method of connecting the distribution pipes via a manifold arrangement taking equal draw-off from both cisterns simultaneously
- siting outlet connections at the opposite end to inlet connections to allow effective water distribution across the cistern
- the method of linking the cisterns using lateral connections between the cisterns.

Frost protection

The major problem associated with freezing temperatures is the dramatic effect it can have on plumbing systems. Be aware that when water freezes it expands by about 10 per cent, causing damage to unprotected pipes and fittings.

When water freezes within the confines of, for example, a copper pipe, it will expand. The water either side of the 'ice plug' cannot be compressed, so the expanding ice will cause the wall of the pipe to split. Once the ice thaws, you have a burst pipe. Water damage in domestic properties can prove to be very expensive. It also causes an undue waste of water, which goes against the requirements of the Water Regulations.

The Water Regulations and BS6700 both deal with frost protection.

Solid and timber floor installations

Where a supply pipe enters the building through a solid floor, it should be ducted, and the ends sealed to prevent moisture entering the pipe. Depending how close the pipe is to the outside wall as it rises to the property, it may need to be insulated as well. If it enters the building through a suspended timber floor, it should be ducted, and the ends sealed, and insulated as well. This is because the space between the suspended timber floor and the oversite concrete is exposed.

Remember

When water freezes it expands by about 10%

Did you know?

The minimum depth for pipes to be laid underground is 750 mm. This is to protect them from frost damage

FAQ

When does a pipe burst due to freezing? Is it when it thaws out?

No, the pipe actually bursts (splits open) when it freezes due to the expansion of water changing to ice, it is only when it thaws out that we realise it has burst by seeing the water leak out.

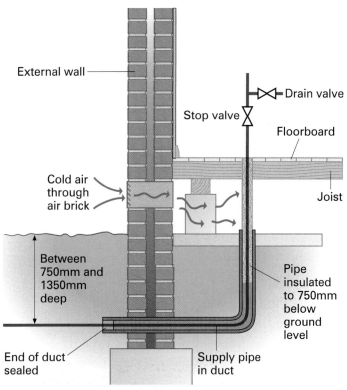

External wall

Drain valve

Stop valve

Floorboard

Joist

Cold air through air brick

Between 750mm and 1350mm deep

Pipe insulated to 750mm below ground level

End of duct sealed

Supply pipe in duct

Figure 6.24 Suspended timber supply pipe entry

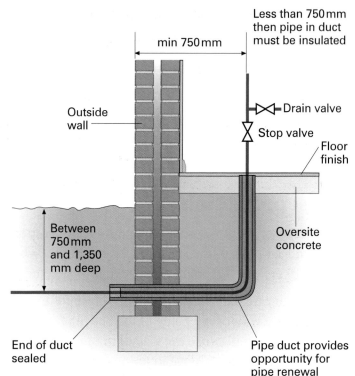

min 750 mm

Less than 750 mm then pipe in duct must be insulated

Outside wall

Drain valve

Stop valve

Floor finish

Between 750 mm and 1,350 mm deep

Oversite concrete

End of duct sealed

Pipe duct provides opportunity for pipe renewal

Figure 6.25 Solid floor supply pipe entry

Other outside applications

Outside taps or standpipes for garden hoses are common. If the tap has a dedicated underground supply, it requires protection.

The diagram shows a typical garden tap installation. Note that a drain valve (see illustration) is located below the pipework. This enables the pipework to be drained once the stop tap is turned off. A double check valve is also fitted to protect against contamination.

Location of pipework

When planning pipe runs, you should try to avoid areas that are hard to keep warm, such as:

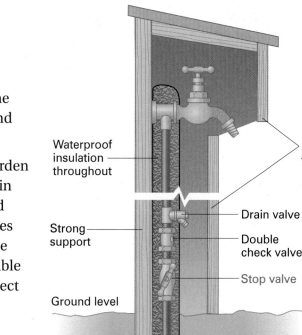

Waterproof insulation throughout

Hinged flap

Strong support

Drain valve

Double check valve

Stop valve

Ground level

Figure 6.26 Installation requirement of externally sited and fed hose union tap

- any outside locations or outside walls
- roof spaces, cellars and under floor spaces
- garages and other outbuildings
- near to windows, air bricks or ventilators.

Obviously it is not always possible to avoid these locations, on these occasions you will need to provide some protection to the pipework.

Protection of pipes and fittings

Where pipework is at risk of freezing, it should be protected using pipe insulation material. The thickness of the insulation used will depend on the type of insulation used and its insulation properties.

The table gives the minimum recommended thickness of insulation in millimetres for 15 mm, 22 mm and 28 mm water pipes, to comply with BSEN 1057 and Water Regulations Schedule 2, paragraph 4.

External diameter of pipe	Thermal conductivity of materials 0°C in W(M.K)				
	0.02	0.025	0.03	0.35	0.04
15	20	30	25	25	32
22	15	15	19	19	25
28	15	15	13	19	22

Figure 6.27 Minimum recommended insulation thickness

The table shows how thick the insulating material must be for products with varying insulation properties. Note that all of the entries in the table require an insulating material that is thicker than the typical 10 mm wall thickness you buy off the shelf at the merchant – so be careful.

Generally, the insulating material must be resistant to, or protected from:

- mechanical damage
- rain
- moisture
- subsoil water
- vermin.

Trace heating

Trace heating involves attaching a low temperature heating element to the outside of the pipe, controlled by a thermostat which activates the heater when

Did you know?

The wrong thickness of insulating material provided is a common fault identified by many water company inspectors when checking systems

the temperature is low, thereby preventing the pipe from freezing. It is more common on industrial installations, but there are domestic products available.

Pipes and cisterns in roof spaces

The illustration shows the correct installation of a cold-water storage cistern and associated pipework in a roof space. Note:

- the provision of a 350 mm gap to the roof surface to allow access for maintenance

- both pipework and cistern are fully insulated

- the space under the cistern is left uninsulated to allow heat from the property to warm the cistern.

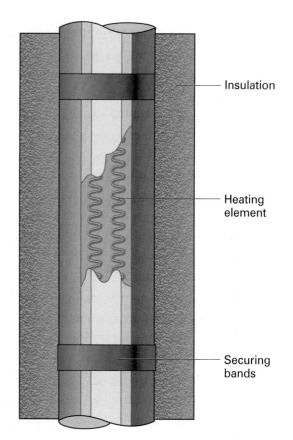

Figure 6.28 Trace heating element protecting pipework against frost

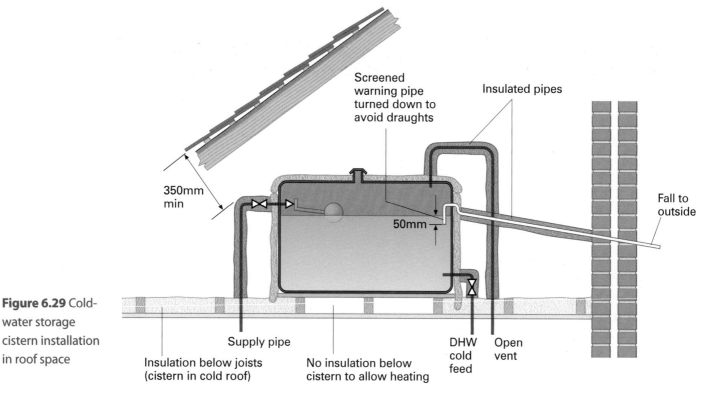

Figure 6.29 Cold-water storage cistern installation in roof space

Water Regulations also require that cold-water systems should not exceed 25°C. The use of insulation on pipework and components acts in hot weather to prevent stored water from becoming over-heated.

FAQs

Why does the water come out of some cold water taps with so much force?

This is due to high pressure in the cold-water supply to the taps which have been fed directly from the main.

The cold-water system in my house is quite noisy, when someone flushes the loo the water screeches through the pipework until the loo cistern is full again.

High water pressure is undoubtedly the cause of the noise; appliances fed from a storage cistern are much quieter in operation although it is possible to fit a pressure-reducing valve to the incoming supply.

Does water in storage cisterns and pipework go stale?

Yes it does, that why storage cisterns coupled together with pipework must be so arranged that all the water be changed as it is used and does not stagnate. When you return from holidays, or if the property has been empty then it is wise to run all taps for a while to replenish any stale water with fresh.

Knowledge check

1   Is untreated water from a deep well usually regarded as wholesome?

2   What compounds are dissolved in hard water?

3   What three methods can be used to treat hardness in water?

4   Water Regulations are enforced by whom?

5   Which of the following is not a requirement of the Water Regulations?
    a. Prevent contamination of the water supply.
    b. Prevent the waste of water.
    c. Prevent air from entering the system.
    d. Prevent misuse of the water supply.

6   State one purpose of the drain and stop valve on the cold-water supply.

chapter 7

Hot-water supply

OVERVIEW

A supply of hot water is an essential requirement for anyone. We use it for personal washing and general cleaning purposes. There is a variety of hot-water systems available. In this chapter you will learn about:

- **Hot-water systems**
  - properties of hot water
  - types of hot-water system

- **Storage heaters**
  - typical storage heaters
  - combination storage systems

- **Instantaneous hot-water systems**
  - types of instantaneous heater

It would be hard to imagine life without hot water: no hot showers, baths or water for cleaning. It is the plumber's job to ensure that hot water is available as required.

A supply of hot water to domestic dwellings only became commonplace after the Second World War, and systems design has been improving since then. Hot water system design is also covered by the Water Regulations and BS 6700.

The design of a hot-water system will depend on the type of building. For example, when installing a toilet in a factory canteen, an instantaneous water heater is a more economical option than a storage system. In a domestic situation, there are instantaneous methods of supplying hot water as well as systems fed from storage.

Properties of hot-water systems

At the end of this section you should be able to:

- explain the basic properties of hot water
- explain the requirements of hot-water systems and connections to service pipework
- explain key difference between hot-water storage and instanteous systems.

Properties of hot water

The intensity of heat is measured by its temperature. In open vented domestic hot water systems, the temperature in the hot water storage vessel should be 60–65°C. The quantity of heat takes into account the volume of a substance and is measured in kilojoules (kj).

The effect of heat on any matter – solid, liquid or gas – causes it to expand. This is a very important principle in plumbing, especially when dealing with the expansion of hot water, and the material in which it is stored or transported, e.g. copper pipework.

Heat travels by **conduction**, **convection** and **radiation**.

Conduction: Have you noticed when you use a blowlamp to make a soldered joint on a copper pipe that, although you apply the flame to the fitting, the pipework either side is always hot for some distance? This is because the heat has travelled along the pipe by conduction.

Convection: The transfer of heat by convection arises from the physical movement of molecules within a fluid or a gas. Liquids and gases expand when heated, and the molecules nearest the source of heat expand more quickly than those further away. On expansion, they become lighter, or less dense, and are consequently pushed upwards by the colder and heavier molecules surrounding them. Convection is the principle behind hot-water circulation between a boiler and a hot-water storage vessel, and for the majority of the heat given off by a radiator.

Radiation: Radiant heat can be described as a form of energy that travels in a straight line through both air and space. Probably the best example of the radiation of heat is that given off by the sun – used to good effect on solar hot-water heating systems.

Did you know?

There are many types of hot-water system but, generally speaking, they are either instantaneous systems (supplying hot water instantly) or storage systems (where water is heated and stored in a vessel until needed)

Safety tip

Catering for the expansion that occurs during the heating process is vital to ensure that the system works correctly and safely

Types of hot-water system

Connection

The connection between main and service pipework will depend on the type of hot water system that is to be installed, but generally will be categorised as direct or indirect.

Direct: Appliances such as instantaneous water heaters, when supplied directly should be fitted with a servicing valve (ball type) as close to the appliance as possible. Direct hot-water storage systems are fed via the cold-water storage cistern (CWSC).

Indirect: Instantaneous water heaters can also be supplied indirectly, and again should be fitted with a servicing valve. Again, on indirect hot-water storage systems the cold water supply is via the CWSC and the installation details were covered in the cold water supply section.

There is a wide range of hot water systems available for domestic properties. The following factors should be considered in their selection and design:

- the quantity of hot water required
- the temperature during storage and at outlets
- the cost of installation and maintenance
- fuel energy requirements and running costs
- any wastage of water and energy
- safety for the user.

Methods of heating hot-water systems

The following energy sources or fuels are used for heating hot water.

- electricity
  - immersion heater
  - instantaneous heater
  - storage heater

- gas
  - boiler
  - water circulator
  - instantaneous heater
  - storage heater

- solid fuel
  - boiler
  - combined cooker and boiler

- oil
  - boiler
  - combined cooker and boiler

- solar (usually used to supplement other systems)
  - solar panel or collector.

Choosing a system

Systems can range from a simple, single-point arrangement supplying one outlet to a more complex centralised boiler system supplying hot water to a number of outlets. BS6700 sets out a number of ways of supplying hot water, as detailed in figure 7.1 below.

The chart is divided into **centralised** and **localised** systems. A centralised system is one where water is heated and can be stored centrally within a building. The heating of the water can be controlled by a thermostat. A system of pipework supplies the heated water to various draw-off points. A localised system is one where water is heated where it is needed – for example a single-point water heater sited over a sink. Localised systems are often used in situations where long distribution pipe runs would involve a waste of water and energy.

> **Definition**
>
> **Centralised storage system** – stores heated water ready to be supplied to a number of outlets
>
> **Localised system** – heats water at the point at which it is required

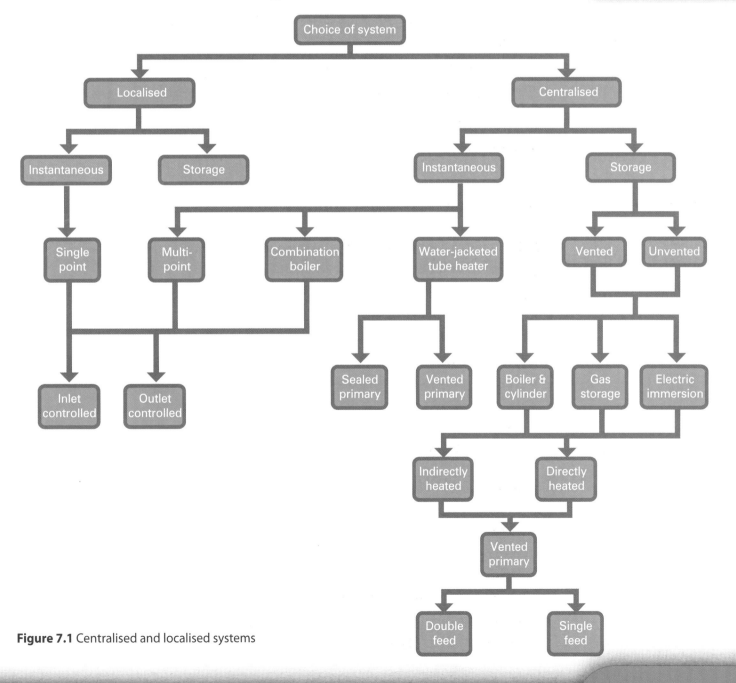

Figure 7.1 Centralised and localised systems

Definition

Gravity circulation – cold water is heavier than hot water and gravity therefore exerts a stronger pull on it, drawing it down and allowing the hot water to rise through the system

Immersion heater – is an electric element fitted inside the hot water storage vessel. It can be controlled by a switch and thermostat

Direct hot-water system (vented)

You are only likely to work on the system shown in the illustration when carrying out maintenance or repair, as it is no longer widely used. This direct system works by water being heated in a boiler (by gas, oil or solid fuel) and then rising due to the principle of convection – otherwise known as **gravity circulation**. It rises through the primary flow pipe and into the hot water storage vessel, heating the contents of the vessel directly. The hot water from the boiler is replaced by the cooler and heavier water moving in the primary return from the lower area of the storage vessel. The system does not always have to be heated by a boiler via primary flow and return pipes. It can be directly heated by means of an **immersion heater**; gas circulators are also used on direct systems, connected directly to the storage vessel.

There are a number of points to remember about direct systems using gravity circulation.

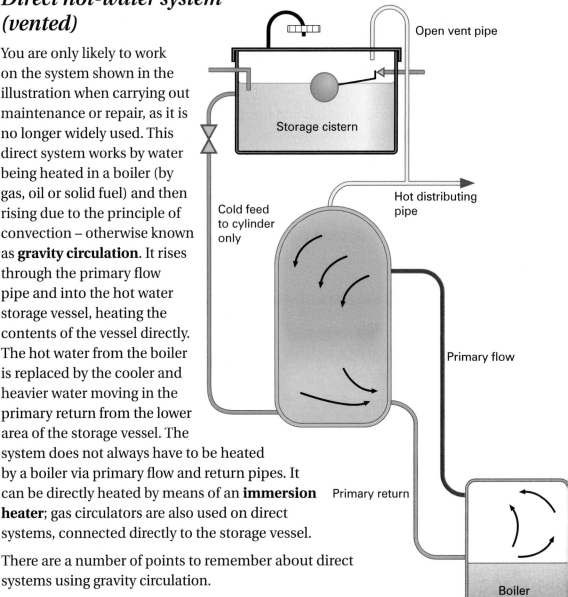

Figure 7.2
Direct system of hot-water

- There are minimum sizes for primary circuits to hot storage vessels:
  - 22 mm for short pipe runs
  - 28 mm for longer pipe runs (or from continuous burning appliances).

- Vent pipes should not be less than 22 mm in diameter.

- All pipes to be laid to falls to prevent air locks and help system drain down.

- The vent route from the boiler, primary flow and open vent should not be valved.

- The cold-feed pipe should be sized in accordance with BS6700 – the cold feed is the key route in which expansion water is taken up from the cylinder when it is heated, i.e. the heated water from the cylinder moves through the cold-feed pipe and the water level rises in the storage cistern.

- The open-vent pipe cannot be taken directly from the top of the hot-water storage vessel. The hot draw-off pipe should incorporate a 450 mm offset between the storage vessel and its point of connection to the open-vent pipe to prevent one pipe circulation.

- Corrosion inhibiters should not be used, as the water in the boiler is fed directly to the appliances.

- No other supplies or draw-offs should be connected to the cold feed.

The next illustration shows a direct system heated by an immersion heater. The immersion heater should be controlled by a thermostat.

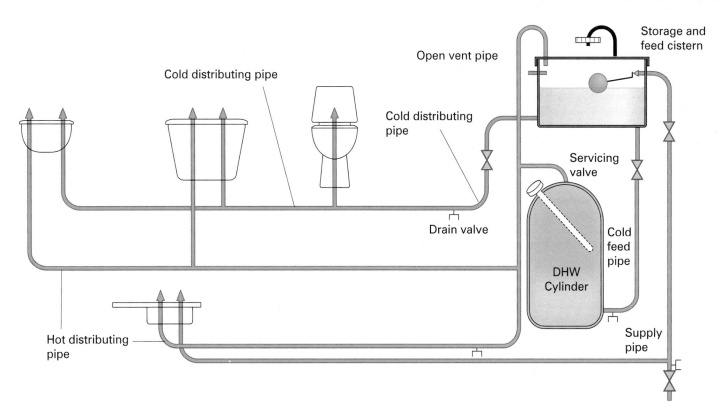

Figure 7.3 Direct hot-water system heated by an immersion heater

The purpose of the vent pipe

The purpose of the vent pipe is to maintain atmospheric conditions in the pipework. It permits any air entering the system to escape and, should the water in the system become over heated, it allows it to expand up the vent pipe and discharge into the cistern. The use of better system controls has reduced the risk of overheating, but it can occur on direct systems, particularly ones using solid-fuel back boilers.

Indirect hot-water system (vented)

This is the most common form of vented domestic hot-water system. It allows the boiler to be used for the central heating circuit as the system permits the use of a variety of different metals because the primary circuit is totally separate from the secondary circuit. The system is called indirect because the water contained in the storage vessel is heated indirectly through a heat exchanger. The diagram shows a vented double-feed indirect system.

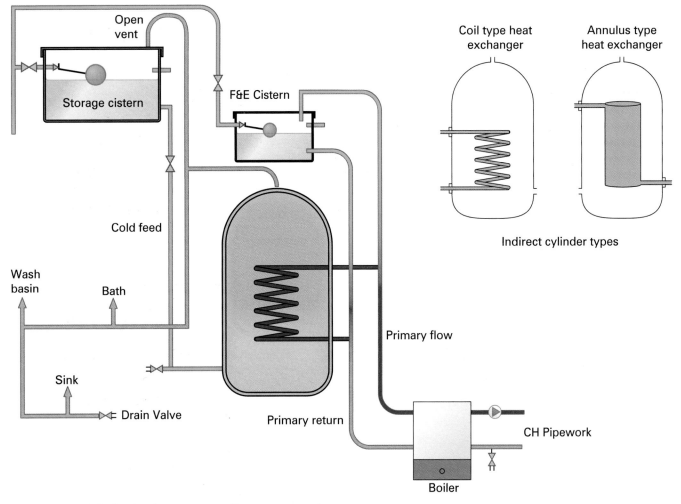

Figure 7.4 Double-feed indirect system of hot-water supply

Key points of the double-feed type of cylinder:

- The open-vent and cold-feed pipes may be connected to the primary flow pipes as shown, or fed separately into the boiler.

- Where the vent pipe is not connected to the highest point in a primary circuit, an air release valve should be fitted.

- A separate feed and expansion cistern needs to be provided to feed the primary circuit – this ensures that where a double-feed cylinder is used, the primary water is kept totally separate from the secondary hot water.

Indirect single-feed system

This system uses a self-venting cylinder and does not require a separate feed and expansion cistern. The water in the primary and secondary circuits is separated by means of an air bubble. The installation of the cylinder needs to be carefully made

Remember

Hot water storage vessels must be insulated. New cylinders are now pre-lagged in the manufacturing process, using expanded foam, but lagging jackets can also be bought

in accordance with the manufacturer's requirements, in order that the air bubble is not dislodged in the cylinder permitting the two waters to mix. It is for this reason that this type of cylinder does not tend to be widely used in modern vented types of hot-water system.

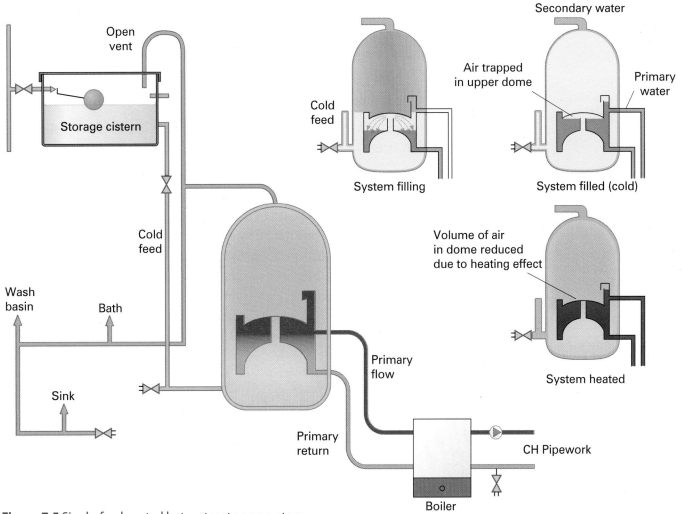

Figure 7.5 Single-feed vented hot-water storage system

Key points of the single-feed type of cylinder:

- Water enters the primary circuit via a number of holes at the top of the vertical pipe immediately under the upper dome.
- The system is self-venting through the air vent pipe while the primary circuit is filling.
- Once the primary circuit is filled, the filling of the secondary supply (the one which feeds the appliances) continues.
- When the secondary supply is full, two air seals are formed and a permanent seal is maintained.
- Once the water is heated, expansion of the water in the primary circuit is taken up by forcing the air from the upper dome.

Remember

The manufacturer will specify a maximum head of water above the base of the cylinder

Definition

Sacrificial anode – a piece of metal low in electro-motive series (electrolytic corrosion) that will be destroyed before all others

Note also the following points:

- The primary circulation system must not be too large. An excessive quantity of water in the system, once expanded, would exceed the amount of space in the dome. The result would be the loss of the air seal.

- Care must be taken when using circulating pumps on the primary circuit to the hot water cylinder, again this could result in the loss of the air seal.

- The loss of air seal would convert the system into a 'direct system'. That is, water from the central heating system would mix with the secondary water.

- Corrosion inhibitors or other additives must not be added to primary circuits.

The disadvantages tend to outweigh the advantages of single-feed vented cylinders, so they are not widely used.

The cylinders used in direct and indirect systems are available in a number of standard sizes – 900 mm × 450 mm, storing approximately 110 litres of water, being the most common. Cylinders are supplied in a number of grades (thicknesses): grade 4 has been commonly used, but it is of an inferior quality, possibly with poor heat exchanger surface area and should no longer be considered. Grade 3 is the preferred minimum option with a heat exchanger and insulation level that will meet the requirements of Part L1 of Building Regulations. The cylinder will usually be labelled to indicate that it complies. Most cylinders will include a **sacrificial anode** inside (made from magnesium) to guard against pitting corrosion and protect the lifespan of the cylinder.

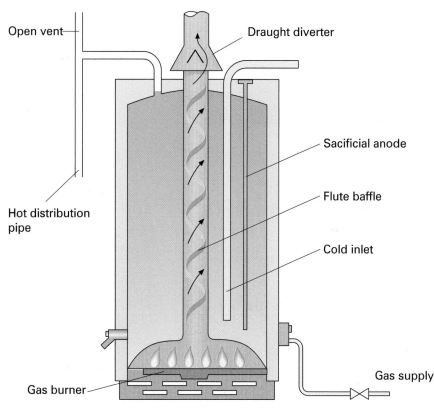

Figure 7.6 Direct-fired gas storage hot-water heater

Storage heaters

At the end of this section you should be able to:

- identify and describe various types of hot-water storage systems

Inlet controlled

These are more common in large domestic or small commercial/industrial buildings.

The diagram shows a typical storage heater. In this case, it is heated by gas, but electrical storage heaters are also available. A gas storage heater is basically a self-contained boiler and storage system. This system also includes an open flue, which must be terminated externally. Often referred to as pressure-controlled

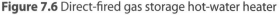

water heaters, they are usually designed to be fed by a cistern (indirect) or mains (direct) supply, in which case it would be included in an unvented system. This type of storage heater is classified as outlet controlled, as the supply is controlled at the appliance outlet, e.g. hot tap. It will also serve multiple outlets.

Outlet controlled

As the illustration shows, these are generally seen as single-point heaters, fitted either above the appliance with a swivel outlet spout, or under the appliance.

The heater is fed from the supply pipe, which has an inlet control. It is important that the outlet must not be obstructed, or any connections made to it, as the open outlet allows for expansion of the water on heating. If an under-sink model is used, then a special tap will be required to allow venting of the water heater. They can be heated by either gas or electricity.

Combination storage systems

Combination storage cylinders include a cold-water storage cistern. The base of the cistern must not be positioned lower than the level of the highest connected water outlet. It should also be high enough to give adequate water flow at the outlets. They are not very widely used, as the relatively low head provided to the hot water outlets in typical installations can be problematical.

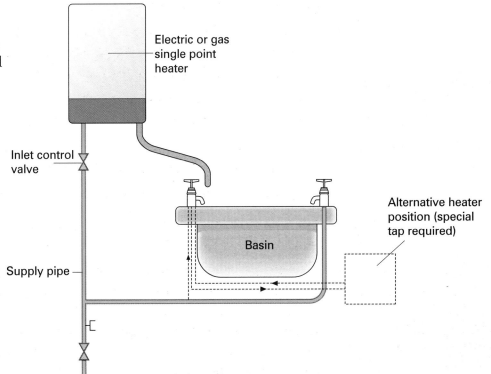

Figure 7.7 Small, inlet controlled single-point storage heater

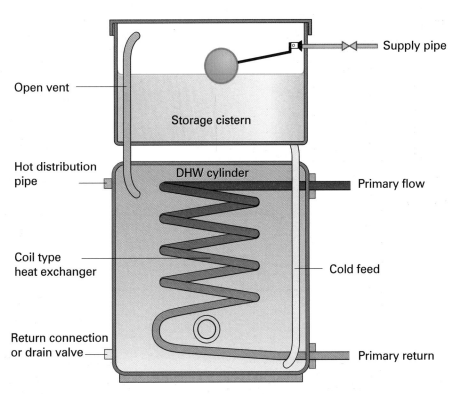

Figure 7.8 Combination storage cylinder

The advantages of combination storage systems are:

- low installation costs
- they are useful for flats where space is limited, providing minimum flow rates can be achieved.

However, there are disadvantages:

- can not be used for showers (without pumping)
- low pressure at hot taps
- cold-water storage space limited.

Instantaneous hot-water systems

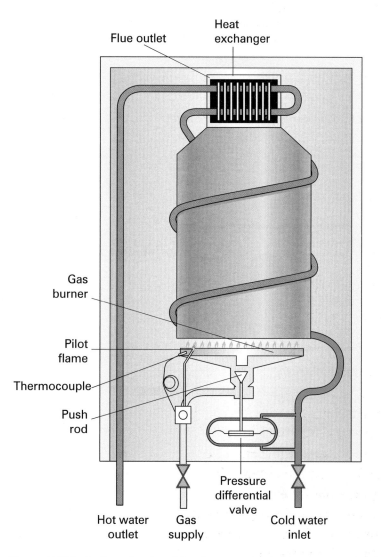

Figure 7.9 Typical gas fired multi-point water heater

At the end of this section you should be able to:

- list the various types of instantaneous water heaters
- describe their operation

Instantaneous systems work by passing cold water from the service pipe through a heat source, which heats the water by the time it comes out at the application end. The heat source can be either gas or electric, or in the case of a water-jacketed tube heater, oil.

The speed that the water can be heated is limited, so the flow rate of the water needs to be controlled so it can be heated properly. Because of the reduced flow rate, it is not possible to supply a large number of outlet points all at once, so they would not be installed in situations where there is high demand. For example, you might find a multipoint in a small property, or a single point in an office kitchen area or WC.

Types of instantaneous heater

Multipoint – gas fired

As you can see from the illustration, this consists of a gas burner sited beneath the heat exchanger.

When the hot tap is opened, it allows water to pass through the heater. This causes the gas valve to open as a result of the drop in pressure in the differential valve. This drop in pressure is caused by water passing through the **venturi**, which creates a negative pressure as it sucks the water from the valve. The diaphragm is connected to a push rod and, as it lifts, it opens the gas line. The gas is then ignited by the pilot light. When the hot-water tap is turned off, the pressure in the differential valve is equalised, the diaphragm closes and the gas supply is turned off.

The water-jacketed heater

This is also known as a thermal storage system.

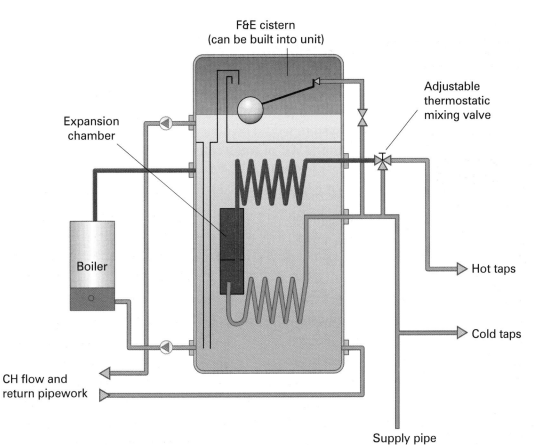

Figure 7.10 Water-jacketed tube heater

When the hot tap is turned on, cold water – either from the mains or storage cistern – passes through a heat exchanger, which is situated in a heat store of primary hot water. The size of this heat store will be calculated based on the volume and rate of flow that can be delivered without an unacceptable drop in temperature. The primary water-flow from the boiler is programmed by the cylinder thermostat. Hot water is pumped to the radiator heating-circuit and is returned to the heat store; the cooler water from the heat store is then returned to the boiler where it is reheated. This is similar to an indirect domestic hot-water system, only in reverse.

Single-point vented electric water heater

Single-point instantaneous water heater

This type of heater uses electricity or gas. Sited directly above the appliance, it is usually inlet controlled, with the hot water delivered via a swivel spout. The electric multipoint is a small tank of water with an electric heating element inside. Because of the low volume of water, it quickly heats up as it is drawn through the heater. The temperature at the outlet will be related to the water flow rate and the kW rating of the heater.

You will find single-point heaters used in situations where a small number of hot-water draw-offs are fed by individual heaters in a non-domestic type building, and where the use of a centralised hot water system would be uneconomical. You might find them in the WC of small cafés, etc.

Instantaneous electric showers

The electric heater shown is designed for mains connection, although some can be fed indirectly. The electrical rating can be in excess of 10 kW, so it is important that the supply is adequate, and wired directly from the mains distribution unit (MDU). For a rating of 9.6 kW, the circuit protection device requirements would be 45 amp and the cable 10 mm². The shower should also be isolated with a switch; this should be located within easy access outside of the shower room. Any pressure variations in the cold water supply to the shower will be handled by the flow governor. Most electrical instantaneous showers are fitted with a flexible hose outlet – they will require a **check valve** to be fitted to protect against possible contamination of the supply, as required by the Water Regulations Schedule 2, paragraph 15. This will usually be provided by the manufacturer.

Instantaneous electric shower

FAQ

Why is it that a hot-water cylinder is always full of water even when the hot tap runs dry?

A hot-water cylinder has the cold-water inlet at the bottom and the hot outlet at the top therefore it always remains full even if the cold supply is disrupted and hot taps are opened until they run dry. Furthermore it is safe for immersion heaters and boilers to remain working because the cylinder is always full.

Knowledge check

1   State 3 of the key factors to be considered when selecting and designing hot-water systems.

2   The water temperature in a hot-water storage vessel should be maintained at:
    50-55°C   65-70°C
    60-65°C   70-75°C

3   Heat transmitted by gravity between the boiler and cylinder in a hot-water system uses the principle of:
    Radiation   Convection
    Conduction   Reflection

4   Which of the following statements is correct?

    A single-feed indirect cylinder –
    a. Should not be filled with corrosion inhibitor
    b. Should be filled with corrosion inhibitor
    c. Requires a feed and expansion cistern
    d. Requires a primary vent pipe

5   Which of the following statements is incorrect?

    A combination storage cylinder –
    a. Is space saving
    b. Provides high pressure hot water
    c. Provides low pressure hot water
    d. Does not require a storage cistern

6   List the main types of instantaneous water heaters.

chapter 8

Common requirements of hot- and cold-water supply

OVERVIEW

In previous chapters, you have looked at some of the specific aspects of hot- and cold-water systems. This chapter considers the common requirements of both systems, including the range of equipment used, installation, soundness testing and maintenance:

- **Taps and valves**
  - the range available
  - their location within systems

- **System installation requirements:**
  - installation of pipework
  - dead legs, secondary circulation and stratification
  - installation requirements of shower mixing valves
  - noise in systems

- **Soundness testing of hot- and cold-water systems:**
  - procedures
  - final checks

- **maintenance and decommissioning:**
  - maintenance requirements
  - decommissioning of hot- and cold-water systems

Taps and valves

At the end of this section you should be able to:

- identify and describe the various taps and valves used in plumbing systems
- state where the taps and valves should be located in a plumbing system
- identify the basic operating principles of taps and valves.

Taps and valves are used to isolate supplies, reduce the flow rate through pipework, permit drainage from systems, and provide an outlet to an appliance, for example sinks, baths, washing machines etc.

Water Regulations require that all fittings must be:

- suitable for their purpose
- made of corrosion-resistant materials
- sufficiently strong to resist normal and surge pressure
- capable of working at appropriate temperatures
- easily accessible to allow the renewal of seals and washers.

Taps and fittings for both hot and cold water should conform to BS1010, parts 1 and 2, BS1552 and BS5433.

Taps and valves are usually made of brass pressings or castings, and are chrome plated when appearance counts. Chrome plating also makes them easier to clean.

The range available

Stop valve

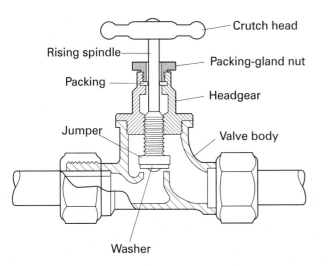

Figure 8.1 Stop valve

Stop valve

Remember

The use of taps and valves applies equally to hot and cold systems.

We dealt with float-operated valves in Chapter 6

Did you know?

Plastic taps and valves are also available. These are manufactured from a thermosetting plastic called acetyl

Safety tip

Always look out for the arrow indicating flow of water to ensure the stop valve is installed in the correct way

The stop valve pictured here is a typical screw-down valve. The washer and jumper fits into the threaded part of the spindle and is raised or lowered by turning the crutch head. When lowered to its maximum extent onto the seating of the valve, the incoming supply is isolated.

The stop valve is usually located on high-pressure pipelines – for example between the incoming mains and rising main – and is therefore used as a main supply isolation valve.

Stop valves used below ground must be made of a corrosion-resistant material such as gunmetal or bronze to avoid dezincification, and marked CR in the UK or DRA in Europe.

Gate valve

Gate valves are sometimes referred to as full-way gate valves, because when they are open there is no restriction through the valve. The gate valve is fitted with a wheel head attached to the spindle. When the head is turned anti-clockwise, the threaded part of the spindle is screwed into the wedge shaped gate, raising it towards the head.

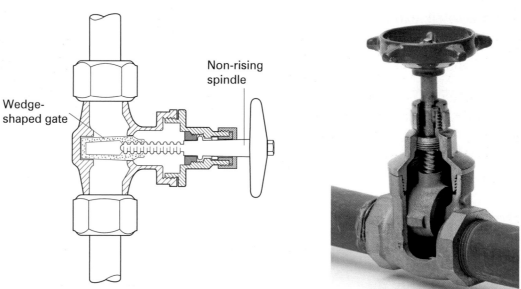

Non-rising spindle

Wedge-shaped gate

Figure 8.2 Gate valve

Gate valves are usually located in low-pressure pipelines, for example the cold feed from the cold-water storage cistern (CWSC) to the hot-water storage cylinder. You might also find them used on supplies to shower valves.

Quarter-turn valves (spherical plug valves)

The 'ball valve'-type quarter-turn valve (spherical plug valve) can be used as a service valve to a cistern, in which case it will have a compression nut and olive on each side of the valve. It can also be used on the hot and cold supplies to a washing machine or dishwasher, when it will have a flexible hose connection. These valves are used to isolate the supply for the service, repair or maintenance of components and appliances within the hot and cold system.

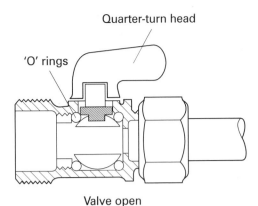

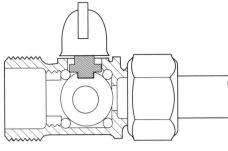

Find out

Why should a quarter-turn valve be closed slowly on high-pressure pipework?

Figure 8.3 Quarter-turn valve

The valves work by the quarter-turn operation of a square head (or slotted head) located at the top of the valve, which aligns the hole in the valve with the hole in the pipe. When the head is in line with the direction of the pipe, the valve is open. The quarter-turn head could be made from plastic and operated by hand – as used on washing-machine supplies – or it could have a slotted head, operated by a screwdriver, as fitted on supplies to cisterns. It can be used on low or relatively high-pressure installations for servicing purposes.

Ball-o-fix

Draw-off taps

Draw-off taps, such as bib taps and pillar taps, work in a similar way to screw-down valves. The more modern type do not work as screw-down valves, but allow two polished ceramic discs to turn and align with two port holes through which the water can pass.

The bib tap

The tap shown overleaf includes a hose union attachment used for garden hosepipes.

You are unlikely to find the bib tap shown below ('supatap') in most modern domestic dwellings, but you may come across it when carrying out maintenance work on other properties.

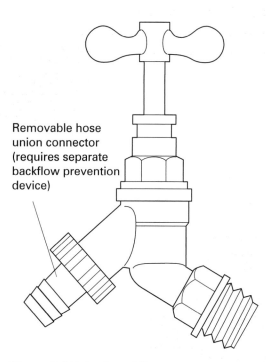

Removable hose
union connector
(requires separate
backflow prevention
device)

Figure 8.4 Union hose bib tap

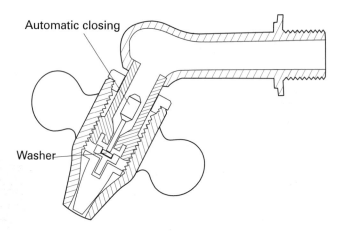

Automatic closing

Washer

Figure 8.5 Bib tap: supatap

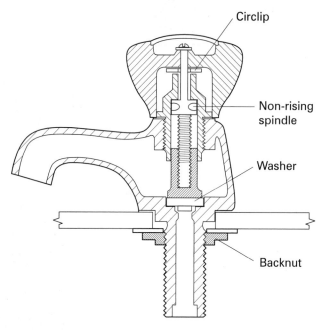

Circlip

Non-rising
spindle

Washer

Backnut

Figure 8.6 Pillar tap

The pillar tap

The pillar tap has a long vertical thread which passes through pre-drilled holes in the appliance. It is held in position with a backnut, and the supply pipe is attached to the thread via a tap connector. Its method of operation is very similar to the stop valve we saw earlier.

The tap shown in figure 8.7 uses quarter-turn ceramic discs. This type of tap is in common use. However, they often have maximum water pressures within which they operate: if you install outside the requirements laid down by the manufacturer, the discs can shatter under operating conditions.

Globe taps

You may come across globe taps when carrying out maintenance work on an old bath installation. Instead of having a male thread connection to the supply, like a bib tap, they use a female thread. The male threaded outlet from the water supply is connected through prefabricated holes in the bath, and the globe tap is then screwed on to the fitting; they can pose a real threat to water-supply contamination, as they discharge below the spillover level of the bath. Their method of operation is similar to the stop valve.

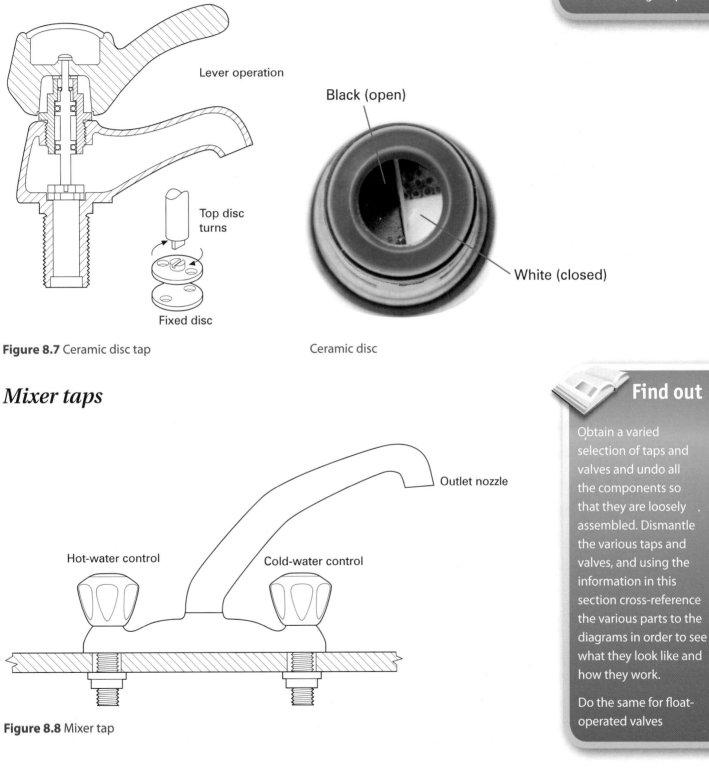

Lever operation

Top disc turns

Fixed disc

Black (open)

White (closed)

Figure 8.7 Ceramic disc tap

Ceramic disc

Mixer taps

Outlet nozzle

Hot-water control

Cold-water control

Figure 8.8 Mixer tap

There are a number of design variations in mixer taps, but they all work on the principle of allowing the water from the hot and cold supply to flow from one outlet.

There are two types of mixer tap:

- single-flow outlet
- twin-flow outlet.

In the case of the twin-flow outlet nozzle, as shown in the illustration, the water is not allowed to mix in the tap swivel outlet. In a single-outlet unit, mixing can take place and cross-contamination of water can occur between the hot and cold supplies.

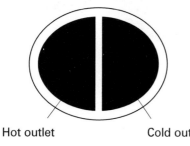

Hot outlet   Cold outlet

Figure 8.9 End view of outlet: twin-flow

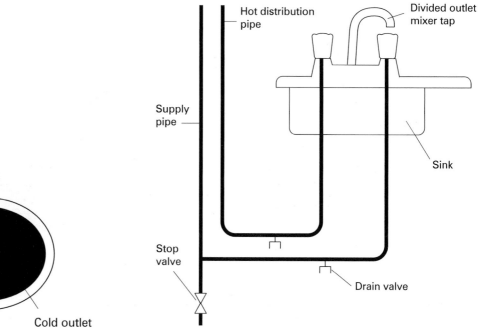

Figure 8.10 Mixer tap pipework

The drain-off tap

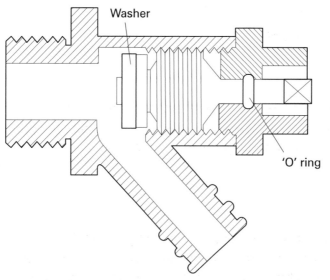

Figure 8.11 Drain-off tap

Conforming to BS2879, the drain-off tap is located at the lowest point of any system and has a 'ribbed' outlet to enable a good grip for a hosepipe connection. The tap is usually fitted with a 'lockshield head', which means it can only be operated by a purpose-made key. Combined stop valves with drain-off taps are also available.

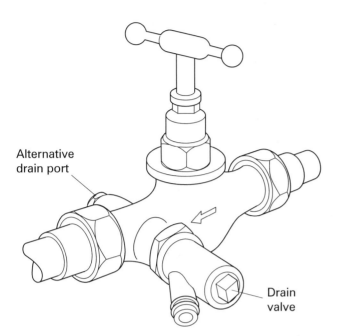

Figure 8.12 Combined stop valve with drain-off tap

System installation requirements

At the end of this section you should be able to:

- explain the requirements for installation pipework
- explain what is meant by dead legs, secondary circulation and stratification
- state the connection requirements for showers and bidets
- state the main cause of noise in a plumbing system.

Installation of pipework

You covered some of the pipework installation requirements in Chapter 5 on Common plumbing processes. This section looks at some of the basic installation aspects of system pipework and the reasons why you need to think carefully about how you route pipework.

Accessibility

Water Regulations require that water pipes and fittings are easy to get at so that they can be inspected and repaired. However, customers and installers are not keen to see exposed pipework, so the tendency is to hide it under floors or behind wall surfaces.

BS6700 gives recommendations about the accessibility of pipes and water fittings.

Find out

The requirements of pipework installed in non-accessible locations in the Water Regulations and British Standards

Hot-water pipework

One of the most common causes of problems in hot-water systems is air locks. Air locks are usually caused when pipework from cold-water storage cisterns, hot-water storage vessels and boilers is not installed correctly. The illustration shows two installation details that highlight the problem.

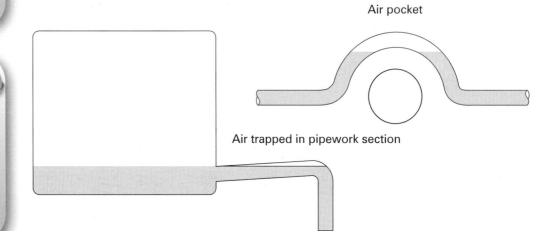

Air pocket

Air trapped in pipework section

Figure 8.13 Air lock

Pipework running horizontally should be level or to an appropriate fall allowing air to escape from the system. Horizontal runs of gravity primary circulation pipework should also be installed to the appropriate falls.

Dead legs, secondary circulation and stratification

Dead legs are long lengths of pipe from the hot-water storage vessel to the appliance (see figure 8.14).

Generally speaking, the length of pipe measured from the hot-water storage vessel or heater to the tap should be as short as possible.

Dead legs should be avoided for two reasons:

1 they waste water by having to run off cold water before it turns hot

2 they waste energy by heating up the volume of water contained in the dead leg, which then cools.

One method of overcoming the problem of dead legs is **secondary circulation**. This can be demonstrated with the typical installation layout using a pumped secondary circulation system, shown in figure 8.15.

A flow-and-return loop is installed, which feeds all the appliances. The water is kept circulating, either by gravity or by a non-corrosive circulating pump (bronze manufacture).

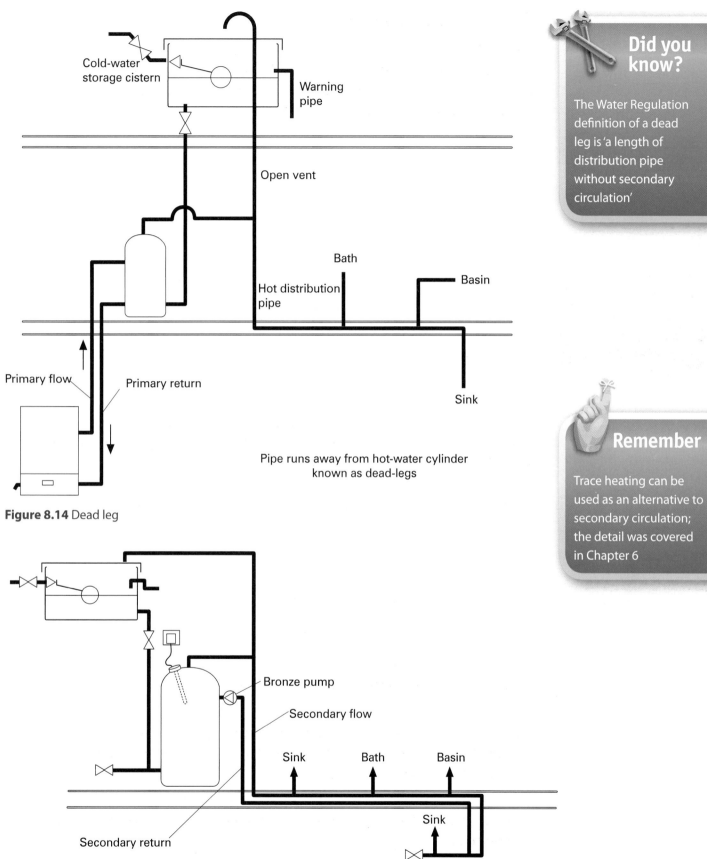

Figure 8.14 Dead leg

Figure 8.15 Secondary circulation

Warning
pipe

Open vent

Bath

Basin

Hot distribution
pipe

Sink

Primary flow

Primary return

Pipe runs away from hot-water cylinder
known as dead-legs

Cold-water
storage cistern

The return pipe is connected in the top third of the cylinder. This prevents cooler water lower down the cylinder from mixing with the hot water.

Did you know?

The Water Regulation definition of a dead leg is 'a length of distribution pipe without secondary circulation'

Remember

Trace heating can be used as an alternative to secondary circulation; the detail was covered in Chapter 6

Stratification

In a hot-water storage vessel, layers of water form from the top of the vessel where it is hottest to the base of the vessel where it is coolest. **Stratification** must take place if the vessel is to function to its maximum efficiency.

Manufacturers of storage vessels build stratification into the design. The following design rules enable stratification to take place:

- the vessel should be cylindrical in shape

- the cylinder vessel should be installed in a vertical position rather than horizontal

- the cold-feed connection should be in a horizontal position.

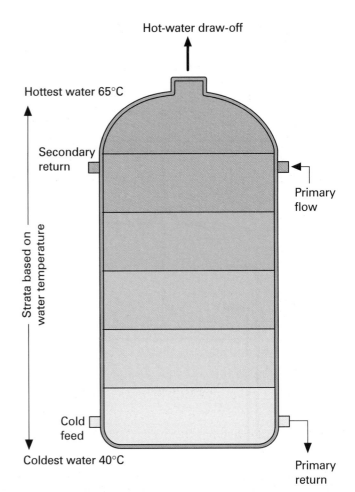

Figure 8.16 Stratification in a hot-water tank

Labels on figure: Hot-water draw-off; Hottest water 65°C; Secondary return; Primary flow; Strata based on water temperature; Cold feed; Coldest water 40°C; Primary return

Connection requirements of showers

Here we will consider the layout and connection requirements of shower mixing valves. A shower can be fed by mains pressure, or it can be storage-fed by gravity or storage-fed using booster pumps.

Mains-fed

Water Regulations require that provision be made to ensure that backflow cannot occur. This is done either by installing a double check valve, or by using a rigid connection to the showerhead. Mains connection can only be used on thermostatic mixing valves, using the manufacturer's valves designed specifically for this purpose.

Shower mixer

On the job: Dilemma for Craig

During alterations to a plumbing system on a house extension, Craig, a final-year apprentice, has been given an ordinary shower mixer by the customer for fitting, although the cold water is mains fed. Craig is aware from his studies that the shower should not be fitted directly to the mains, and he explains this to the customer. The customer accepts the explanation that it is not good practice and against the Water Regulations to fit it, but insists that this is his house and that it is acceptable to him.

'Anyway,' he says, 'thermostatic showers are three times the price – no chance, just fit it.'

1. What should Craig do now?

2. Who do you go to for expert advice?

Storage-fed

The hot and cold supplies need to be of equal pressure, and the showerhead a minimum of 1 m below the bottom of the cold-water storage cistern to ensure that adequate pressure from the showerhead is achieved.

FAQs

Why is the water flow from my shower poorer than it used to be?

Often the cause is just a build-up of debris in the holes of the shower rose/spray head. Removing it and cleaning it out is sometimes all that is needed to rectify the problem.

My friend complains that the water temperature in their inexpensive shower is very difficult to control and to set at a comfortable level. Why might this be?

The symptom is typical of a manual shower that has been installed with unequal pressures – that is, the cold from high mains pressure and the hot from low-pressure storage. Only special thermostatic mixer-type showers can cope with different pressures, and even then there are regulations to comply with to prevent backflow risks.

Safety tip

For non-thermostatic shower mixers, the cold-feed connection to the hot-water storage vessel should be above the cold-water supply to the shower. This is to prevent scalding should the water supply to the cold-water storage cistern be accidentally turned off, and the content of the cistern allowed to drain down

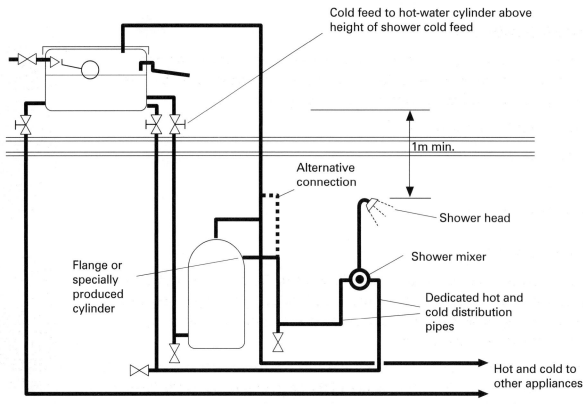

Cold feed to hot-water cylinder above height of shower cold feed

1m min.

Alternative connection

Shower head

Flange or specially produced cylinder

Shower mixer

Dedicated hot and cold distribution pipes

Hot and cold to other appliances

Figure 8.17 Pipework feeding a shower

Boosted supply

In the system illustrated below, water pressure is increased using a booster pump. The diagram shows a single-impeller booster pump and a double-impeller booster pump. The pump increases the pressure, which means that the minimum static head of 1 m is no longer needed. However, a minimum of 150 mm head is necessary to allow the flow switch to open when the supply is turned on.

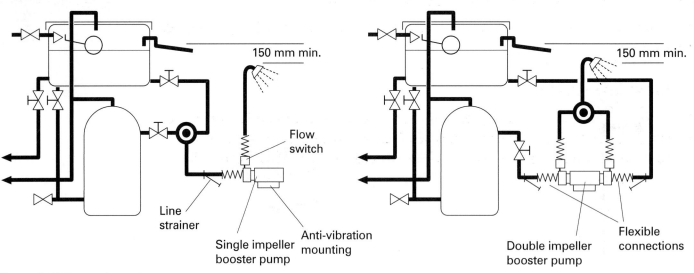

150 mm min.

150 mm min.

Flow switch

Line strainer

Anti-vibration mounting

Single impeller booster pump

Double impeller booster pump

Flexible connections

Figure 8.18 Boosted supply

Noise in systems

Noise in systems is usually caused by vibration. Not only is vibration a source of annoyance to the occupier of the building, but in severe cases it can also cause damage to pipework and fittings, eventually causing leaks.

Noise in systems is categorised as:

- water hammer
- flow noise
- expansion noise.

Water hammer

This is probably the most common cause of complaint from customers. When a valve is closed suddenly, shock waves are transmitted along the pipework, making a loud hammering noise. The problem is made worse where cold-water supply pipework is not adequately clipped.

Defective float valves and tap washers on cold taps can cause water hammer, so it can be cured through regular maintenance of the valve or taps. The velocity of the cold-water supply will further affect the problem; reducing the flow rate by turning down the stop valve will therefore help to reduce the incidence of water hammer.

Flow noise

Pipework noise becomes significant at velocities over 3 m/s, so the system should be designed to operate below 3 m/s even if this means increasing the diameter of the pipe or installing a special device known as a pressure-reducing valve.

Flow noise is also sometimes heard from cisterns. This can be caused by splashing noises as the incoming water hits the water surface as it fills.

Silencer tubes on float valves were once used to cure this, but are no longer allowed, except for the collapsible type.

Expansion noise

This usually occurs in hot-water pipework. As the system expands and contracts, it causes creaks and cracking sounds. The use of relevant pipe clips, brackets or pads between pipes, fittings and pipework surfaces should help to deal with expansion and contraction.

Did you know?

Connections to bidets are covered in detail by Water Regulations; this subject will be dealt with at Level 3

Definition

Water governor – pressure-reducing valve

Find out

What are the main causes of air locks in a hot-water system? How can these be avoided?

Soundness testing of hot- and cold-water systems

At the end of this section you should be able to explain the procedures for carrying out soundness testing on hot- and cold-water systems, including customer liaison.

Soundness testing on completed hot- and cold-water systems is essential to ensure that there are no leaks. Even the most competent plumber can make the occasional mistake, so testing is always required.

BS6700 provides the standard for soundness testing on hot- and cold-water systems.

Procedures for carrying out soundness testing

Soundness testing of hot- and cold-water systems includes:

- visual inspections
- testing for leaks
- pressure testing
- final checks.

Visual inspection

This includes making sure all pipework and fittings are thoroughly inspected to ensure that:

- they are fully supported, including cisterns and hot-water cylinders
- they are free from jointing compound and flux
- all connections are tight
- terminal valves (sink taps etc.) are closed
- in-line valves are closed to allow stage filling
- the storage cistern is clean and free from **swarf**.

It is useful at this stage to advise the customer or other site workers that soundness testing is about to commence.

Testing for leaks

When testing for leaks you should follow this checklist:

> ## *Testing for leaks checklist*
>
> 1. Slowly turn on the stop tap to the rising main
>
> 2. Slowly fill, in stages, to the various service valves, and inspect for leaks on each section of pipework, including fittings
>
> 3. Open service valves to appliances, fill the appliance and again visually test for leaks
>
> 4. Make sure the cistern water levels are correct
>
> 5. Make sure the system is vented to remove any air pockets prior to pressure testing.

Pressure testing

Pressure testing of installations within buildings is done using hydraulic pressure-testing equipment. BS6700 has separate procedures for testing rigid pipes and plastic pipes.

The procedure for testing rigid pipes (e.g. copper)

- Make sure any open-ended pipes are sealed, e.g. vent pipes

- Once the system has been filled, it should be allowed to stand for 30 mins to allow the water temperature to stabilise.

- The system should be pressurised using the hydraulic testing equipment to a pressure of $1\frac{1}{2}$ times the system working pressure.

- Leave to stand for 1 hour.

- Check for visible leakage and loss of pressure. If sound, the test has been satisfactory.

Hydraulic pressure tester

- If not sound, repeat the test after locating and repairing any leaks.

The procedure for testing plastic pipes

BS6700 has two test procedures for plastic pipes – procedures A and B. See Water Regulations for more details.

Test A procedure:

- Apply test pressure, which is maintained by pumping for a period of 30 minutes and visually inspect for leakage.

- Reduce pressure by bleeding water from the system to 0.33 times maximum working pressure. Close the bleed valve.

- Visually check and monitor for 90 minutes. If the pressure remains at or above 0.33 times working pressure, the system can be regarded as satisfactory.

Test B procedure:

- Apply test pressure ($1\frac{1}{2}$ times maximum working pressure) and maintain by pumping for a period of 30 minutes.

- Note the pressure and inspect visually for leakage.

- Note the pressure after a further 30 minutes. If the pressure drop is less than 60 KPa (0.6 bar), the system can be considered to have no obvious leakage and then visually check and monitor for 120 minutes. If the pressure drop is less than 20 KPa (0.2 bar), the system can be regarded as satisfactory.

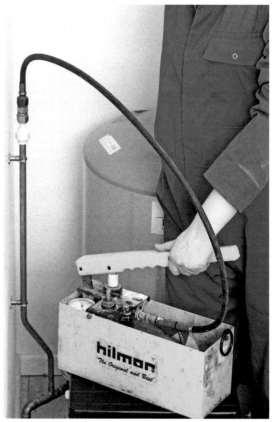
Using a hydraulic pressure tester

Final system checks

After the system tests have been completed, the system should be thoroughly flushed out to remove any debris or swarf before carrying out a final visual check for leaks. Advise the customer and/or other site workers that testing is complete. At this point in the test procedure a performance test of the system is conducted, plus a hand-over, both of which will be covered at Level 3.

Maintenance and decommissioning

At the end of this section you should be able to:

- describe the maintenance requirements for a:
  - stop tap
  - hot- and cold-water tap
  - float-operated valve
  - shower mixer valve

- describe the procedures for decommissioning hot- and cold-water systems.

Hot- and cold-water system maintenance at Level 2 requires you to understand the principles of maintaining some of the key components within a hot- and cold-water system. Maintenance is important as it ensures that the system components operate as they were designed to. It also reduces the chance of wasting water and energy by preventing dripping taps, float-operated valves and showers.

Maintenance work usually includes turning off hot- and cold-water supplies and then turning them on again. You need to ensure the customer is notified when you are doing this, so that they can fill kettles and turn off washers and dishwashers.

The principles of maintenance/servicing are quite simple:

- find out what the fault is with the component to be serviced

- isolate the supply

- strip the components

- repair or replace the defective part

- reassemble the components

- turn on the supply and test the component for correct operation.

Maintenance requirements

Here are maintenance check lists for the following:

- stop tap

- hot- and cold-water tap

- float-operated valve

- shower mixer valve.

Stop tap

Common faults

- The stop tap won't turn off the supply properly. This is usually due to a worn-out washer or a defective seating.

- The stop tap is stiff and hard to turn. This is due to infrequent use and lack of regular maintenance. The spindle and packing-gland nut oxidise over time which has the effect of 'bonding' the two together.

- The stop tap is leaking. If the stop tap is a compression type, the nut and 'O' ring joint could be leaking. Similarly, with a capillary fitting, the solder joint could be leaking. It is more likely, however, that the packing-gland nut will be leaking. Another possibility is that condensation has been mistaken for a leak. The incoming main is very cold, and the stop tap is usually located in a kitchen, where moisture levels are high. If the pipe is not insulated, the moisture will condense on the cold surface of the pipe and trickle down the pipe to the stop tap, giving the impression of a leak.

Tap reseating kit

On the job: The wrong taps

Ben and Zoë are both apprentices working for a large firm on a maintenance contract for a housing estate. They are carrying out routine servicing of sink mixer taps that are fed directly from the incoming cold-water supply. Zoë discovers that the type of sink mixer tap is not permitted by Water Regulations and suggests to Ben that they should inform the supervisor. Ben says 'Don't bother, it will only cause a problem as the firm is already behind schedule, and anyway, it is nothing to do with us.'

1   Who is right?

2   What should be done?

Checklist

1   **Isolate the supply**

- Make sure you have a stop-tap key in readiness.
- Locate the external stop tap. It should be at the boundary of the property (footpath at the front of the house). On other property types, it may not be easy to locate. In some cases you may need assistance from the Water Company to isolate the supply.
- Once you have located the external stop tap, lift the cover and locate the external stop-tap head.

- Be prepared for the stop-tap box to be filled with silt or other rubbish. You may have to clear that first to find the head.
- Once you have located the head, use the stop-tap key to turn off the supply. If it is stiff, DO NOT FORCE IT – you may snap the head. Try twisting it both ways to get some movement. Again, if it is impossible to move it, you may have to call the Water Company.
- Turn off the tap clockwise.
- Return to the property and drain down the cold water from the lowest point (usually the kitchen sink). Hopefully there should be a drain-off tap close to the stop tap. On older properties, drain taps are not always fitted.

2 Strip the component

- You will need to remove the headgear from the body. This is not always easy, as it may not have been stripped for years. It often helps to apply heat to the joint between the headgear and body. Remove the headgear using adjustable grips/spanners. If a drain-off tap was not fitted, get ready for a rush of water from the pipework above the stop tap.

3 Repair the component

To fully repair the stop tap:

- Strip the headgear by removing the **crutch head** and **packing-gland nut**.
- Clean the rising spindle and the inside of the packing-gland nut using fine emery cloth (or a file in some cases).
- Replace the defective washer.
- Check the seating for defects. If the seating is severely pitted, it might be worth replacing the stop tap. Alternatively, you could re-seat the stop tap using a re-seating tool which 'grinds' the seating flush again.

4 Reassemble the component

- Fit all the component parts together, making sure that the spindle and inside of the packing-gland nut are lubricated.
- Repack the gland nut using PTFE tape.
- Refit the crutch head.
- Before refitting the headgear, make sure the washer between the headgear and body is in place and intact.
- Refit the headgear and tighten.
- If not already fitted, install a drain-off tap using adjustable grips/spanners.

5 Turn on and test

- Make sure the internal stop tap is turned off, as well as the drain tap.
- Turn on the external stop tap.
- Return to the building, and make sure the cold tap on the sink is open.
- Slowly open the stop tap and allow water to flow from the sink tap. This will help in removing any bits of debris that may have entered the pipework.
- Open the stop tap fully and assess the pressure at the sink tap.
- Turn off the sink tap and allow the supply to charge the system.
- Check that all the appliances fed by the supply are operating properly; check the cistern water levels.

Safety tip

Remember that the water draining off may be hot if you used heat to remove the headgear

Definition

Crutch head – The handle of a stop tap

Packing-gland nut – Nut used to compress packing to make valve spindles watertight

Hot- and cold-water tap – kitchen sink

The servicing requirements of hot and cold taps are generally the same, but the isolation of the supply and turning it on, differs.

Common faults

Depending on the type of taps fitted, faults tend to be:

- dripping taps
- leaking taps (glands)
- tap heads that are difficult to turn
- noise when the cold tap is turned on.

On some properties, servicing valves may have been fitted to the hot and cold supply under the sink. This is good practice but is not a requirement of the Water Regulations. If servicing valves are fitted, isolation is achieved by turning them off.

The checklist below assumes servicing valves are not fitted.

Checklist

1 Isolate the supply
- On a direct cold-water system, turn off the cold-water supply to the system and drain through the sink tap.
- On an indirect cold-water system, turn off the service valve to the cold-water distribution pipe and drain through the sink tap.
- Turn off the gate valve to the domestic hot-water cylinder.
- Drain down the system pipework through the sink tap.

2 Repair or replace the defective part
This procedure will depend on the type of taps fitted. Most taps fitted to kitchen sinks are chrome-plated brass, but some may be plastic.
- Remove the cover from the tap head assembly; these are usually screwed through the top of the tap cover and concealed behind a chrome or plastic cap. Lift the cap with a small flat-headed screwdriver; the cover screw (which is usually cross-headed) can then be removed.

You are then left with a headgear similar to the stop tap dealt with above. The process for re-washering the tap is the same.

3 Reassemble the component
- This is the same procedure as for the stop tap, but you will also need to replace the cover.

4 Turn on and test
- Make sure the hot and cold taps to the sink are turned off.
- Turn on the stop tap if it is a direct system.
- Turn on the service valve on the cold-water distribution pipe if it is an indirect system.

Did you know?

You may come across sink taps using ceramic discs rather than the more traditional washer and seating methods of control. The ceramic discs do wear and allow leakage. These discs can be replaced, but you may have difficulty locating the parts. It may be more cost effective to replace the headgear completely

Remember

If the cover is plastic, don't overtighten the fixing screw, or you could split the plastic

- Turn on the gate valve to the hot-water storage cylinder.
- Return to the sink taps and make sure the supply pressure to the cold-water tap is satisfactory (direct) and that the cold-water (indirect) and hot-water supply is flowing smoothly.
- Test the operation of the taps.

Float-operated valve

There are three types of float-operated valve:

- diaphragm ball valve
- Portsmouth ball valve
- equilibrium valves:
  Portsmouth type
  diaphragm type.

The trend on new installations is towards the use of plastic diaphragm ball valves both on CWSC and WC cisterns, but on maintenance work you could come across a range of different types.

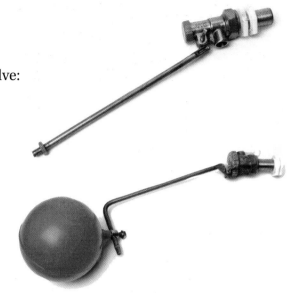

Diaphragm ball valve

It is a Water Regulation requirement that servicing valves are fitted as closely as possible to the cistern in order to isolate the float-operated valve for maintenance work.

Common faults

- Probably the most common fault is a running warning or overflow pipe. This is caused by the float-operated valve not closing off the supply properly.

- No water in the cistern, toilet will not flush, no hot water. This is usually because the hole in the seating is blocked with tiny bits of debris that got into the system when it was installed.

- Excessive noise in the system, particularly when the WC cistern is flushed or hot/cold water is drained off from the cold-water supply cistern.

Checklist

1 Isolate the supply
- If a service valve is fitted then you can simply turn this off.
- If a service valve is not fitted, you will have to turn off the supply at the stop tap.
- If you are working on a CWSC, run the level down to enable you to work easily on the float-operated valve. If it is a WC cistern, pull the flush to allow the float to drop. You might think this wastes water, but you need to check the correct operation of the float valve by watching it working as it fills up again.

2 Repair or replace the defective part (when dealing with overflows)
- If the float valve is plastic, then it is either the washer in a Portsmouth valve that is defective, or the diaphragm in the diaphragm ball valve.

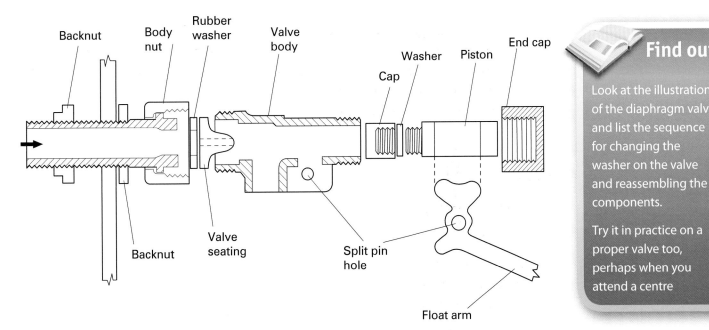

Figure 8.19

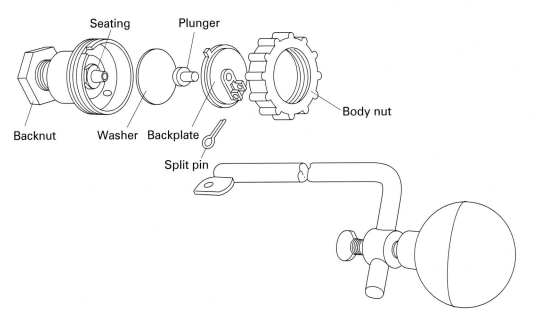

Figure 8.20

Brass Portsmouth ball valve

You may come across this type of ball valve on maintenance work.

Common faults

Most problems are caused by defective washers. Others are caused by oxidisation or a build-up of scale between the lever arm and piston.

Checklist

Once you've split the body from the body nut, follow this sequence:

1 **Remove the end cap**.

2 **Remove the split pin** from the lever arm, this frees the piston.

3 **Remove the piston cap** and take out the defective washer and replace.

4 **Reassemble the components**:
   - Refit the cap.
   - Make sure the piston, and the end of the lever arm that locates the piston, are cleaned thoroughly.
   - Replace the piston, relocate the arm and split pin.
   - Replace end cap.
   - If a service valve is fitted, turn it on; if not, turn on at the stop tap.
   - For both the WC cistern and CWSC, allow the cistern to fill naturally. Make any final adjustments using the float adjustment.

5 Service valves should be fitted in installations where these do not exist.

Other maintenance points

WCs that are not flushing or hot-tap outlets with no hot water can often be caused by a float-operated valve problem. This is usually because small deposits, such as solder or grit, have got into the system and eventually become lodged in the throat of the seating, thus preventing the water from getting through. This is rectified by cleaning out the seating.

System noise known as 'water hammer' (see earlier) can also be a problem with float-operated valves, particularly when fed directly from the mains. You need to check the inlet pressure is not too high by adjusting it with the stop tap. Fitting an equilibrium valve, which distributes pressure equally on both sides of the washer, will also help to solve the problem.

Shower mixer valve

There are a number of shower mixer valves available. It would be impossible here to give detailed maintenance requirements for all the units on the market. We will, therefore, cover the general maintenance requirements here.

Common faults

- Constantly dripping showerhead – this is caused by the deterioration of seals within the main body of the shower on the outlet side of the showerhead.

- Erratic temperature control, particularly on thermostatically controlled showers – this is due to failure of the thermostatic control mechanism or the entry of dirt or grit into the shower unit.

Remember

Poor supply could also be caused by poor maintenance of the shower head itself. Strip down the shower head or rose first, clean and replace

If servicing valves have not already been fitted, they should be installed during the maintenance work

- Manually controlled showers rely on the hot- and cold-water supplies being at the same pressure. If one becomes partially blocked (sometimes pieces of debris find their way into systems), flattened or raised above the other, this could affect the temperature.

Checklist

1 **Isolating the supply**. Hopefully, the hot and cold supply to the shower from the CWSC and HWSC can be isolated using service valves. Otherwise you will have to drain down 225 litres of water. Some shower manufacturers include isolating check valves in the actual shower unit, which improves the ease of servicing. Once the supply is turned off, maintenance can begin.

2 **Strip the component**. Again, much will depend on the manufacturer, but it means gaining access either to the thermostatic or to the manual cartridge. This is usually done by removing the on/off control cap. Behind this should be screws holding the temperature-control housing.

3 **Repair or replace components**. This could involve replacing a set of 'O' ring washers (for leaks) or replacing a thermostat assembly for control (erratic temperature). It could be as simple as cleaning out the internal ports.

4 **Turn on and test**. Once the mixer has been reassembled, the supply is turned on and the mixer shower tested.

Decommissioning of hot- and cold-water systems

Decommissioning domestic systems means turning off the supply to the system, removing system pipework and components, and making sure that the hot or cold supplies are sealed or left so that they can not be turned back on.

This might be necessary:

- where an old system is to be completely stripped out of a domestic property and replaced with a new, or alternative system, for example a direct hot-water system being replaced with an indirect hot-water system

- where a system is to be stripped out permanently, for example prior to the demolition of a building.

FAQ

How is it that quarter-turn taps are associated with water hammer?

The sudden closing from full-bore flow to closed in a fraction of a second can cause severe shock waves and even cause damage. It is necessary to close these types of valves slowly so as not to cause the flow to come to an abrupt stop.

FAQs

The cold water supply has been turned off by the water authority for about an hour and my hot taps have now run dry. Is it safe to leave my immersion heater switched on?

Yes. Although the hot taps have run dry, the hot-water cylinder always remains full in short-term situations like this, so it is perfectly safe.

Why do I sometimes get an air lock in my hot-water system?

It is simply poor installation of pipework, where the pipe is not laid with sufficient fall to allow air to escape to the highest point. The most common problems are found in cold feeds and gravity circuits, where air becomes trapped. Forcing the air out with a hose is only a temporary measure as air will accumulate over and over again. The only real solution is to alter the pipework to give correct falls and allow air to escape naturally.

Knowledge check

1   What are the main causes of air locks in a hot-water system, and how can these be avoided?

2   What is meant by the term 'dead leg'?

3   What is meant by the term 'stratification'?

4   List the three main causes of noise in systems.

5   Which of the following BS specifications covers soundness testing
     BS6600
     BS6700
     BS7600
     BS7700

6   What are the 4 main stages when carrying out soundness testing?

7   Describe the procedure for carrying out a pressure test on a hot- and cold-water system. (The installation pipework is copper.)

chapter 9

Central-heating systems

OVERVIEW

The Level 2 Certificate is limited to pipework systems and solid fuel boilers only. This nevertheless allows us to look at the various controls and components used on central-heating systems, and to think about the other types of oil and gas boilers on the market.
This chapter will cover:

- Types of central-heating system
- Building Regulations and central heating
- Central-heating system components
- Heat emitters
- Central-heating system controls
- How the controls work together
- Soundness testing of central-heating pipework systems
- Maintenance and decommissioning of central-heating systems

Although we have discussed cold- and hot-water systems (both direct and indirect) separately, in reality, in a new domestic dwelling, they would be installed together.

In this chapter, we are going to add central-heating systems, which are closely linked to the hot-water supply as one boiler is usually used to heat both the hot water and the central-heating system.

Types of central-heating system

At the end of this section you should understand:

- the difference between gravity-fed and pumped heating systems.

The various types of central-heating system can be summarised with a chart.

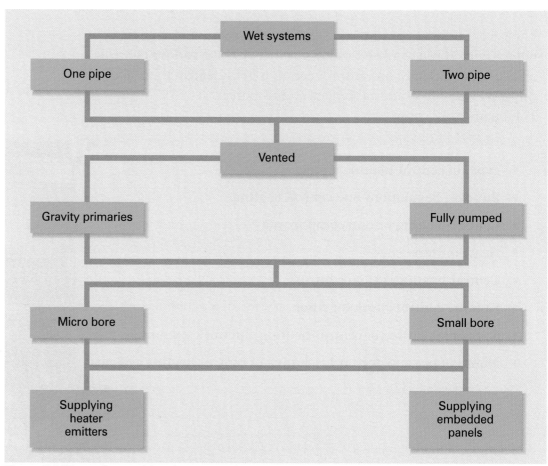

Figure 9.1 Types of central-heating system

Sealed systems or combination boilers will be fully covered at Level 3.

Full gravity heating systems

These are no longer installed, but you might come across an old system during maintenance work in buildings such as old village halls. They were sometimes referred to in domestic properties as background heating: one radiator, for example in the bathroom, which was fed off the primary flow-and-return pipe from the boiler. As with any gravity system, it worked on the principle of having the pipework at the correct fall to aid circulation. Flow-and-return to the radiator was usually 22 mm. The key feature of these systems was large-diameter pipes and no pump in the system for either heating or hot water.

Pumped heating and gravity hot-water systems (semi-gravity)

These can be either one-pipe or two-pipe systems.

One-pipe system

Again, you are unlikely to install these from new, but you may come across them.

Remember

Gravity systems were never really suitable for domestic properties because of the excessively large-diameter pipes that were required to get only the smallest amounts of heat to the radiators

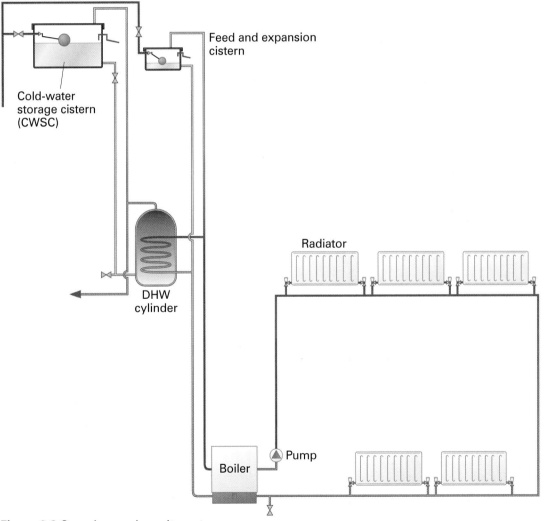

Figure 9.2 One-pipe semi-gravity system

This system works on the basis of a one-pipe run, with a flow and return from the boiler. The disadvantages with the one-pipe system tend to outweigh the advantages, so they are no longer installed in domestic properties.

Advantages	Disadvantages
Lower installation cost compared to a two-pipe system	The heat emitters on the system pass cooler water back into the circuit. This means that the heat emitters at the end of the system are cooler
Quicker to install	The pump only forces water around the main circuit and not directly through the radiators. This means it is important to select radiators that allow minimum resistance to the flow of water
Lower maintenance costs	The 'flow' side to the radiator is usually installed at high level to improve circulation, creating additional unsightly pipework

Did you know?

The key to recognising whether a system is a one-pipe system is that there will be a complete ring of pipework from the flow on the boiler back to the return

Careful balancing of the radiators is a must and – for a given heat requirement in a room – the radiators at the end of the circuit will need to be larger than those at the beginning.

To ensure that radiators operate correctly they must be sited as close to the pipework ring as possible. If you try to cut a couple of connections 100 mm apart into the pipework and run it for 5 m to the radiator, it will not work. The full ring has to be extended to run under the radiator with short tail connections on to it.

Two-pipe system – semi-gravity system

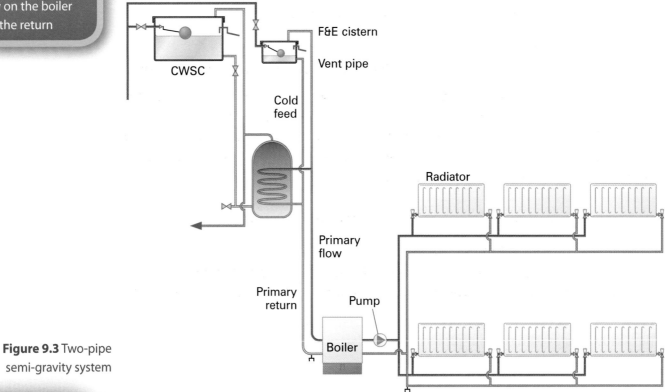

Figure 9.3 Two-pipe semi-gravity system

This was a popular choice of system particularly in the 1970s. It is no longer permitted on new properties (other than with solid-fuel boilers) without additional controls being installed. Nor do they meet the requirements of the Building Regulations for extension or boiler replacements to existing oil- or gas-fired systems.

In a two-pipe system, water is pumped around both the circuit and the radiators. This improves the speed with which radiators heat up. The system can easily be balanced by adjusting the lock shield valve on each radiator.

Did you know?

The two-pipe system is the most common method of feeding radiator circuits

Building Regulations and central-heating

By the end of this section you should understand:

- why there are Regulations for domestic heating systems
- how systems can be installed to comply with the Regulations
- how to upgrade existing installations.

The Building Regulations 2000 deals with the conservation of fuel and power in an approved document *Approved Document L1: Conservation of fuel and power*.

This document sets out requirements for:

- space-heating systems controls
  - zone control
  - timing controls
  - boiler control interlocks
- hot-water systems
- alternative approaches for space heating and hot-water systems
- commissioning of heating and hot-water systems
- operating and maintenance instructions for heating and hot-water system controls.

The legislation sets out methods for improving energy conservation in dwellings by improving the control of a system using temperature control to individual rooms, permitting only the most energy-efficient boilers to be installed, and prohibiting the installation of gravity circulation systems in new properties. It is covered in more detail in the section on environmental awareness.

Meeting the minimum requirement of the Building Regulations

The next system we will look at is designed to bring controls up to Building Regulation specifications for work on existing systems such as boiler replacements.

Pumped heating and gravity hot water (semi-gravity)

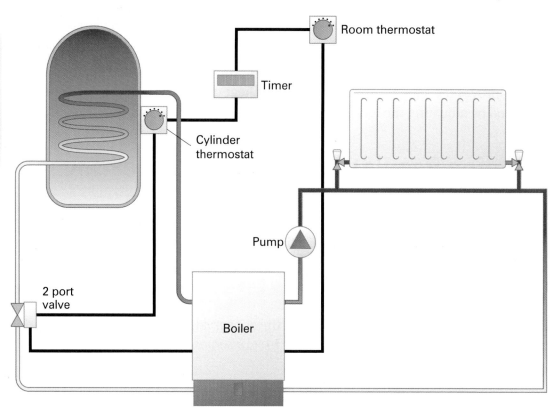

Figure 9.4 Semi-gravity system with two-port valve

This semi-gravity system uses a two-port valve and provides independent temperature control of both the heating and hot-water circuits in a pumped heating and gravity domestic hot-water heating system. The pump and the boiler are switched off when space and hot-water temperature requirements are met. Time control can be with either a time switch or a programmer. Thermostatic radiator valves (TRVs) could also be fitted to provide overriding temperature controls in individual rooms. The pump and circulation to the cylinder can be turned off independently by the thermostats, and there is an interlock to ensure that when the final circuit turns off so do the pump and boiler, thereby avoiding wasting energy.

Fully pumped systems

This type of system is often used on small and micro-bore heating systems, which are covered in detail at Level 3.

In this system, the hot water and the heating circuits are operated completely by the pump. Therefore, because there is no requirement for gravity circulation, the boiler can be sited above the height of the cylinder, giving more design options.

Installations are controlled by motorised valves. There are a number of system designs incorporating two-port zone valves or three-port valves (two-position and mid-position) that meet the requirements of the Building Regulations Approved Document L1.

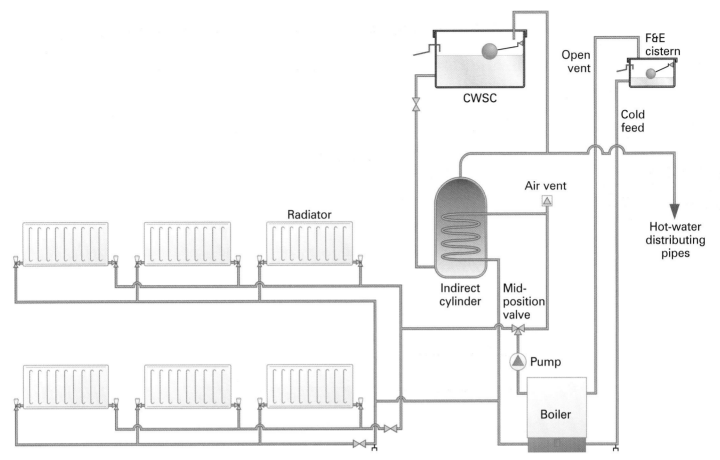

Figure 9.5 The fully pumped system

Fully pumped system using a two-position three-port diverter valve

This was one of the first fully pumped systems to be installed in domestic properties. However, it is no longer widely used.

This system is designed to provide independent temperature control of the heating and hot-water circuit in fully pumped heating systems. The design, when used with a programmer, satisfies the Building Regulations. The two-position three-port diverter valve is usually installed to give priority to the domestic hot-water circuit – that is, it can only feed either the

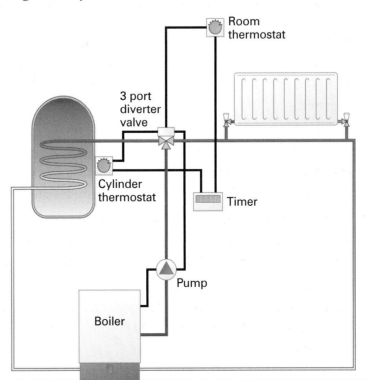

Figure 9.6 Fully pumped system with two-position diverter valve

hot-water or the central-heating system at any one time. Because the system is a priority control system – to produce domestic hot water – it should not be used where there is likely to be a high hot-water demand during the heating season: the heating temperature could drop below comfort level when the demand for hot water is high.

Because the system feeds only one circuit at a time, you will find that this system tends not to be widely installed.

Fully pumped system using three-port mid-position valve

This type of system is commonly installed in new domestic properties.

The system is designed to provide separate time and temperature control of the heating and domestic hot-water circuits. To fully meet the requirements of the Building Regulations, time control must be via a programmer and TRVs, and an automatic bypass valve must be fitted (where required).

The mid-position valve permits both hot-water and heating circuits to operate together.

On the job: Installing central-heating

Ray qualified as a plumber last year. He now has his own apprentice and needs to explain to her the installation requirements and system operation for a fully pumped system using a two-position diverter valve.

List all the points Ray needs to cover.

Fully pumped system using 2 × 2 port valves

This type of system is commonly installed in new domestic properties, particularly larger ones.

This system is recommended for use in dwellings with a floor area greater than 150m². The main reason for this is the limited capacity of a three-port valve installation to satisfy the heat demands of a larger system. The use of the 2 × 2 port valves also gives greater flexibility in system design, with additional valves being added to the system to zone separate parts of the building. The system provides separate temperature control for both heating and hot-water circuits. Again, the features are similar to the other systems shown earlier.

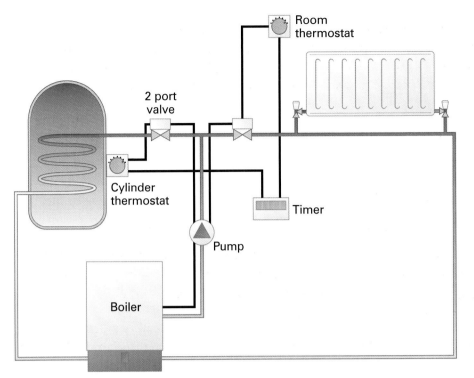

Figure 9.7 Fully pumped system with 2 × 2 port valves

FAQ

Can I use a small boiler with a diverter valve for individual control of hot water & heating?

Yes, but it is not widely done because while the cylinder is being heated up (HW Priority) the heating system is off. The smaller boiler is calculated on the max load of the heating system only. Once the cylinder is hot all heat is diverted to the heating system.

Central-heating system components

By the end of this section you should be able to:

- identify the various components of heating systems:
  - cisterns
  - air separators
  - circulating pumps.

Feed and expansion cistern

As you have seen from the diagrams in this chapter, the feed and expansion (F&E) cistern is used on all open-vented central-heating systems. While the cistern allows the system to be filled up, its main purpose is to allow water in the system to expand. The water level should therefore be set low in the cistern when filling the system. The cold feed to the system in an average domestic property is usually 15 mm minimum, and this pipe should not include any valves. This is to ensure that, in the event of overheating, there is a constant supply of cooler water to the system to prevent the dangerous condition of boiling. The servicing valve to the system should be located on the cold-water inlet pipework to the cistern.

The F&E cistern is located at the highest point in the system, and it must not be affected by the position and head of the circulating pump. To avoid any problems with gravity primaries, a minimum height can be obtained by dividing the maximum head developed by the pump by three. In fully pumped systems, the level of water in the F&E cistern should be a minimum of one metre above the pumped primary to the direct hot-water storage cylinder. If a valve were fitted in the cold feed it may be inadvertently closed. This could have disastrous consequences if the open vent also became blocked.

Space for expansion of water

The cistern and float valve must also be capable of resisting a temperature of 100°C. The system volume expands by about 4 per cent when heated, so a system containing 100 litres would expand by 4 litres. Space must be allowed in the F&E cistern to take up the additional volume when heated.

Primary open safety vent

In a fully pumped system, this should usually rise to a minimum height of 450 mm above the water level in the F&E cistern to allow for any pressure-surge effects created by the pump.

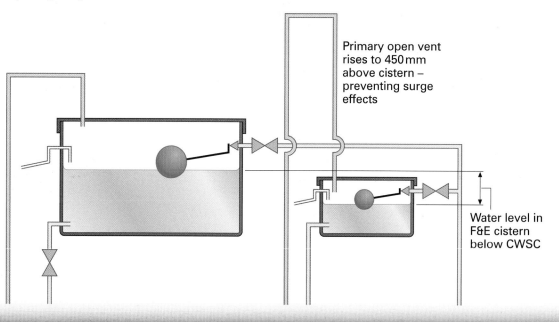

Primary open vent rises to 450 mm above cistern – preventing surge effects

Water level in F&E cistern below CWSC

Figure 9.8 Open safety vent to F&E cistern in a fully pumped system

The purpose of the primary open safety vent is to:

- provide a safety outlet should the system overheat due to component failure
- ensure that the system is kept safely at atmospheric pressure.

The minimum diameter of the safety vent is 22 mm, and the pipe should never be valved.

Air separators

The purpose of the air separator is to enable the cold feed and vent pipe to be joined closely together into a correct layout to serve the system. The grouping of the connections causes turbulence of water flow in the separator, which in turn removes air from the system. This reduces noise in the system and lowers the risk of corrosion.

The diagram shows a fully pumped system containing an air separator.

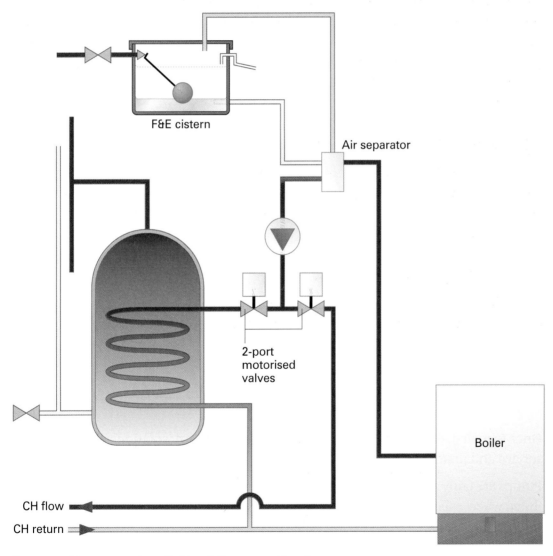

Figure 9.9 Air separator installed in a fully pumped heating system

Domestic circulating pumps

Domestic circulating pump

Ball-type pump valve

Gate-type pump valve

Pump stripped down to show the impellor

Once installed, domestic circulating pumps are fitted with isolation valves to permit service and maintenance.

Pumps are fairly simple. They consist of an electric motor which drives a circular fluted wheel called an impeller; this 'accelerates' the flow of water by centrifugal force.

It is the pump's job to circulate water around the central-heating system, ensuring that the water is delivered at the desired quantities throughout all of the system components. Most pumps have three settings, and pump manufacturers provide performance data for each, which shows flow rate in litres per second and pressure in k Pa and m head.

Flow rate should not exceed 1 litre per second for small-bore systems and 1.5 litres per second for micro-bore systems; anything higher can create noise in the system. Most pumps deliver 5 or 6 m head. This is usually enough to overcome the flow resistance of the whole heating circuit.

It is good practice to position the pump so that it gives a positive pressure within the circuit. This ensures that air is not drawn into the system through microscopic leaks.

The pump position is even more critical in a fully pumped system because of its position in relation to the cold feed and vent pipe.

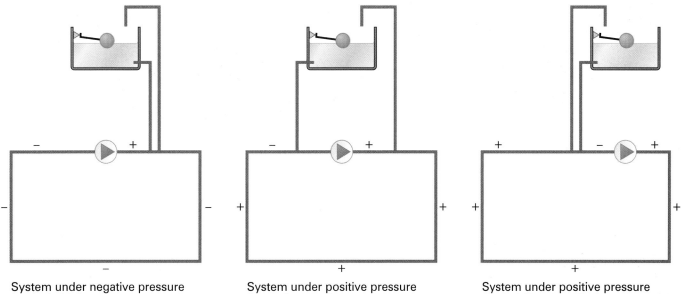

System under negative pressure (undesirable)

System under positive pressure "pumping over" may occur

System under positive pressure (desirable connection)

Figure 9.10 Pump position in relation to cold feed and vent pipe

Heat emitters

By the end of this section you should be able to:

- identify various types of radiator and heater
- explain how each type works and for which spaces they are appropriate
- understand the basics of solid-fuel and gas boilers.

We use the term 'heat emitter' because it describes all types of devices used to heat the rooms that we live in. These include:

- cast-iron column radiators
- skirting heaters
- fan-assisted convector heaters
- panel radiators.

Cast-iron column radiator

Cast-iron column radiators

Sometimes called hospital radiators, these are mostly found on older installations in such buildings as schools or village halls. However, some are now being installed in domestic properties as part of a 'designer' décor.

Skirting heaters

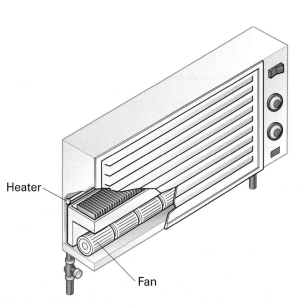

Figure 9.11 Skirting heater

These work on the principle of natural convection; the fins provide a large surface area for heat output. The radiator is heated up by conduction from the heating pipe, cool air then passes through it and as the air is heated it rises and passes from the panel via louvres at the top.

Fan-assisted convector heaters

These work on the basis of forcing cooler air through the heating fins (heat exchanger) using an electrically controlled fan, and they therefore require connection to the central-heating control system.

Figure 9.12 Fan-assisted convector heater

Panel radiators

Despite the name 'radiator', about 85 per cent of the heat is given off by convection. The heat output of a standard panel radiator can be further improved by the addition of fins welded onto the back. These

Typical panel radiator

have the effect of increasing the surface area of the radiator, as they become part of its heated surface. The design of the fins will also help convection currents to flow.

Types of panel radiator

Radiator designs have developed dramatically over the last 60 years. Manufacturers aim to provide radiators that are efficient and also offer the maximum choice of styles. The most common types of steel panel radiators are shown in the illustration.

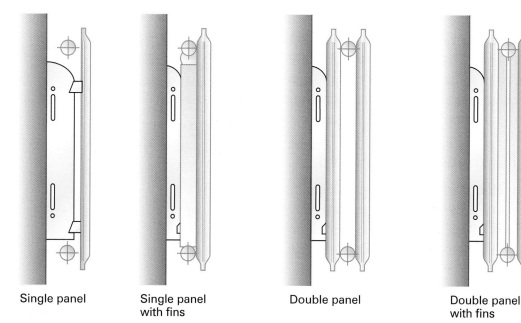

Single panel   Single panel with fins   Double panel   Double panel with fins

Figure 9.13 Types of panel radiator

Manufacturers will provide a range of at least four height options, from 300 mm to 700 mm. Width measurements are from 400 mm, with increments of 100–200 mm, to a maximum of around 3 m. The recommended height from the floor to the base of the radiator is 150 mm (depending on the height of the skirting board). This allows adequate clearance for heat circulation and valve installation.

Outputs will vary depending on design. You must ensure that the output will be sufficient to heat the room the radiator is going to be in. Radiator sizing will be covered in Level 3.

Radiator styles

Seamed top panel radiator

This is currently the market leader and is the most commonly fitted radiator in domestic installations. Top grilles are also available for this radiator.

Seamed top panel radiator

Remember

When choosing radiators from catalogues, you should take note of manufacturers' fixing positions. It is often said that a radiator should be positioned beneath a window to reduce drafts. Curtains must finish 10cm above the radiator

Compact radiators

These have all the benefits of steel panel radiators, with the addition of 'factory fitted' top grilles and side panels, making them more attractive to the consumer.

Rolled-top radiator

Rolled-top radiator

As the popularity of the compact has increased, the market for rolled-top radiators has declined. The method of manufacture means that some of the production seams can be seen following installation, making these radiators less attractive to the customer.

Combined radiator and towel rail

This product combines a towel rail and radiator in one unit. It allows towels to be warmed without affecting the convection current from the radiator. These are generally only installed in bathrooms.

Tubular towel rail

Often referred to as designer towel rails, these are available in a range of different designs and colours. They can also be supplied with an electrical element option, for use when the heating system is not required. They tend to be mounted vertically on the wall.

Low surface-temperature radiators (LSTs)

These were originally designed to conform to Health Authority requirements, where surface temperature of radiators was not allowed to exceed 43°C when the system was running at max. LSTs are now becoming popular in children's nurseries, bedrooms and playrooms, and in domestic properties where occupants are disabled.

Compact radiator

Combined radiator and towel rail

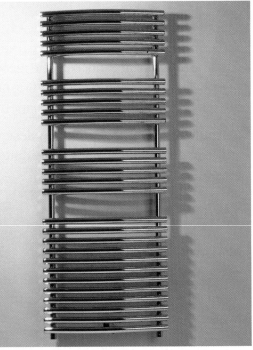

Tubular towel rail

Safety tip

You should not wedge the back of radiators over skirting boards or tightly under window sills, as this will prevent air circulation across the back panel reducing the output. Radiator shelves have a similar effect

Panel radiator accessories

Tappings

Radiator plug and manual air vent

Radiators are usually provided with $\frac{1}{2}$" BSP tappings, into which you attach the radiator valves and a plug and air-bleed valve. Domestic installations have the valves placed at the bottom of the radiator at opposite ends. Alternatively the connections could be placed top and bottom at opposite ends where a person may be elderly or disabled, making access easier.

Radiator brackets

A number of brackets and fixing styles are available, and are usually provided with the radiator. The bracket shown has deep hanging slots and corresponding lug positions. These provide greater stability. Plastic inserts are used to seat the radiator precisely and help to minimise expansion and contraction noises.

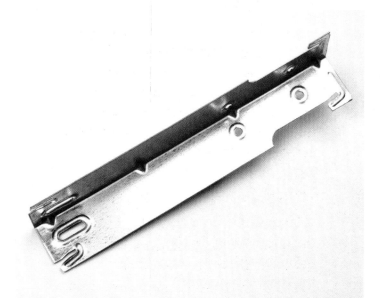

Typical radiator bracket

Radiator valves

There is a wide selection of valves available from manufacturers. Building Regulations require thermostatic radiator valves (TRVs) to be installed on new systems or for there to be some other means of controlling an individual room's temperature. A thermostat and motorised valve may be used, but these are usually too expensive to include in all rooms. You may be required to install manual radiator valves on repair jobs.

Wheel-head radiator valves

These enable the occupier of the building to control the temperature of the radiator manually by turning it on or off.

Rotating the plastic 'wheel head' anti-clockwise will raise the spindle through the body of the valve, lifting the valve and opening the flow to the radiator.

Wheel-head radiator valve

Lock-shield radiator valves

These valves are intended to be operated not by the occupier of the building but by the plumber. The plastic cap conceals a lock-shield head which can only be operated with a special key or pliers. The plumber will use this valve either to isolate the supply if removing the radiator, or to balance the system when commissioning. The valve illustrated shows an added feature – an in-built drain-off facility.

Lock-shield radiator valve with drain-off facility

Thermostatic radiator valve

Thermostatic radiator valves (TRVs)

These control the heat output from the radiator by controlling the rate of water flowing through it. The user adjusts the valve to maintain a desired temperature in the room; the valve then works automatically to maintain that set temperature.

TRVs are fitted with a built-in sensor which opens and closes the valve in response to room temperature. Liquid, wax or

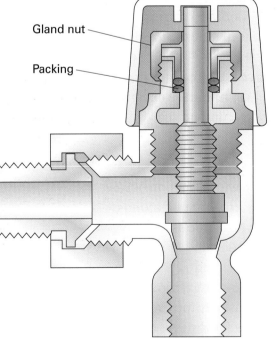

Gland nut

Packing

Figure 9.14 Section through a wheel-head radiator valve

gas expands into the bellows chamber as the sensor heats up. As the bellows expand they push the pin down, closing the valve. There are a number of settings on the head of the valve to enable a range of room temperatures to be selected.

You need to install a bypass valve to prevent the boiler and pump working against a closed system should all the TRVs close down and the water flow rate in the system fall to a low level.

Solid-fuel boilers

The only boiler you cover in the Level 2 Certificate in Plumbing is the solid-fuel boiler, although we will look briefly at the basics of other boilers.

Types of solid-fuel boiler

In domestic installations you could come across:

- solid-fuel room heater with back boiler
- independent free-standing boiler.

Convector room heater

This type of appliance is available in two ratings: 10 kW and 13 kW. The 10 kW version will heat the room in which it is located, plus the domestic hot water, and will provide a total radiator surface of up to 13.8 m²; the 13 kW can serve up to 20.5 m² radiator surface. It fits a standard fireplace opening and is thermostatically controlled. This boiler requires a flue of 175 mm. It has two flow tappings and two returns for the domestic hot water and heating, all 1″ BSP.

Temperature adjusting head

Heat sensor

Bellows chamber

Pressure pin

Union tail to radiator

Valve

Figure 9.15 Section through a thermostatic radiator valve

Convector-type solid-fuel room heater

Independent free-standing solid-fuel boiler

The model shown in the illustration is a gravity-fed boiler.

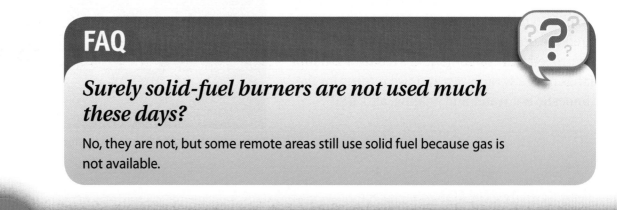

Figure 9.16 Section through an independent gravity-fed solid-fuel boiler

This type of boiler is available in a range of outputs from 13.5 kW to 29 kW. The model shown has a range of 13.5 kW, 17.6 kW, 23.5 kW and 29.3 kW. There is a built-in fan to boost heat output on demand. The hopper is easy to fill from the top, and a typical unit will hold enough fuel to burn continuously for 14 hours.

These boilers are designed for installation with a class 1 flue system.

FAQ

Surely solid-fuel burners are not used much these days?

No, they are not, but some remote areas still use solid fuel because gas is not available.

Oil-fired boilers

The following ranges of oil-fired boilers are available:

- combination
- condensing
- traditional
    - wall-mounted internal – room sealed or open flues
    - wall-mounted externally sited
    - floor-mounted internal – room sealed or open flues
    - floor-mounted external.

Outputs vary, but are typically as shown in the table.

Floor-mounted oil-fired boiler

Type of boiler	Capacity	Comment
Combination	Available in a range of capacities	
Wall-mounted internal	20 kW, 22 mm flow and return connection	Suitable only for fully pumped systems
Wall-mounted external	14.6 kW and 19 kW, 22 mm flow and return connections	
Floor-mounted internal	Outputs up to 70 kW (240,000 Btu/h)	
Floor-mounted external	Outputs range from 14.7 kW to 33.7 kW (50,000 to 115,000 Btu/h)	

Figure 9.17 Outputs of oil-fired boilers

Gas boilers

Gas boilers are covered in detail at Level 3. There are three basic types of gas boiler:

- traditional wall mounted or floorstanding
- combination boilers
- condensing.

In all cases, new boilers have to meet minimum efficiency standards. This is set out as a **SEDBUK rating**. There's more about this in Chapter 13 on Environmental Awareness.

Definition

SEDBUK rating – a measure of the seasonal efficiency of a boiler installed in typical domestic conditions in the UK

Traditional or regular boilers

These include:

- boilers with cast-iron, high-alloy steel, copper or aluminium heat exchangers

- system boilers.

All of the above can be floor-standing or wall-mounted, and are usually designed to fit in with kitchen unit installations. Gas-fired back boilers are also available; these are concealed in a chimney opening at the back of a gas fire.

Traditional or regular boilers supply hot water indirectly via a domestic hot-water storage cylinder, which is usually sited at first-floor level.

Cast-iron heat exchangers

The water in this type of boiler is heated by hot gases passing though the heat exchanger. The heat exchanger on a cast-iron boiler is a close network of waterways or, for the alloy type, a tube that passes through a series of fins. Boiler outputs for general domestic installations range from 9 kW to 29 kW.

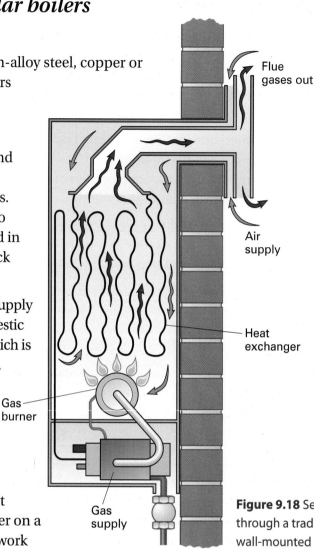

Figure 9.18 Section through a traditional wall-mounted boiler

Condensing boilers

Condensing boilers are the most energy-efficient boilers currently on the market.

The condensing boiler has to include a fan to make the process work, and the heat exchanger is larger than that in non-condensing boilers. The boiler is able to extract more heat from the combustion process than traditional boilers, making it more efficient. As the flue gases are cooled, the water vapour they contain turns to liquid (condensates), which has to be drained from the boiler to a drain or soakaway.

System boilers

System boilers supply domestic hot water indirectly. The key system controls – pre-wired, pre-plumbed and pre-tested – are all built-in to the boiler unit, including the expansion vessel, the pump, and the system bypass.

Combination boilers

This type of boiler is designed to heat up a cold-water supply instantaneously for domestic hot water, so it is classed as direct hot-water supply. When required, it will also supply hot water for the central-heating system. Combination boilers are covered in greater detail in Level 3.

Central-heating system controls

By the end of this section you should understand:

- why it is important to control heating systems
- the different ways to control heating systems.

Everyone has a moral obligation to do what they can to conserve energy. Energy conservation is also a high priority of government. For this reason, Building Regulations have been amended to improve energy efficiency in buildings.

The revised Regulations have had the effect of enforcing fixed specifications for system design and, in particular, its control. We will therefore now look at mechanical and electrical controls used on central-heating systems, beginning with the basic system controls.

Time switch

A time switch is an electrical switch operated by a clock to control either space heating or hot water, or both together.

Full programmer

This can be used on gravity primary and fully pumped systems. It allows the time settings for space heating and hot water to be fully independent. More sophisticated programmers can include separate time controls for multiple space heating circuits.

Gas-fired combination boiler

Central-heating programmer

Room thermostat

This device measures the air temperature within the building or room where it is sited and controls the operation of the heating circuit (depending on whether the timing device is calling for heat). It can be used directly to switch on a circulating pump or boiler, or to operate a motorised valve.

Room thermostat

Programmable room thermostat

This is a combined time switch and room thermostat. It enables the user to set target temperatures for different periods for the central-heating. This can be done on a daily or weekly cycle: for example, a higher temperature may be required for a certain period in the evening or at weekends.

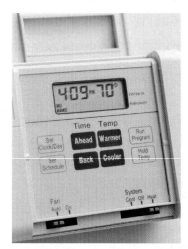

Programmable room thermostat

Cylinder thermostat

The cylinder thermostat measures the temperature of the water stored in the hot-water storage vessel and, when it reaches a pre-set temperature, switches off the hot-water circuit to the vessel. This is done by connecting to the circulating pump or boiler, or by the operation of a motorised valve.

Frost thermostat

These give automatic frost protection to boilers and pipework that have to be located in areas at risk from severe cold. External boilers are a typical example.

Frost thermostat

Automatic bypass valves

These are mechanical devices used to make sure water can flow through the boiler to maintain a minimum water-flow rate should a system using zone valves or TRVs become closed. Once set, the valve opens automatically as the TRVs or zone valves close and the system pressure goes over the pre-set limit.

Use of the bypass valve reduces system noise and increases pump life by preventing it working against a 'dead head'.

Automatic bypass valve

Automatic air vents

These are used on central-heating systems in order to remove air from the system automatically. A vacuum break on the bottom of the valve prevents an air lock forming and encourages air to be released from the water.

Vents should always be installed on the positive side of the system and positioned where air is likely to get trapped.

Automatic air vent

Motorised valves

The type of valve used will depend on the system design, but the following are available:

- three-port diverter valve
- three-port mid-position valve
- two-port motorised zone valve.

Three-port diverter valve

This valve has been designed to control the flow of water on fully pumped central-heating and hot-water systems, on a selective priority basis, which is normally for the domestic hot water.

Three-port diverter valve

Two-port motorised zone valve

Three-port mid-position valve

This valve looks very similar to the diverter valve. It is used on fully pumped hot-water and central-heating systems, in conjunction with a room and cylinder thermostat, to provide full temperature control of both the hot water and heating circuits, which can operate independently of each other or both at the same time.

Two-port motorised zone valve

A single zone valve is used on gravity domestic hot water and pumped central-heating systems to enable separate temperature control of both the heating and hot-water circuits. Motorised valves are also used in fully pumped systems to provide separate control of both heating and hot-water circuits. They can be used to zone different parts of a building, for example upstairs and downstairs.

Junction box or wiring centre

These provide the connections between the electrical system components and the mains electricity supply.

Most manufacturers supply their controls in packs, and these can include a wiring centre which is designed to simplify the wiring of a particular system pack. Packs include control valves, programmers and thermostats. All the terminal connections are clearly marked, and full instructions are included with the wiring centre.

Wiring centre

Boiler control interlocks

This is a term used in Building Regulations Approved Document L1. It is not an actual control device but an interconnection of controls – such as room and cylinder thermostats, programmers, zone valves etc. – designed to ensure that the boiler does not fire up when there is no demand for heat.

In a system with a traditional boiler, this would require the correct wiring of the room **stat**, cylinder stat and motorised valve(s). It may also be achieved by more advanced controls such as a boiler energy manager.

How the controls work together

By the end of this section you should understand:

- how the components discussed in this chapter operate with each other for the following systems:
  - pumped heating and gravity hot water using a two-port valve
  - a fully pumped system using a three-port mid-position valve
  - a fully pumped system using 2 × 2 port zone valves.

So far, we have looked at system layouts, components and controls. We now need to think about how they work together. We will focus on the following system types, as they are the ones commonly installed.

For existing system modifications such as boiler replacements:

- pumped heating and gravity hot water – two-port valve.

For fully pumped systems (the only option for new installations and the preferred option for any existing system modifications or upgrades):

- fully pumped system using three-port mid-position valve
- fully pumped system using 2 × 2 port zone valves.

Pumped heating and gravity hot water, two-port valve

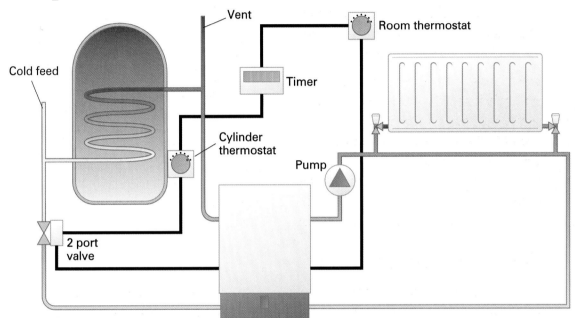

Figure 9.19 Components of a pumped heating and gravity hot-water system

Installation

- Two-port valve
  - The two-port valve should be installed in the domestic hot-water circuit, so that when it is closed the cold feed or vent is not affected
  - It is good practice to fit the valve on the primary return to the boiler
- Room thermostat
  - Locate this where it will not be unduly affected by excess heat gains (e.g. directly above a heat source), direct sunlight or draughts – it is usually positioned in a hallway at a height of 1.5 m from the floor
- Cylinder thermostat
  - This should be located between a quarter and a third of the distance from the bottom of the hot-water storage vessel and should be easily accessible
- Time control
  - A twin channel programmer is the preferred way to provide independent switching of hot-water and central-heating circuits to meet Building Regulations requirements.

How this installation works

- **Heating only**. When the room stat calls for heat, the pump and boiler are switched on. The zone valve on the gravity primary remains closed.

- **Hot water only**. When the cylinder stat calls for heat, the zone valve is opened. Just before the valve reaches its fully opened position, the auxiliary switch closes and switches on the boiler only.

- **Heating and hot water**. When both stats call for heat, the pump and boiler are switched on and the zone valve is opened.

Fully pumped system using three-port mid-position valve

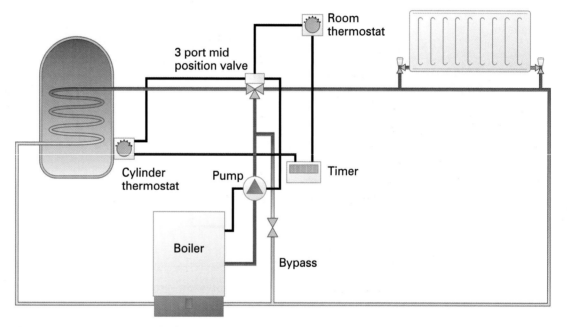

Figure 9.20 Components of a fully pumped system using a three-port mid-position valve

Installation

- Three-port valve
  - water flow must be into port AB
  - port A is connected to the heating circuit
  - port B is connected to the hot-water storage vessel primary circuit

- Room and cylinder stat as for the previous system

- Time control
  - must be provided by a programmer that permits the selection of heating without hot water

- Bypass
  - an automatic bypass should be installed immediately after the pump, between the flow and return; pipework should be 22 mm, and flow should be in the direction of the arrow marked on the valve body.

How this installation works

- **Heating only**. When the room stat calls for heat, the valve is activated so that the central-heating port only is opened and the pump and boiler are switched on.

- **Hot water only**. When the cylinder stat calls for heat, the valve remains open to the hot-water storage vessel; the pump and boiler are switched on.

- **Heating and hot water**. When both stats call for heat, the valve plug is positioned so that both ports are opened. The pump and boiler are switched on; they're switched off once the stats are satisfied. The valve remains in the last position while the programmer is in the 'on' position.

Fully pumped system using 2 × 2 port zone valves

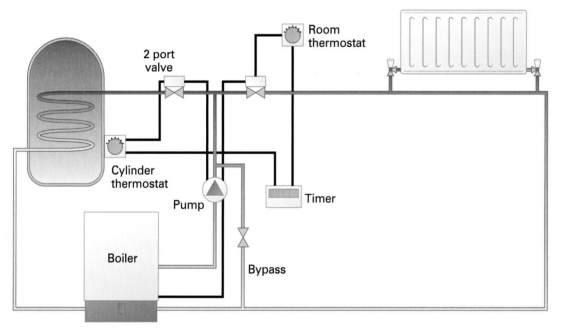

Figure 9.21 Components of a fully pumped system using 2 × 2 port valves

Installation

- Two-port valves
  - water flow must be in the direction of the arrow on the valve body

- All other components are the same as the previous system. A programmer must be used to satisfy Building Regulations.

This system provides greater flexibility than the others. Additional zone valves and thermostats can be added so that the heating circuit can be separated into several zones. This improves temperature control and efficiency.

How this installation works

When either stat calls for heat, the respective motorised valve will be opened. Just before it is fully opened, the auxiliary switch sited in the motorised valve head will be closed and will switch on both pump and boiler. Once both stats are satisfied, the valves are closed and the pump and boiler switched off.

Soundness testing of central-heating pipework systems

By the end of this section you should be able to conduct:

- visual inspections
- testing for leaks and flushing the system
- pressure testing
- final checks
- flushing of pipework
- anti-corrosion procedures.

You have already looked at the testing requirements for hot and cold systems; these aspects apply equally to central-heating systems. In this section, we will also look at flushing central-heating system pipework and using additives to protect against corrosion.

Soundness testing procedures

Visual inspection

In the same way as for hot and cold water pipework, testing includes making sure that all pipework and fittings are thoroughly inspected to ensure:

- they are fully supported, including F&E cisterns and hot-water storage cylinders
- they are free from jointing compound and flux
- that all connections are tight
- that in-line valves and radiator valves have been properly tightened and are closed to allow stage filling
- that the inside of the F&E cistern is clean
- that all the air vents are closed.

Before filling, it is a good idea to remove the pump and replace it with a section of pipe. This will prevent any system debris entering the pump's workings.

Testing for leaks

- Turn on the stop tap if the installation was for a complete cold-water, domestic hot-water and central-heating system, or the service valve to the F&E cistern if only the central-heating circuit has been installed.
- Allow the system to fill.
- Turn on the radiator valves fully and bleed each radiator.
- Visually check all the joints for signs of leaks.

- Drain down the system, flushing out all debris, wire wool, flux etc.

- Reinstall the pump.

- Refill and test for leaks again.

- Make sure the water level in the F&E cistern is at a level that allows for expansion of water in the system when hot.

- Heat system and flush while hot.

Pressure testing

On larger jobs, testing could be part of the contract specification. It is usual to test the system to $1\frac{1}{2}$ times the normal working pressure or 3 bar, whichever is the greater, for a period of one hour; the test pressure is achieved by using hydraulic test equipment.

Hydraulic test equipment

A safe mains connection (one that will not contaminate the supply) or a hydraulic test kit can be used to pressure test the system. All open ends in the system must be temporarily sealed prior to the test and a test point provided to connect the hydraulic kit or mains supply. The system is then increased to $1\frac{1}{2}$ times its normal working pressure and left to stand for one hour. Larger systems must be tested section by section, e.g. first floor, second floor, or by zones.

Final system checks

Once the pressure test has been completed, a final check is made to ensure everything is sound. Advise the customer or other site workers when system testing is complete. The requirements for full commissioning, water treatment of the system and performance testing are covered at Level 3.

Hydraulic test kit

Corrosion protection

Some corrosion will occur in any domestic central-heating system. Just how severe it is will depend on:

- the types of metal in the system

- the nature of the water supply

- the degree to which air can be drawn into the supply.

It is good practice to install a corrosion inhibitor or protector to maximise the working life of the system. It should be applied in accordance with the manufacturer's instructions (there may be several stages to the process of applying the inhibitor), which may be done while filling the system after flushing, or may be injected through a radiator vent.

Corrosion protection

- prevents a build-up of 'black sludge', the major cause of central-heating problems

- cuts fuel bills

- stops frequent venting

- has a non-acidic neutral formation, so is safe to use

- is harmless to the environment.

Maintenance and decommissioning of central-heating systems

By the end of this section you should be able to:

- isolate and drain down sections of the heating system to
  - replace defective valves
  - replace a defective pump

- conduct general maintenance

- decommission a heating system.

Fault diagnosis will be fully covered at Level 3. In Level 2 you are required to understand the principles of maintaining the components within the system (excluding the boiler). It also requires you to be able to change a defective radiator valve and central-heating pump, so we will focus on these two components.

By now, you should have realised how important energy efficiency is. Maintenance plays an important role here: it ensures that the system components continue to operate as they are designed to. For example, a pump that is not performing to its specification could make the boiler work harder, thus wasting energy.

Maintenance procedures consist of:

- finding out what the fault is

- isolating the supply

- stripping the component (if possible)

- repairing or replacing the defective part

- reassembling components (if applicable)

- turning on supplies and testing the component for correct operation.

Replacing a defective radiator valve

Valves could be one of three types:

- TRVs
- lock shield
- manual radiator valve.

Replacement of any of these valves will follow the same procedure. There are four ways to do this job:

1 Create a vacuum in the system to hold the water while replacing the valve – this is usually the preferred option as it is quicker.

2 Drain down the central-heating system – this is usually a last resort.

3 Use pipe-freezing equipment to isolate the supply to the radiator valve.

4 Some manufacturers have produced specialist servicing tools and valve inserts for both manual valves and TRVs. These enable the valve to be serviced without draining down.

The first two methods are the most commonly used, so we will focus on them.

Creating a vacuum in the system

This is a quick and simple solution to changing a component such as a radiator valve as it saves draining down the system. It works by completely plugging and closing all outlets to the system such as vent pipes, automatic air vents etc. It only works where you are working on one open pipe end (you cannot use this procedure for cutting into pipework) and it can only be used for vertical sections of pipework – making it ideal for radiator-valve replacement.

Replacing a defective radiator valve

1. Isolate the radiator valve

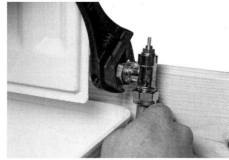
2. Drain the radiator

3. Relieve system pressure

4. Saw through compression ring (optional)

5. Replace valve

6. Finish the job

Isolate the system

Make sure the electrical supply to the heating system is isolated. Turn off the boiler. Remove the fuse from the spur outlet to the wiring centre or junction box for the controls. Advise the customer of what you are doing and not to touch the controls. Only work on the system when the water has cooled. If you speak to the customer beforehand to arrange the job, it is a good idea to ask them to turn off the system so that it is cold when you get there.

Close all outlets

Turn off the service valve to the F&E cistern; close any automatic air vents. The open vent pipe and cold feed now need plugging – manufacturer devices are available, to carry this out.

Drain down the radiator

Before progressing any further, the radiator that is serviced by the valve to be replaced needs to be drained down. This is carried out by isolating both radiator valves and the loosening of the union nut on the radiator valve to be replaced. Catch the water from the radiator in a small bowl, and have a bucket handy to empty the bowl from time to time. Air needs to be let in to the radiator through its air vent to ensure that the water in the radiator fully drains out.

Relieving the system pressure

The radiator valve to be replaced should now be 'cracked open' at its control mechanism. A small amount of water should be discharged into the bowl until the pressure in the system reduces to atmospheric pressure. After a short period, the water flow should stop. If it does not, or the flow is excessive, this suggests that you have not closed all of the outlets or there is a leak in the system. If there is a leak, there is no option but to try another method of isolation, such as draining down. When the flow stops, you can move on to replacing the valve.

Replacing the valve

Check that the new valve body will fit the existing valve tail screwed into the radiator. If it does not fit, change the tail. If it is the type with an external nut, this can be done with an adjustable spanner. If not, then you will need a radiator spanner/**allen key** which goes inside the tail to remove it. When fitting the new tail, wrap PTFE tape around the threads five or six times in the direction of the thread.

It is at this point that you will need to establish whether modifications to the existing pipe are required. If you are fitting a valve by the same manufacturer, it will probably be a straight swap; otherwise, if there is some lift in the pipe, you may be able to remove the old compression ring and nut by sawing and lifting the pipe to the required height of the new valve. It is also possible that the length of pipe may have to be extended. Once the pipe has been prepared, the valve should be installed and all the nuts tightened.

Finishing the job

You are now ready to put the system back into operation. Remove the plugs from the cold feed and vent pipe, and open any automatic air vents in the system. Open the radiator valves and refill the radiator, checking for any leaks to the replacement valve. Replace the fuse and operate the system. Advise the customer that the work has been completed.

Draining down the system

This turns the relatively simple task of changing a defective radiator valve into a long job. It is not the draining down but the refilling and making sure all the system is working correctly that takes time.

Isolate the system

Make sure the electrical supply to the heating system is isolated. Turn off the boiler. Remove the fuse from the spur outlet to the wiring centre or junction box for the controls. Advise the customer of what you are doing and not to touch the controls. Only work on the system when the water has cooled. If you speak to the customer beforehand to arrange the job, it is a good idea to ask them to turn off the system so that it is cold when you get there.

Drain down the radiator

Locate the drain valve at the lowest point below the radiator valve to be changed. If you are working on an upstairs radiator drain only the first-floor level, not the whole system. You can do this by slightly loosening the connection from the valve to the radiator (have a bowl and cloth ready) and checking it until the water has stopped dripping. Drain-off is done by connecting a hose pipe to the drain valve and running the other end to an outside drain.

Fitting the valve

Once the system has been drained, one of two situations will exist:

1 The valve can be fitted without altering the pipework. This is because the valves are the same size, or there is sufficient 'play' in the pipe to move it up or down to fit.

2 The valve dimensions are different, so the pipework has to be altered.

If the valve can be changed without altering the pipework, remove the valve, leaving the existing nut and compression ring on the pipe.

Check that the new valve body will fit the existing valve tail screwed into the radiator. If it does not fit, change the tail. If it is the type with an external nut, this can be done with an adjustable spanner. If not, then you will need a radiator spanner/allen key which goes inside the tail to remove it. When fitting the new tail, wrap PTFE tape around the threads five or six times in the direction of the thread.

Remove the nut and compression ring from the new valve and place the new radiator valve body in position. If it is a TRV, remove the thermostatic head first.

Safety tip

Always use the electrical safe isolation procedure of removing the fuse

Remember

Always use dust sheets or proprietory drip catchers to protect floor coverings from dirty radiator water!

Because the compression ring has been used, some plumbers wrap PTFE tape around the joint between the compression ring and the valve body.

If you have to alter the pipework – either shortening by cutting, or extending by use of a fitting – complete the process, fit the new nut and compression ring to the pipe, and follow the next steps.

Once the valve has been secured to the pipe, fit the valve body to the radiator tail and tighten using adjustable spanners/grips. Make sure you use opposite force on the valve body to stop the valve twisting and possibly affecting the nut and compression ring joint.

Finishing the job

Turn off the drain tap, turn on the supply to the F&E cistern and begin the air-release process. This will involve:

- bleeding all the radiators, so do not forget your air key
- bleeding the pump
- bleeding the air valve (if manual type) on the pumped primary on fully pumped systems.

Using pipe-freezing equipment

There are a number of kits on the market. They vary in specification and cost, and you will have to decide what is the best kit to suit your needs and your pocket. The kits illustrated can generally be used on copper and plastic, and all are suitable for use on central-heating systems.

- Make sure you can use your kit competently.
- Only carry out this process on cold pipework with no water flow.
- If you try to use a pipe-freezing kit on warm pipes, you cannot predict how long the plug of ice will last and so you may get wet and the house may be flooded.
- Make sure the supply to the opposite end of the radiator is turned off.
- Remove the floorboards from above the pipe run to allow plenty of access to the pipe to be frozen.

Pipe freezing

- Following the manufacturer's instructions, place the freezer head or jacket at the required distance from the valve.

- Most kits will freeze the supply within 2 minutes and will last for 30 to 45 minutes.

- Removing and replacing the valve is the same as the draining-down job, but remember that here you will have to drain the excess water out of the radiator. When you're doing this, loosen the air vent to allow the water to drain more quickly. Have a bowl and cloth available to deal with any excess water.

- Once the process of replacing is complete, fill the radiator and test.

Replacing a defective heating pump

In the same way as valve replacement, advise the customer what you are doing, both in advance and during the actual job. On maintenance jobs, a pump replacement could be on any type of central-heating system, and the pump could be located almost anywhere in the building. However, it is more than likely that it will be located close to the domestic hot-water storage vessel. On solid-fuel systems or gravity primaries, pumps can be located near the boiler; on system boilers, pumps are located inside the boiler casing.

Carry out some preliminary checks:

- First, make sure that the pump really is faulty – the fault could be in the electrical control, or something as simple as a blown fuse to the fused spur to the pump.

- Check to see whether the replacement pump will fit without any pipework alterations. If it doesn't, go and get a pump that will! Check that the pump valves are in good condition (which they rarely are!). If the answer is yes to both these checks, you will not have to drain the system down. If the answer is no to either, then you will.

Isolate the supply

Depending on the type and the age of the system, there may be a fused spur to the pump, or a fused isolating switch. The pump could also be part of a system control package, in which case you would have to isolate the fused spur to the wiring centre or junction box.

In all cases, remove the fuse to isolate the supply. If a wiring centre or junction box has not been used (i.e. older systems), turn off the boiler as well. If you need to alter the pipework, you will have to drain down the system in the same way as for replacing a radiator valve.

Safety tip

Always follow the safety instructions that come with the equipment

Safety tip

Always use a temporary continuity bond before breaking any pipework connections

Remember

Because the electrical system is involved in this job, the safe isolation procedure must be followed at all times

Don't forget: some manufacturers produce pump extensions which will allow you to avoid draining the system, so check for these first

Replacing a pump

1. Disconnect the cable and close the valves

2. Undo union connections

3. Remove old pump

4. Clean the union surfaces

5. Fit the new pump

6. Final electrical connections

Disconnect the pump cable

Disconnect the electrical supply. First remove the plastic cover. This will reveal wiring that is the same as on an ordinary plug.

Remove the plastic cable retainer and disconnect the cables, making sure there is enough bare wire ready for reconnection to the pump fittings, and that the wires are in good condition. Cut back the wire and re-strip if not, or replace the cable, and wire the new pump before it is put in position. This makes the job easier.

Remove the old pump

Assuming that you do not have to drain down, turn off the isolating valves and remove the pump. If it has been in place for a long time, this may be difficult. You may have to apply heat to the pump valve nuts to loosen them. You will also need large **stillsons** or grips on both valve nuts to stop the pump rotating. Be careful when doing this not to disturb the joint between the pipe and the pump valve.

Cleaning the union surfaces

Once you have got the pump out, make sure the pump valve faces are free of any left-over valve washers.

> ## Remember
>
> Remember your plug wiring: blue to neutral, yellow and green to earth and brown to live. On older systems the wiring could be red to live, black to neutral, green to earth

Fitting the new pump

Fit the new pump in accordance with the manufacturer's instructions and making sure the new washers are fitted. If you altered the pipework to fit the new pump, ensure the pump and the new valves are fitted. Turn the water supply back on. If you drained the system, refill it as you did for the radiator valve replacement. If not, turn the pump valves back on. Bleed the pump through the slotted head in the body of the pump or pump bleed screw.

Final electrical connections

Reconnect the pump cable to the cable terminals, carry out any electrical checks to the electrical supply and test the pump. Advise the customer that the system is back in service and complete any commissioning records that may be required.

General maintenance requirement of systems

Prevention is better than cure, and more and more plumbing firms offer their customers planned maintenance packages for the full domestic system, not just heating. This will include:

- servicing of boiler and other heating appliances at least once a year (a legal requirement for gas appliances in tenanted properties)
- regular servicing of taps and valves
- cleaning and flushing of central-heating systems.

Maintenance contractors report that the majority of maintenance and repair jobs in domestic properties are caused by:

- failure of central-heating controls, boiler parts, motorised valves and thermostats
- leaks or burst pipes
- leaks caused by householders themselves, a typical example being the removal and replacement of radiators for decorating.

What is of concern is that contractors receive calls from consumers who think their system is faulty, when in fact they do not understand how to work it properly. This is why it is vitally important to make sure that your customer fully understands everything about the system you install.

Motorised valves are usually very reliable, but will eventually come to the end of their working life. At one time this meant draining the system. However, it is now possible to replace defective motors or the actual power head.

Replacement two-port motorised valve motor

On older systems, it may be necessary to flush pipework out when carrying out maintenance work. A power flushing kit is designed for this purpose.

Approved cleaning agents should be used with this equipment, which comes complete with hoses and valved connections.

Decommissioning central-heating systems

As with the decommissioning of hot- and cold-water systems covered earlier, this could include the following situations:

Power flushing kit

- An old system is going to be completely stripped out of a domestic property and replaced with a new system, for example a gravity solid-fuel system being replaced with a fully pumped condensing boiler system.

- A system is to be stripped out permanently, for example prior to the demolition of a building.

Make sure you:

- keep customers or other site workers informed of what work is being carried out

- make sure all supplies are isolated and capped off (where necessary) and that electrical supplies are effectively isolated

- use notices on taps or valves, for example 'not in use', 'do not use' etc.

FAQs

Why do I need to know about "one pipe" systems, hardly anyone installs them these days?

You will no doubt end up working on a one pipe system one day so it is wise to understand a little about the principles of operation.

FAQs

Is it good practice to use an inhibitor in heating systems

Yes, tests have proven that an inhibitor correctly added to a system will prolong the life of boiler, radiators and other components.

Are correct settings of thermostats really that important? What difference does a few degrees make?

Well, quite a lot over time in terms of wasted fuel and money. By setting thermostats wisely we can save on fuel bills and help the environment too.

Knowledge check

1   What Building Regulations document sets out energy conservation requirements for central-heating systems?

2   List three specific areas that the Regulations cover.

3   What's the main advantage of a two-pipe central-heating system over a one-pipe system?

4   What is the main purpose of an F&E cistern?

5   What two performance details would you expect a pump manufacturer to supply with their product specification?

6   Radiators give out 85% of their heat by radiation and 15% by convection. Is this statement true or false?

7   Name three types of radiator valve.

8   List three types of solid fuel boiler and three types of oil fired boiler.

9   What's the difference between a time switch and a full programmer?

10 List the four main stages of carrying out soundness testing on a central-heating system.

11 State two factors that affect corrosion in a central-heating system.

12 In what circumstances would you need to drain the central-heating system to replace a defective pump?

Electrical systems

OVERVIEW

At Plumbing Level 2, electrical work is restricted to basic practical skills on components that are not connected to the supply – that is, non-live work. On completion of Level 2 you *must not* regard yourself as competent to work on electrical systems. You will cover electrical work in much more detail at Level 3.

The job of a plumber involves installing and maintaining a wide range of appliances and components, many of which are powered by electricity. You need to be able to work safely and competently on the electrical supply to items such as heating controls, immersion heaters, showers etc., including being able to inspect and test electrical systems, and connect wiring from an outlet to the appliance or component. This chapter will provide the underpinning knowledge for this by covering:

- **Electricity supply systems installation (Part1):**
  - what you have learned so far
  - safety requirements when working with electricity
  - safe isolation
  - testing
- **Electricity supply systems installation (Part 2):**
  - types of cable and cords
  - installing pvc cables
- **Electricity supply systems installation (Part 3):**
  - installing fixed electrical appliances

- **Testing and decommissioning (knowledge only):**
  - different tests for electrical supply and earth continuity systems installations, including polarity, insulation, resistance, the use of multimeters and methods of preventing the unauthorised or inadvertent use of the electrical supply
  - procedures for advising customers or line managers when carrying out testing procedures for decommissioning system.

Electricity supply systems installation (Part 1)

By the end of this section you will:

- have revised the coverage of electricity in previous chapters

- be able to describe the safety requirements for working on electrical systems.

Plumbers carry out work on electrical systems, which is mainly restricted to connecting supplies to appliances and controls. You are unlikely to be required to install electrical circuits in either new or existing dwellings, but it is important that you have a basic understanding of how circuits work.

Anyone working on electricity supplies *must be competent*. This chapter aims to give you sufficient knowledge to do your part of the job competently.

What you have learnt so far

In Chapter 4, Key plumbing principles, you looked at the principles of electricity and electricity supply and control. Here's a checklist of what you should now be able to do:

- state the basic principles of electricity

- demonstrate an understanding of electrical units and terms

- explain the basic principles of electrical power and resistance

- state the function of fuses and fuse rating

- state the basic types of electrical circuit

- describe the difference between direct and alternating current

- state the main principles behind the generation and supply of electricity

- explain the main features of basic domestic circuits

- state the basic principles of earth continuity, bonding and temporary bonding.

In Chapters 7, 8 and 9 you have also covered hot-water supply, including instantaneous showers and central heating systems, and controls and pumps. Go back over these chapters to revise this material before continuing with the rest of this chapter.

Safety requirements when working with electricity

The two most important factors are:

- circuit protection

- safe isolation (avoidance of electric shock).

Circuit protection

The Electricity at Work Regulations 1989 require that an electrical current is disconnected automatically if a greater current flows through a conductor (cable) than the circuit was designed for. This should happen within 0.4 seconds for anything with a socket outlet and 5 seconds for fixed appliances, e.g. an immersion heater.

Faults in circuits are caused by:

- overloading
- short circuits
- earth faults.

These faults are usually down to badly designed circuits, faulty appliances and/or poor installation. Circuits are protected against overload and short circuits by either fuses or circuit breakers. Circuit protection against earth faults includes **fuses**, **circuit breakers** or a **residual current device** (RCD) – this provides the greatest level of protection.

Electric shock

There are two ways in which a person can receive an electric shock:

- by direct contact with a live supply
- by indirect contact with a live supply.

Direct contact

There are ways of preventing direct contact. For example, the conductors in a cable or flex are protected with a plastic coating that insulates the live parts of the system that carry the current. Components such as pumps and valves using electric motors are also insulated internally.

Another method is to ensure that, under normal circumstances, live components are placed within an enclosure, such as a junction box, which means the exposed cabling cannot be touched.

Indirect contact

Indirect contact is prevented by a method called **earth equipotential bonding**, which ensures that all exposed metalwork in a building is bonded together and connected to the main earthing terminal. Should any exposed metal become live due to an electrical fault, the current will be discharged to earth. The fault current will also be detected by the protective device, which will automatically cut off the supply.

Safe isolation

Before working on any electrical supply you must make sure that it is completely dead and cannot be switched on accidentally without you knowing. Not only is this a requirement of the Electricity at Work Regulations 1989, but it is *essential* for your personal safety and that of your customer or co-workers.

The proper way to test if a circuit is live is to use an approved voltage-indicating device, similar to the one shown in the photograph on page 285.

Voltage-indicating devices may use either an illuminated lamp to indicate the presence of a voltage or a meter scale. Test lamps are normally fitted with a 15 watt lamp and must be so constructed that they are not dangerous if the lamp is broken. They must also be fitted with protection against excess current, either by a fuse not exceeding 500 mA or a current-limiting resistor and a fuse. The test leads should be held captive and sealed into the body of the voltage detector. Test lamps and voltage indicators should be clearly marked with the maximum voltage that may be tested by the device and the maximum voltage that the device will withstand.

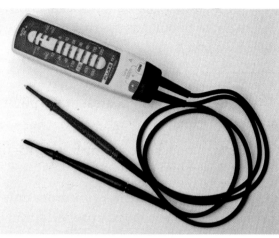

An approved voltage indicating device

Testing

Most fatal accidents involving electricity occur at the isolation stage. This is when you must be really careful and be aware of what you are doing, as you may have no idea of the type of supply you are confronted with. Do not take any risks, and if you are not sure, seek assistance. We will look at how to:

- identify sources of supply
- isolate
- secure isolation
- test that the equipment or system is dead
- begin work

Standardised procedures have been drawn up by the Electrical Contracting Industry and these are also used as the standard for safe working in the plumbing industry.

Identify sources of supply

It is important that you identify both the *type* of supply and the *source* of supply. In domestic dwellings, this will be a single-phase 240 volt supply to the circuits.

In most cases there are two types of circuit:

- 32 amp ring main for 13 amp socket outlets
- 6 amp lighting circuit.

In addition to these, you may also encounter:

- 16 amp supplies to fixed components such as immersion heaters
- supplies to other fixed components such as cookers and central heating systems and showers.

As a plumber you are unlikely to work on lighting circuits, but you need to be able to recognise the circuit layout and cable sizes. In older properties you may also find a 16 amp radial circuit. These supply the socket outlets but, unlike the ring main, the circuit terminates at the last socket and is not returned to the consumer unit. They are not as popular as the ring main, because the number of sockets that can be installed is limited.

To identify the source of supply:

1   Locate the plug socket nearest to the point where you intend to work on the component or appliance.

2   Make sure the socket is live by testing it – this can be done by plugging in a power tool to check that it works.

3   If it is sound, a check at the consumer unit should clearly indicate which circuit breaker (or fuse on an older property) will isolate the supply.

Isolate

Now that the type and source of supply have been identified, it needs to be isolated. Regulations require that a means of isolation must be provided to enable skilled persons to carry out work on or near parts that would otherwise normally be energised (live).

Isolating devices (fuses, miniature circuit breakers, RCDs) must comply with British Standards, and the isolating distance between the contacts must comply with the requirements of BS EN 60947-3 for an isolator. The position of the contacts must either be externally visible or be clearly, positively and reliably indicated.

Secure isolation

To prevent the supply being turned on accidentally by the customer or other co-workers, the fuse or circuit breaker should be removed and kept in your pocket, or the isolator locked off, with a special locking-off device. As an extra precaution, a sign saying 'work in progress and system switched off' must be left at the consumer unit or the area in which you are working.

Test equipment and checking system is dead

Any circuit you work on *must* be tested to ensure it is dead. Test equipment must be regularly checked to make sure it is in good and safe working order. Your test equipment must have a current calibration certificate, indicating that the instrument is working properly and providing accurate readings. If it is not calibrated, test results could be inaccurate. Before starting work:

* Check the equipment for any damage – look to see if the case is cracked or broken, indicating a recent impact, which could result in false readings.

* Check that the insulation on the leads and the probes is not damaged, and that it is complete and secure.

Remember

Supplies to cookers and showers are in 6 mm^2 to 10 mm^2 cable, immersion heaters are normally 2.5 mm^2, and are generally provided with a 16 amp over-current protection device

The approved method of testing whether a circuit is dead is to use a voltage-indicating device

If you have any doubts about an instrument or its accuracy ask for assistance. These instruments are very expensive and any unnecessary damage caused by ignorance should be avoided.

Test the voltage indicator on a proven supply before you start; this will confirm that the kit is working. The best piece of equipment for doing this is a **proving unit**.

Only now can you use the voltage-indicating device to establish that the circuit you are to work on is dead. You should check phase (live)-to-neutral conductors, phase-to-earth conductors and neutral-to-earth conductors to make sure all connections are dead.

You are not quite ready to begin work – you should again check the test equipment on a known supply to make sure it is working correctly and has not become damaged during the testing procedure.

Only if all the above procedures have been followed correctly should you consider the circuit to be dead and safe to work on.

Begin work

Make use of warning notices: 'Plumber at work'. It may be helpful to put your name and contact number on the notice, so that if you have to leave the job while the customer is out, they can contact you to find out why the power has been turned off.

The flow chart above shows the procedure for isolating an individual circuit or item of fixed equipment.

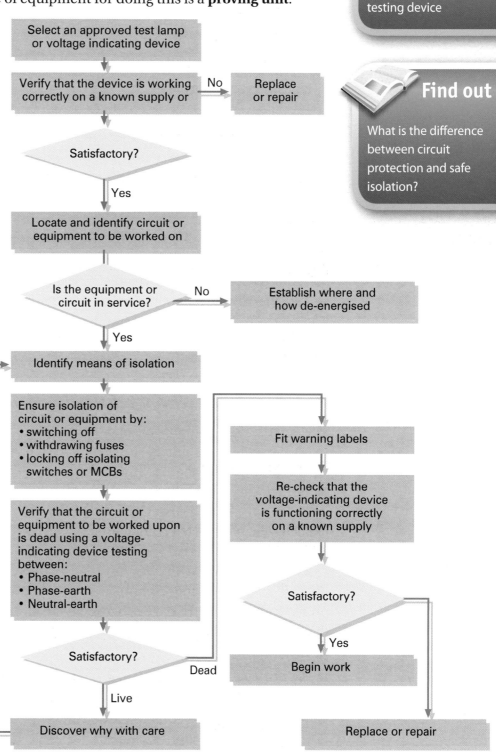

Figure 10.1 Flow chart – isolating a circuit

Definition

Proving unit – Low-voltage, inverted d.c. testing device

Find out

What is the difference between circuit protection and safe isolation?

Electricity supply systems installation (Part 2)

By the end of this section you should be able to describe the practical requirements for installing appliances and controls on electrical systems.

Before we look at a specific installation, we will consider some aspects of installation good practice.

In this section we will look at PVC cables and cords (flex), including:

- types of cable and cords
- installing PVC cables
- terminating cables and flexible cords
- connecting to terminals.

Types of cable and cords

PVC (Polyvinylchloride) insulated and sheathed cables are the most popular type of cables in current use and are employed extensively for electrical installations in domestic dwellings. PVC is versatile, tough, cheap and easy to work with and install, and is generally the most economical material for this type of work. However, its level of insulation is limited in conditions of excessive heat and cold, although this is not a problem in domestic properties. It can also suffer mechanical damage unless you apply additional mechanical protection in certain situations.

PVC insulated and sheathed flat-wiring cables

This cable is used for domestic and industrial wiring. It is suitable for service wiring where there is little risk of mechanical damage. It is available as a two- or three-core. Two or three plain copper, solid or stranded conductors are individually insulated with PVC and sheathed overall with PVC. An uninsulated plain copper circuit protective conductor (CPC) lies between the cores.

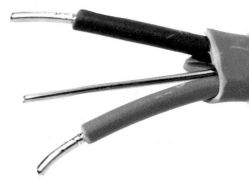

Flat-wiring cable

The core colours for two-core cables are brown and blue plus earth. For three-core cables, they are brown, black and grey plus earth. The sheath colours are normally grey or white. The construction of three-core cables is exactly as mentioned above with the inclusion of an uninsulated plain copper circuit protective conductor (CPC) between the coloured cores.

Multi-strand PVC-insulated and sheathed cable

This cable is suitable for surface wiring where there is little risk of mechanical damage. It is normally used for 'meter tails' to connect the consumer unit/distribution board to the PES (Public Electricity Supply) meter, and as single core for conduit and trunking runs where conditions are difficult. The construction of this cable is PVC-insulated and PVC sheathed solid or stranded plain copper conductors. The old core colours are normally black and red, new colours are blue and brown. Sheath colours are normally black, red or grey, although other colours are available.

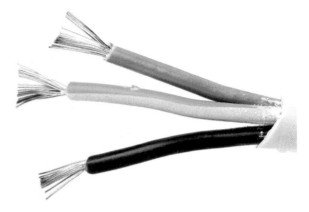

Multi-strand cable

Single core PVC-insulated and sheathed cable

This cable is used for domestic and general wiring. It comprises a PVC-insulated copper conductor with a PVC sheathed overall. The old core colours are red or black, new colours are brown or blue and the sheath colour is normally white or grey.

Single-core cable – old colour

Heat-resisting PVC-insulated and sheathed flexible cords

These flexible cords are suitable for use in ambient temperatures up to 85°C. They are not suitable for use with heating appliances. Their construction comprises plain copper flexible conductors insulated with heat-resisting (HR) PVC and (HR) PVC sheathed.

The core colours for single- and twin-core are brown or blue; for three-core, brown, blue and green/yellow; and for four-core, blue, brown, black and green/yellow. The sheath is white.

Installing PVC cables

Cables are fixed at intervals using plastic clips, which incorporate a masonry-type nail.

Where PVC cables are installed on the surface, the cable should be run directly into the electrical accessory, ensuring that the outer sheathing of the cable is taken inside the accessory to a minimum of 10 mm. If the cable is to be concealed, a flush box is usually provided at each control or outlet position.

Cable-fixing clips

Remember

When bending PVC cable around corners, the radius of the bend should be such that the cable or conductor does not suffer damage

Installing and clipping PVC cable

- In order to ensure a neat appearance, PVC cable should be pressed flat against the surface between the cable clips

- The cable can be straightened by running the thumb or fingers over it before clipping. The palm of the hand can be used for bigger cables.

Flattening cable before clipping

This sequence of forming the cable should be carried out after inserting the last cable clip and before fixing the next cable clip.

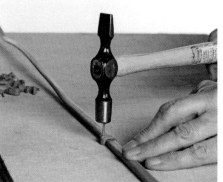

Fix the first clip

Fix the last clip

Fix the intermediate clips

- When a PVC cable is to be taken around a corner or changes direction, the bend should be formed using the thumb and fingers as shown.

- Care must be taken to ensure that the bend does not cause damage to the cable or conductors and that cable clips are spaced at appropriate intervals.

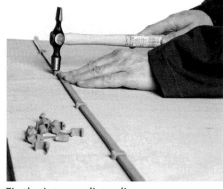

Clipping round a curve

Cable runs

Cable runs should be planned so as to avoid cables having to cross one another, which would result in an unsightly and unprofessional finish. When installed in cement or plaster, they should be protected against damage. This is done by covering the cable with a plastic or metal channel or by installing them in an oval PVC **conduit**.

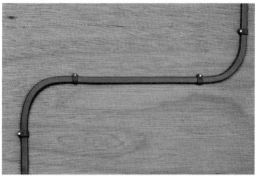

The finished result

PVC-sheathed cables should not come into contact with polystyrene insulation, as a chemical reaction takes place between the PVC sheath and the polystyrene resulting in the migration of polymers from the cable, known as 'marring'.

Terminating cables and flexible cords

The entry of the cable end into a fitting is known as a **termination**. In the case of a stranded conductor, the strands should be twisted together with pliers before terminating. Care must be taken not to damage the wires. BS7671 requires that a cable termination of any kind should securely anchor all the wires of the conductor. This is to prevent any undue stress on the terminal or socket, causing it to overheat.

The cable, under working conditions, will expand and contract due to heating and cooling, again putting pressure on the terminal connection. This may result in increased resistance and probably overheating. So, when terminating the flex, it should be gripped with a flex clamp, tightened down onto the protective sheathing.

Stripping PVC cables

Pulling cable apart

Removing the sheathing

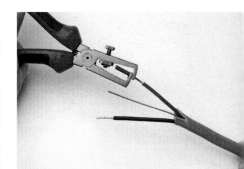

Stripping the insulation

- Nick the cable at the end with a knife and pull apart.

- When the required length has being stripped, remove the surplus sheathing with a knife or snips.

- The insulation can also be stripped from the conductors with a knife or purpose-made strippers. Examine the conductor insulation for damage.

- An alternative method of stripping the insulation from the conductors is with a pair of purpose-made strippers.

Connecting to terminals

There is a wide variety of conductor terminals. Typical methods of securing conductors in fittings are:

- pillar terminals
- screw heads
- nuts and washers
- claw washers
- strip connectors.

Pillar terminals

A pillar terminal has a hole through its side into which the conductor is inserted and secured by a setscrew. If the conductor is small in relation to the hole it should be doubled back. When two or more conductors are to go into the same terminal they should be twisted together. Care should be taken not to damage the conductor by excessive tightening.

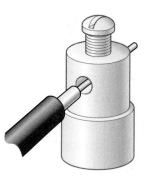

Figure 10.2 Pillar terminal

Screw-head, nut and washer terminals

When fastening conductors under screw heads or nuts, it is best to form the end of the conductor into an eye, using round-nosed pliers; the eye should be slightly larger than the screw shank but smaller than the outside diameter of the screw head, nut or washer.

The eye should be placed in such a way that rotation of the screw head or nut tends to close the joint in the eye. If the eye is put on the opposite way round, the motion of the screw or nut will tend to untwist the eye and will probably result in a bad contact.

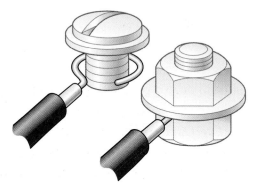

Figure 10.3 Screw terminal

Claw washers

In order to get a better connection, claw washers can be used. Lay the looped conductor in the pressing. Place a plain washer on top of the loop and squeeze the metal points flat using the correct tool. The diagram illustrates this method.

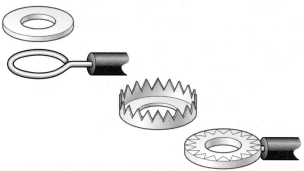

Figure 10.4 Claw washer

Strip connectors

The conductors are clamped by means of grub screws in connectors, which are usually made of brass and mounted in a moulded insulated block. The conductors should be inserted as far as possible into the connector so that the pinch screw clamps the conductor. A good, clean, tight termination is essential in order to avoid high resistance contacts resulting in overheating of the joint.

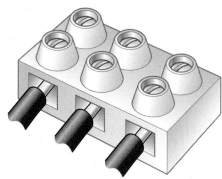

Figure 10.5 Strip connector

Electricity supply systems installation (Part 3)

Installing fixed electrical appliances

In this section we will look at how the electrical appliances and components are connected into a domestic ring-main circuit. This will also cover basic practical installation requirements for:

- immersion heaters
- dishwashers
- washing machines
- macerator-type WCs
- waste-disposal units.

Immersion heaters

Immersion heaters are not wired directly from the ring main but from the consumer unit in 2.5 mm twin and earth PVC cable, and have an overcurrent protection device rating of 16 amps. They are fed to a double pole switch, which has to be located a maximum of 1 m from the connection to the immersion heater. The cable between the pole switch and the immersion heater is 1.5 mm heat-resistant flex, usually butyl. A double-pole switch is safer than a single-pole version, because when it is turned off both the live and neutral are isolated. It is most likely that you would be required to disconnect and reconnect the supply when replacing a defective immersion heater.

Immersion heater flex to double pole switch

Dishwashers and washing machines

These are normally connected via a three-pin plug into a 13 amp socket outlet. While most appliances come with pre-moulded plug heads, you will occasionally need to change a plug. This is best learnt by a practical demonstration from a qualified person.

Macerator-type WCs and waste-disposal units

Like central heating control units, these are connected to the system using a spur outlet. A spur outlet is a bit like a hidden plug connection; it has the advantage that it cannot be unplugged accidentally thus turning off the supply to a control or appliance. It is also less likely to get damaged.

You will probably wire a component or an appliance from the fused outlet. On some occasions, however, you may be required to connect into the ring main.

Remember

We discussed the installation of pumps, electric showers (or water heaters) and central-heating controls in Chapters 8 and 9.

Find out

What should you do to ensure that a section of pipework can be safely removed without any risk of electric shock?

Did you know?

Spur outlets are also used on ring mains where it is inconvenient to fit an outlet within the ring-main loop, such as in an isolated location

Remember

The connection to a macerator-type WC must be an unswitched fused spur

Installing a fused spur outlet

Before you start, make sure the cables and fittings are correct for the job. Check for any damage to the cable, and that it is the right size and type, and that the fittings have the correct amp rating. **Remember the procedure for isolating the supply**. The spur outlet should be located close to the appliance or component it is intended to serve. Check the manufacturer's instructions for any specific requirements.

You need to connect to the ring main as close as possible to the spur outlet. This can be done by locating the nearest socket outlet, lifting a floorboard and then tracing the wire until it is at the closest point to the spur.

The cable is cut. It should be tested to make sure that it is the ring-main circuit and that the cable is sound (see the next section on testing). The joint on the ring main will be made using a junction box.

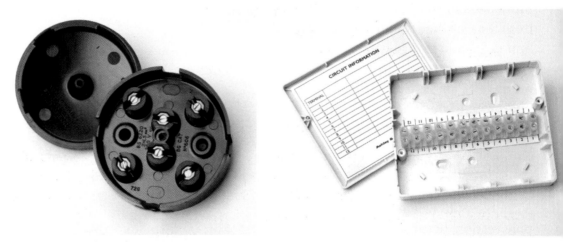

Old and new junction-box designs

Definition

Chasing out – cutting a recess into plaster or block work for the socket or cable

Running the cable to the spur outlet

The spur outlet can be surface-mounted or flush-fitting. If using a flush fitting, the plaster and block work has to be **chased out** to take the box; keep this as neat as possible to reduce the mess and the amount of making good. A chase is cut from the hole for the box to the floor level; make sure this is wide enough to receive plastic capping, which you will have to fix to protect the cable before making good. In some cases, you may be able to thread the cable under the floor. Much will depend on the position of the joists, but in some cases you may have to drill a hole in the floor.

Fit the spur box by drilling and screwing to the wall. If it is a metal box, you must include a rubber grommet in the hole that the cable will pass through. Push the cable through the hole in the spur and under the floor, and pull the cable to the junction box.

If you are running the cable through the joists, the hole should be drilled in the centre to prevent any weakening. Alternatively it can be clipped to the side or the underside of the joists.

The ends of the ring-main cables and the cable to the junction box are prepared, and connected live-to-live, neutral-to-neutral and earth-to-earth.

The cable from the junction box to the spur, and the outlet from the spur, are connected to the back of the spur fitting. Again, if it is a metal box, you need to take an earth wire from the box to the fused spur. The cable is now ready to connect to the component or appliance.

When you have completed the work, carry out a visual inspection to make sure all the connections are sound, the cable routes and fixings are still where they should be, and the cable has not been damaged during installation.

The installation is tested while still switched off (we will look at testing in the next section), and, assuming everything is in order, the supply can be turned on.

Equipotential bonding

By the end of this section you should be able to:

- explain what is meant by the terms:
  - main equipotential bonding
  - supplementary bonding
- state when and why temporary earth-continuity bonding is required
- describe how to carry out temporary earth-continuity bonding.

This is a very important area of electrical wiring, as most of the systems we install require some form of bonding protection, so you should be able to tell if the earthing and bonding have been installed correctly. You may also have to carry out work on these systems, but should do so only if you are competent.

Safety tip

Never carry out work of this nature if you are not sure. If this is the case, contact an approved electrician

On the job: A bonding problem

Callum, a final-year apprentice, is required to carry out a repair job on the copper pipework of an existing heating system, and this involves removing a section of pipework. When he checks the earthing/bonding arrangements, he discovers that none exist. His supervisor has gone to a boiler breakdown several miles away and is not due back for several hours.

1 What should Callum do now?

2 What would you do in this situation?

3 What should you check to make sure that equipotential and supplementary bonding was correctly installed, and how?

Main equipotential bonding

The method explained here of protecting against indirect contact using equipotential bonding and automatic disconnection of supply meets the requirements of the Wiring Regulations (BS 7671:2001). Metal pipework can provide a route for stray electric current to go to earth. This could cause an electric shock for someone touching the live pipework unless the metal is properly earthed. It can also cause the pipework to corrode.

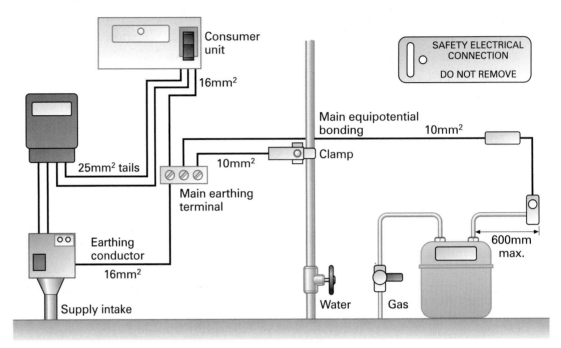

Figure 10.6 Earth-bonding conductors

In the illustration, note the size of the bonding conductors, and the need for the clamp to be labelled. The bonding conductor from the main terminal to the earthing clamp is 10 mm. The bonding to gas, water or other services should be as close as possible to the point of entry, and for the gas supply within 600 mm of the meter.

Bonding the pipework as shown in the diagram will provide a safe route to earth and is described as main bonding.

Earth bonding to pipe

Supplementary bonding

You need to be able to tell if a plumbing installation is correctly bonded. You can see from the previous diagram that the equipotential bonding connects only to one point of the pipework.

There are other exposed metal parts within the domestic hot-water, cold-water and central heating systems, which may not be protected because they have been isolated from the earth by plastic fittings, cisterns etc. used in the system.

This affects the conductivity of the pipework as an earth. To maintain the earth, supplementary cross bonding is used; a typical layout is shown in the diagram. The bathroom is the biggest area of risk.

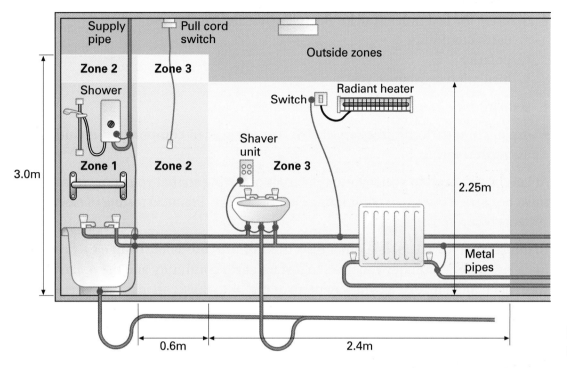

Figure 10.7 Zones in bathrooms

The zones represent the degree of risk – proximity to water services. The bonding, wiring and connections are shown by the blue points and lines.

It is important that, on completion of a job, the services as shown in the illustration – including any supplementary bonding requirement – are carried out. This work will normally be done by a qualified electrician and your company should have advised the customer that the work is necessary or should have made arrangements to carry out the electrical work at the same time as you are completing your plumbing systems installation.

Temporary bonding

If you are required to work on a repair or maintenance job, it is important that you check that the pipework you are going to work on is bonded correctly. If you are going to remove a section of metal pipework or fittings, it is essential that earth continuity is maintained before any cutting or disconnection takes place.

A typical piece of kit consists of crocodile clips and a 10 mm conductor, 250 V rating minimum. It is suitable for up to 28 mm metal pipe.

The clips should be placed in a position to bridge the gap of the pipe or fitting that is going to be removed. Work can then safely take place. Only remove the clamps once the job is complete.

Remember

Plumbers are in constant contact with metal pipework and metal surfaces when carrying out their job – customers can be too!

Safety tip

Remember: by removing the section of pipework, you have removed a section of the system from the main bonding conductor and may, therefore, have exposed yourself to real danger – this one can kill!

Testing and decommissioning

By the end of this section you should be able to:

- describe the test procedure for the following electrical tests:
  - earth continuity
  - polarity
  - insulation resistance
- state the type of electrical testing equipment used
- explain how to decommission electrical appliances or components, including associated wiring.

At Level 2 you need to understand the basics of testing so that you know what must be carried out when completing electrical work. Practical testing of electrical systems is covered at Level 3.

In this section we will look at other tests that confirm the safety of an electrical installation. As a plumber you need to test for earth continuity and the polarity of wires at an appliance or a component. In addition to these, an insulation-resistance test should be carried out which will tell you if the cabling used is in good condition.

Remember

If these tests are not properly carried out then the law is being broken

Electrical tests

The systems we deal with include:

- continuity of earthing conductors
- polarity
- insulation resistance.

Whatever the test, you must use the correct type of test equipment. For electrical work associated with plumbing, the testing and the proper equipment required includes:

- continuity – low-reading ohmmeter
- polarity – low-reading ohmmeter or insulation and continuity tester
- insulation resistance – insulation-resistance tester.

Your training will include showing you examples of the above test equipment.

Remember

The good news is that the tests are carried out while the system is dead

Earth continuity

Earth continuity is making sure that should there be an electrical fault, all exposed metalwork in a building is bonded together and connected to the earthing block in the consumer unit, leaking the current to earth and automatically disconnecting the supply. An earth continuity test will verify that exposed metalwork in a building is bonded together and connected to the earthing block in the consumer unit. The ohmmeter leads are connected between the points being tested, between simultaneously accessible extraneous conductive parts – pipework, sinks etc. – or

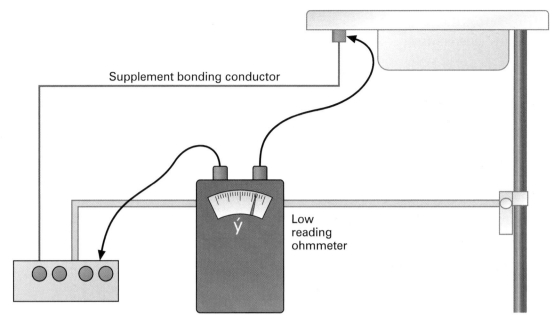

Figure 10.8
Testing for continuity

between simultaneously accessible extraneous conductive parts and exposed conductive parts (metal parts of the installation). This test will verify that the conductor is sound. To check this, move the probe to the metalwork to be protected. This method is also used to test the main equipotential bonding conductors. There should be a low resistance reading on the ohmmeter.

Polarity

Testing for polarity makes sure that phased conductors are not crossed somewhere – neutral from mains connected to live and vice versa (reversed polarity). In this situation, the system might still function as expected; however, when isolated from a switch, the system would be in dangerous mode.

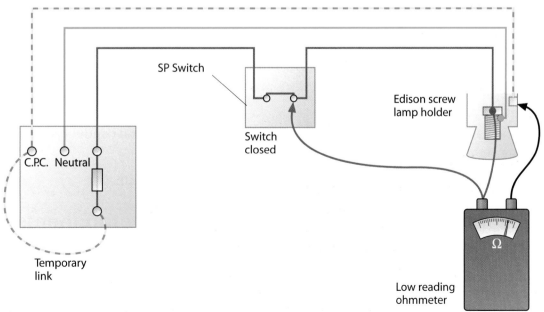

Figure 10.9 Testing polarity

Testing across the conductors, as detailed above, is carried out to make sure that no wires have been crossed.

Insulation resistance

Insulation-resistance tests make sure that the insulation of conductors, electrical appliances and components is satisfactory, that electrical conductors and protective conductors are not short-circuited, or do not show a low insulation resistance (which would indicate a defective insulation).

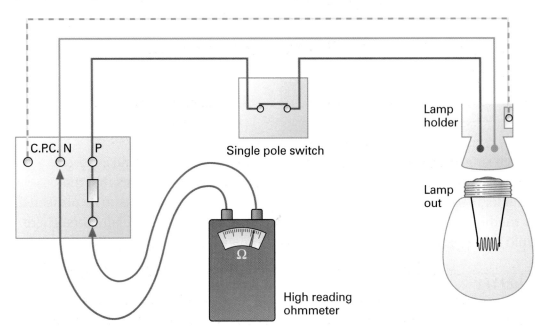

Figure 10.10 Insulation-resistance check

Before testing, ensure that:

1 Pilot or indicator lamps and capacitors are disconnected from circuits to avoid an inaccurate test value being obtained.

2 Voltage-sensitive electronic equipment such as dimmer switches, delay timers, power controllers, electronic starters for fluorescent lamps, emergency lighting, Residual Current Devices etc. are disconnected so that they are not subjected to the test voltage.

3 There is no electrical connection between any phase or neutral conductor (e.g. lamps left in).

An insulation resistance of no less than 0.5 M ohms should be achieved.

Decommission electrical appliances or components, including associated wiring

Your job will require you to take out components and appliances, and that will include the electrical supply. Generally speaking, the work will be a repeat of what you've covered earlier, in terms of safety, isolation and testing, because you're working on an electrical supply.

In some cases you'll decommission an appliance or a component to repair or replace it, so it's disconnect, reconnect. There may be instances, however, where an appliance is going to be removed altogether, which would mean the supply would need to be made safe. This would be done by taking out the cable back to its isolation point, e.g. immersion heater isolating switch, fused spur outlet. All of this should be done following the safety procedures that we have already learned.

FAQs

Why is electricity included in a plumbing qualification?

It is not intended to make plumbers into electricians but plumbers need to know quite a bit about safe methods of working and isolation in the installation and maintenance of plumbing components with electrical connections. There is a wide range of plumbing/electrical components that have to be tested for correct operation, and plumbers often remove sections of pipework which may affect earthing arrangements.

What should I do if I am unsure about the correct action to take in order to safely isolate a component?

The simple answer is DO NOT try to do the work at all, seek help and guidance from someone who is competent in this area of work. Do not take any risks, You are a long time dead!

Surely locking an mcb in the off position while working on a circuit is an overreaction. Is it really necessary?

Yes, it is absolutely necessary, how else can you be sure that the mcb will not be switched on by someone while you are working.

Do I have to be qualified in electrical work in order to carry out any testing of a system controls circuit for correct operation?

No, but you do have to be competent. The rules on this may change in the future, however, so keep up to date.

Knowledge check

1   The main requirement for a plumber working on an electrical system is that they must be:
    a. competent
    b. over 21 years old
    c. Corgi registered
    d. confident

2   Briefly summarise the difference between circuit protection and safe isolation.

3   State how people are protected against indirect contact from an electrical supply.

4   What piece of equipment would you use to check if an electrical circuit was dead?

5   What size cable, ring-main amp rating and socket amp rating would you expect to see on a ring main?

6   You're completely removing a hot-water storage vessel which has an immersion heater installed. You've done all the safety checks and tests. What are your next actions?

7   How would you make sure a system had been safely isolated?

8   Give examples of the type of cable you are likely to find in a typical domestic dwelling.

9   Give three example of good practice when installing cable runs.

10 What three electrical tests are likely to be carried out by a plumber?

chapter 11

Above ground discharge systems

OVERVIEW

Above ground discharge systems (AGDS) include sanitary appliances, sanitary pipework and fittings, and rainwater systems. This chapter will look at the various appliances on the market, how to fit them and pipe them up, and how they are tested and maintained. This chapter also covers rainwater drainage – that is, guttering and rainwater-pipe systems. AGDS are essential to keep our environment clean and hygienic.

This chapter will cover:

- **Sanitary appliances:**
  - British Standards and Regulations
  - types of appliance and materials
  - working principles of sanitary appliances
  - other types of appliances.

- **Traps**
  - types of trap
  - access
  - why traps lose their seal.

- **Sanitary systems:**
  - primary ventilated stack system
  - stub stacks
  - general discharge stack requirements
  - trap seal loss.

- **Pipework installation, access, materials and testing:**
  - preparing to install AGDS pipework
  - the installation and jointing of AGDS pipework
  - connections to drains
  - connecting the soil and vent pipe.

- **Soundness testing:**
  - soundness testing of AGDS.

- **Rainwater and guttering:**
  - materials
  - gutter and rainwater systems.

- **Maintenance and decommissioning**
  - basic maintenance and cleaning
  - decommissioning systems.

Sanitary appliances

By the end of this section, you should be able to:

- state the British Standards and regulations relevant to sanitary appliances
- state the types of sanitary appliances used in AGDS and the materials they are made from
- explain the working principles of sanitary appliances
- describe the installation requirements of sanitary appliances, including preparatory work.

Sanitary appliances can be divided into two main groups: those used for washing purposes and those used for the removal of human waste.

Sanitary appliances play a very important role in our lives: they mean we can keep ourselves and our eating utensils clean, and we can use the toilet in privacy and comfort.

Although we are mainly concerned here with domestic installation we will extend coverage to urinals and flushing cisterns which you may come across when working in public toilets or larger buildings.

British Standards and Regulations

British Standards

British Standard 8000 is the Code of Practice covering workmanship on building sites. Part 13 covers above ground drainage and sanitary appliances, including:

- materials handling
- site storage of components
- the preparation of work, materials and components
- the installation of sanitary appliances
- inspection and testing.

BS8313 gives guidance on the holes, chases and ducts required for pipework.

Water Regulations

Water Regulations do not refer specifically to sanitary appliances, but do deal with how they perform by covering undue consumption, waste of water and protection against backflow.

The regulations governing WC cisterns, urinals and automatic flushing cisterns are limited to the capacity of water for flushing.

Backflow prevention is achieved either through mechanical methods (using valves) or non-mechanical means (using air gaps between outlet and appliance

Remember

Backflow prevention has a bearing on sanitary appliances, which are categorised by the Water Regulations in terms of risk. This covers WC suites, washbasins, bidets, baths, shower trays and sinks

Siphonic action uses differences in the atmospheric pressure in two parts of a pipe to draw water through the pipe. Check back in Chapter 4 if you cannot remember how this works

water levels) to prevent the possibility of water being drawn back into the supply pipework and into the main. This could occur if an open tap outlet was allowed to become submerged in a sink or bath, and if the main outside stop valve was turned off and water drained from the system. As the water drained down from the system, water from the sink or bath would be drawn into the pipework due to siphonic action. Water Regulations will be covered in greater detail at Level 3.

Types of appliances and materials

There is a vast range of styles and designs of sanitary appliances available. However, British Standards ensure that there is standardisation in terms of dimensions,including the size of waste outlets.

Baths

Manufactured to BS4305 (EN198), baths can be supplied in acrylic sheet, heavy-gauged enamelled steel, and vitreous enamelled cast iron. Victorian-style free-standing designs give a traditional 'period look'. Vitreous enamel provides a smooth hardwearing surface that is corrosion resistant and easy to clean.

Baths come in a huge variety of styles, and taps and waste can be either end-mounted or centre-mounted. Space-saving styles include corner baths and compact baths.

More expensive options include enamelled baths manufactured from heavy-gauged porcelain enamelled 3.5 mm thick onto heavy gauge steel. Steel sheet is pressed into shape and then coated in layers of vitreous enamel which is fired at high temperatures. The enamel and steel bond during firing to give the enamel a glass-like high gloss finish.

> **Did you know?**
>
> Acrylic sheet baths are vacuum-formed to the desired shape

This type of bath is particularly suitable for installation in hotels, or in housing association and local-authority housing where durability is important.

The jet-system bath is a popular choice in the 'luxury' market. It is like a standard bath but contains a number of jets (generally eight) in the side of the bath to provide a massage effect. In some cases additional outlets are sited in the floor for foot massages.

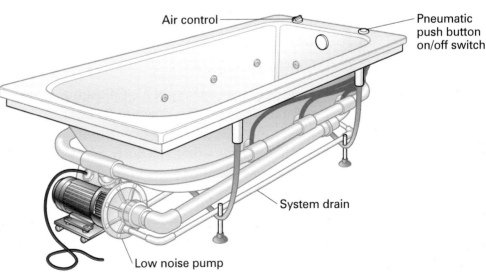

Figure 11.1 Typical jet-system bath

The water is circulated through pipework located around the outside of the bath and forced through the jet nozzles by a pump.

Baths are manufactured with two holes for pillar taps, or single holes for monobloc fixings. Most come with an overflow hole, and waste hole that will take a $1\frac{1}{2}''$ threaded waste. Combined overflows and waste fittings are also provided, as shown.

Water closets (WCs) and cisterns

Again, there are a lot to choose from. They are categorised as:

- back to the wall
- close-coupled
- low-level
- high-level
- concealed.

Figure 11.2 Combined waste and overflow fitting

In the main, WCs are manufactured from **vitreous china** conforming to BS3402. Vitreous china is made from a mixture of white burning clays and finely ground materials. These are fired at high temperatures, and even before glazing the material cannot be contaminated by bacteria, so it is totally hygienic.

Vitreous china is coated with an impervious non-crazing vitreous glaze in either a white or coloured finish. The material is stain proof, burn proof, rot proof, rustfree, non-fading and is resistant to acids and alkalis.

WCs used in public places may also be manufactured in stainless steel. These are more resistant to vandalism than vitreous china.

WC cisterns

Prior to 1993 the capacity of a WC flushing cistern was 9 litres. The Water Regulations brought this down to 7.5 litres, and from January 2001 reduced it further to 6 litres. In addition, Water Regulations also permit the use of dual-flush cisterns; these deliver 6 litres for a full flush, and 4 litres for a lesser flush. The dual flush cistern is now specified more often that the single flush.

There are two types of cistern:

- siphonic
- dual flush valve.

With the siphon type, when the lever is pressed the water in the bell of the siphon (1) is lifted by a disc (or diaphragm) up and over into the leg of the siphon (2) (see figure 11.3). This creates the siphonic affect, which continues until the water level in the cistern has dropped to a level which allows air to enter the bell.

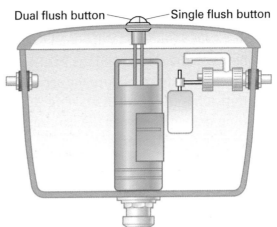

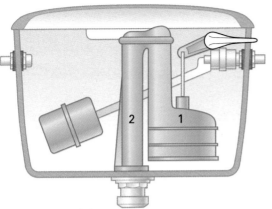

Single flush – siphon type

Dual flush – valve type

Figure 11.3 Different types of cistern

The dual-flush valve-type cistern is operated by pressing either the full flush or lesser flush buttons, which have to be clearly marked on the cistern. These operate a valve which releases the water into the WC pan.

Another type of fitting (see figure 11.4), is also available for 6 litre capacity cisterns. This works on the principle of lifting a hinged flap and is suitable for full flush only.

Cisterns of 9 litres and 7.5 litres are still available. Water Regulations permit these to be installed as replacements for existing cisterns. This is because the WC pan was designed to work on those capacities.

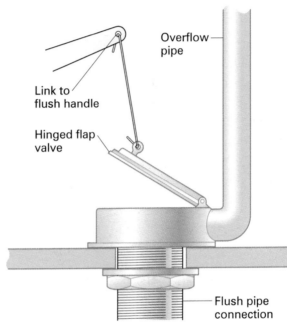

Figure 11.4 Cistern fitted with hinged flap mechanism

The overflow pipe can now discharge into the WC pan using an integral overflow, eliminating the need to provide an external overflow pipe. This pattern of cistern is now supplied with most WCs.

WC pans

There are a number of designs for WC pans, but there are only two main types:

- wash-down WC pan
- siphonic pan.

The wash-down pan uses the force of the water from the cistern to clear the bowl. The principle of the siphonic pan is to create a negative pressure below the trap seal.

With the single-trap pan, this is done by restricting the flow from the cistern and is achieved by the design of the pan.

Remember

WC pans and cisterns are manufactured as a single unit and therefore replacement cisterns to existing pans must always be the same capacity as the original. It is not possible to 'mix and match' as each component has been designed carefully in conjunction with the others to operate properly as a 'suite'

The double-trap closed-coupled pan uses a pressure-reducing device between the cistern and the pan. As the water is released into the second trap, it has the effect of drawing air from the void between the two traps, and siphons the content from the bowl.

Figure 11.7 shows what the pressure-reducing valve looks like between the cistern and pan.

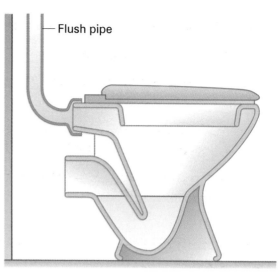

Figure 11.5 Wash-down pan

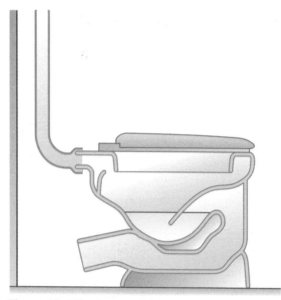

Figure 11.6 Single trap siphonic pan

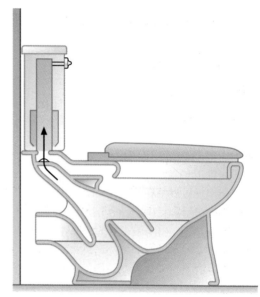

Double trap siphonic pan

Popular WC types include:

Concealed – where the cistern is hidden

Close-coupled – where cistern and pan are joined in one unit

Low-level – where the cistern is no more than one metre above the pan

High-level – where the cistern is more than one metre above the pan necessitating flushing by a chain

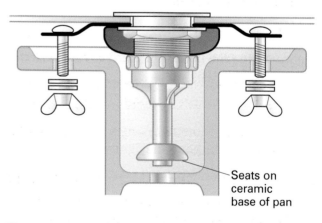

Seats on ceramic base of pan

Figure 11.7 Typical pressure-reducing valve between cistern and pan

The choice of WC suite depends on a number of factors:

- cost – the customer's budget
- location – for public toilets, factors such as durability and ease of cleaning are a consideration
- aesthetics (how good it looks) – Victorian versions of the high-level cistern are now considered very desirable.

On larger housing contracts, or in public buildings, the choice of WC may have already been made by the client's architect as part of the contract specification document.

Some cisterns are reversible – that is, the handles, overflow outlet and water inlets can be either left- or right-handed. In some cases, the cistern is plastic rather than vitreous china.

Kits are usually supplied for the soil outlets, and will be either a 'P' trap or an 'S' trap.

Washbasins

These fall into two main types:

- fixed to the wall
- countertop basins.

Fixed to the wall

These are secured by using brackets or are supported by a pedestal. Larger washbasins, fixed using brackets, are mostly found in non-domestic installations. The pedestal type is probably the most popular choice in homes. Appearance is the main factor as all the pipework and fittings can be seen with the bracket-mounted basin.

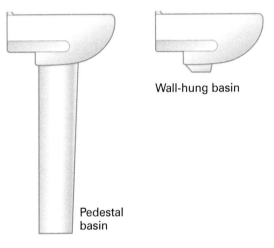

Wall-hung basin

Pedestal basin

Figure 11.8 Fixed to wall

Countertop basins

There are three types:

- countertop – sits proud of the work surface
- semi-countertop – used where space is tight as the front of the basin projects clear of the top
- under the countertop – has the flange surface on top of the basin.

All basins are available with either single tap holes for a monobloc mixer tap, or dual tap holes for pillar taps. Basins are also available with three holes for fittings with an independent spout.

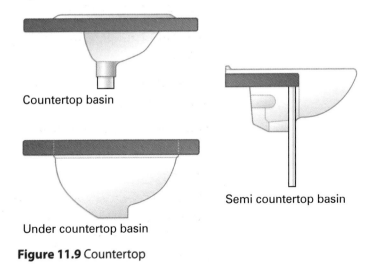

Countertop basin

Under countertop basin

Semi countertop basin

Figure 11.9 Countertop

Countertops are installed in all sorts of buildings, from homes to pubs and restaurants. They are manufactured from either vitreous china or a high-impact plastic called acetyl.

The basin outlet is designed to take a $1\frac{1}{4}"$ slotted waste fitting. Waste fittings can be the standard slotted waste or a pop-up type, which is operated by lifting and pressing a knob on the top of the basin.

Hand-rinse basins

Called hand-rinse basins due to their size, these are used in cloakrooms with WCs. They can be a corner design or traditional, either countertop, pedestal or bracket-mounted.

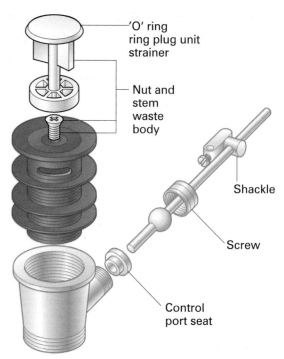

Figure 11.10 Pop-up waste fitting

Countertop range

There are three types:

Standard countertop – this fits into a pre-cut hole and is secured by fixing brackets that secure the underside of the basin to the countertop; the joint is sealed using waterproof sealant

Semi countertop – can be made from vitreous china or acetyl

Under countertop – this appliance is made from acetyl.

Shower trays

Shower appliances were covered in Chapters 7 and 8. Here we will concentrate on shower trays. As with other sanitary appliances, there are a number of designs. Trays are available in reinforced cast acrylic sheet, fireclay for heavy-duty applications or resin-bonded. The outlet is designed to take a $1\frac{1}{2}"$ threaded waste. Corner trays are used where space is limited.

Bidets

Made from vitreous china, there are two types:

- over-the-rim supply
- ascending spray.

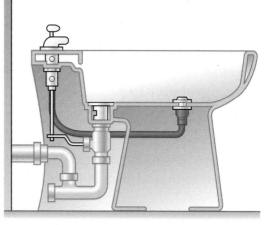

Figure 11.11 Ascending spray bidet

The ascending-spray bidet has to be piped up correctly to avoid risk of contamination to the supply and cannot be used on mains supply, for example with combination boilers and unvented hot-water systems. This is to meet the requirements of the Water Regulations and will be dealt with at Level 3. The bidet can be supplied for use with pillar taps, monobloc fittings and pop-up waste.

Sinks

Belfast sinks and London sinks

These are used in both domestic and commercial situations. They are usually fed by bib taps, but can also be supplied by either pillar taps or monobloc taps in domestic installations.

London sinks are similar, but without an overflow. Both of these sinks are made of heavy-duty fire clay, which is manufactured in a similar way to vitreous china. They are available in white only and are designed to take a $1\frac{1}{2}$' threaded waste.

Kitchen sinks

Again, a vast range of designs is available. Sinks are produced in stainless steel, plastic-coated pressed steel, fire clay and plastic in the form of acetyl. They can be single drainer, double drainer; single and basket (suitable for connection to a small waste-disposal unit) and double sink with single drainer. They are designed to take a $1\frac{1}{2}$' slotted waste. This can be slotted for use with a sink overflow, or plain. Taps can be monobloc or pillar.

Urinals

Usually installed in non-domestic buildings, urinals are manufactured in vitreous china for urinals and stainless steel for bowls fixed to the wall, and stainless steel, plastic or fire clay for slab urinals.

The automatic siphon's operation is very simple:

- as the cistern fills, and the water rises, the air inside the dome of the automatic siphon is compressed
- the increased pressure forces water out of the 'u' tube, which reduces the pressure in the dome
- the reduction in pressure causes siphonic action to take place, flushing the cistern
- when the cistern has emptied, the water in the upper well is siphoned into the lower well.

Did you know?

The water supply to urinals is controlled by an automatic flushing cistern or a pressure flushing valve; this is covered at Level 3

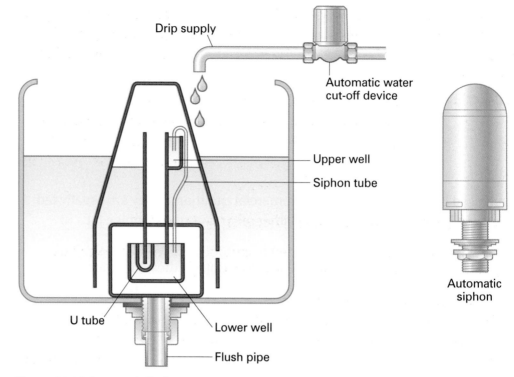

Figure 11.12 Automatic flushing cistern

Water Regulations state that auto-flushing cisterns must not exceed the following maximum volumes:

- 10 litres per hour for a single urinal bowl, or stall

- 7.5 litres per hour, per urinal position, for a cistern servicing two or more urinal bowls, urinal stalls or per 700 mm of slab.

The flow rate can be achieved using urinal flush control valves which allow a small amount of water to pass into the system.

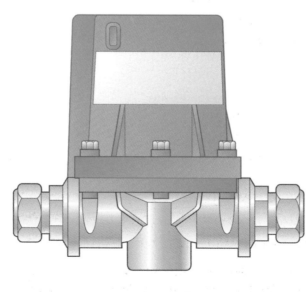

Figure 11.13 Hydraulic flush-control valve

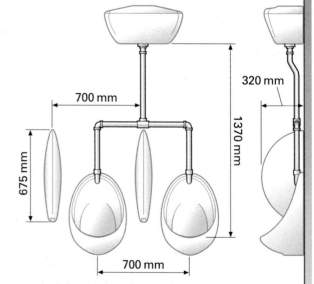

Figure 11.14 Typical two bowl urinal layout

Timed flow control valves can also be used. These have the additional advantage of being switched off when the building is not in use, for example evenings, weekends, factory shutdowns, school holidays etc. They can be set for 'one-off' hygiene flushes, say once every 24 hours.

Figure 11.14 shows a typical example of a urinal set-up showing individual wall fixed urinals (bowls). These are vitreous china. The picture also shows the cistern and flush-pipe arrangements. On slab urinals the flush pipe is often mounted on the face of the urinal and has a number of holes which are used to wash the face of the slab and the channel. These are called **sparge pipes**.

Working principles of sanitary appliances

The main working principles that you need to consider with regard to sanitary appliances apply on the whole to WC pans and cisterns.

Some WC pans clear their waste by the wash-down action of the water as it is released into the bowl. Other designs use siphonic action: siphonic WC pans, siphons in cisterns and automatic siphons. We covered the 'science' aspect of siphonage in Chapter 4, Key plumbing principles, and we looked at how siphonage is used earlier in this chapter.

Installing sanitary appliances

Preparation

It is important to prepare properly before you start installing any appliance. This should ensure there are no hold-ups later owing to missing fittings, or pipe chases not having been prepared as part of the specification.

Storage of materials on site

This is particularly important on larger jobs where you might take delivery of a number of items of sanitary ware. However, some aspects also apply to 'one-off' jobs. Sanitary appliances are expensive items, so it is important that they are stored properly to prevent damage or theft.

Here is a storage checklist:

- Before a delivery takes place, make sure there is somewhere suitable to store the appliances – a lockable materials cabin, with adequate room, is ideal. The storage surface should be raised off the ground of the cabin (use pallets or similar), and it should be clean and dry.

- Before accepting delivery, check the appliance carefully for signs of damage. This does not mean removing the wrappings – these should be left on as long as possible. Production defects are rare, but keep an eye out for these as well.

- Check all traps, taps, water plugs, brackets and seats are with the appliances. There is nothing worse than starting a job and then finding that a waste fitting or bracket is missing.

> **Definition**
>
> **Sparge pipe** – a flush pipe mounted on the face of a urinal connected to a number of holes, used to wash the face of the slab and channel

- When you are happy that everything is complete, check off all the items ordered against the delivery note. If you have the authority, sign for the delivery. You will receive a customer's copy of the delivery note. If anything is not correct, note the missing items and contact the supplier immediately. If you are desperate for the delivery as it stands, sign for the items and get the person making the delivery to countersign that some items are missing.

- Take care when handling the material. Some sanitary items are heavy: remember your lifting technique from the Health and Safety module.

- Store the materials in the materials cabin. As mentioned, these should be clean and dry. The appliances should also be stored away from plaster and concrete, and should also be kept away from areas where other materials could fall onto or into them (brick stacks etc.)

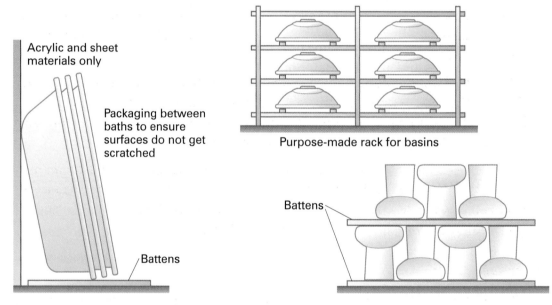

Acrylic and sheet materials only

Packaging between baths to ensure surfaces do not get scratched

Battens

Purpose-made rack for basins

Battens

Figure 11.15 Storage patterns for baths, WCs and basins

The illustrated ways of storing appliances are simple but effective. WC pans should be stacked no more than four high. Make sure the storage surfaces (battens/pallets) are clean: battens are used for all sorts of things on site, and if they are covered in grit, storing materials on them will defeat the object.

Site checks

Check access to the work area, and check that the work area itself is free from obstructions and potential hazards. This should ensure that the risk of accidents is kept to a minimum.

- Make sure you have any job details or specifications/drawings showing the fixing positions of the sanitary appliances and any specific fixing requirements. You will also need the manufacturer's fixing instructions.

- If you find any installation problems, or defects in the appliance as you are working, report them to your supervisor.

- On some jobs you will cut holes or chases or install ducts for pipework. On other jobs, this preparatory work is done by other trades in advance, so you will need to check that it has been done. If not, report back to your boss or the site supervisor.

- When you're at the stage of fixing the appliances, the 'first-fix' pipework must be installed. This may have been done by you, particularly on a single-dwelling job, or it could have been done by another plumber on a large housing contract.

- Make sure servicing valves have been installed where required. Good practice would see them put on every appliance that requires regular service and maintenance.

- On first-fix jobs, open-ended pipes should have been capped. This is to prevent debris getting into the system.

- On jobs where the appliances are being fixed against tiles, baths and showers should be installed before tiling, and concealed bidets, washbasins and WCs after tiling. This makes the job easier for the tiler as they won't have to cut around the shape of the appliance.

- Edges of baths should be set into plaster or plaster board. Careful planning and liaison with other trades can ensure that plasterwork is not taken to full floor level on the bath-edge wall, and in the case of stud-partitioned walls, extra struts can be fixed to support the edges of the plaster board.

- Finally, as you are carrying out the work, avoid standing on any of the appliances. This is sometimes unavoidable with showers, or baths with a shower appliance installed above it. In each case, make sure the surface of the appliance is fully protected with dust sheets and cardboard packaging.

Safety tip

Always use the correct lifting and carrying techniques when transferring materials from the store to the job

Remember

It could be a few days between first and second fix, and in the mean time other trades, such as plasterers, may have been in – it is easy for debris to get into uncapped pipework

Installation

For the purpose of this topic, we are going to work through the installation of a typical new bathroom suite, including:

- bath
- pedestal washbasin
- close coupled WC – wash-down pattern
- shower tray and enclosure
- bidet.

Preparing the appliances

This is often referred to by plumbers as 'dressing the suite' – it means installing the taps, wastes and, in the case of the bath, the cradle frame or feet. It will also include installing float valves, overflows, siphons or flushing valves and the handle assembly to the WC cistern.

Fitting the taps

Fitting the taps is a relatively simple process. Tap manufacturers' designs have to comply with regulations, so your installation will also satisfy the regulations. Remove just enough of the protective coating to make sure a clean joint can be achieved between the fittings and the surface of the appliance. Also, double check to make sure the surface is clean and dry. You will not have to use any other jointing compound, the washers supplied by the manufacturer are all you need. There may be slight differences in the washer kits supplied by manufacturers, but the illustrations here show typical details.

Pillar tap to a basin

The grip washer goes on the underside of the tap. Sometimes a thicker washer is also supplied, which goes between the back nut and the underside of the basin. Tighten the back nut using a purpose-made basin wrench. You can test if it is tight enough by checking by hand that the tap does not move. Make sure both taps are pointing outwards in parallel; some plumbers turn the taps slightly inward.

Monobloc taps

These are often supplied with an 'O' ring, which should be fitted between the base of the tap and the basin. The rubber washer goes underneath as shown. The assembly is completed using a metal washer held in place by a fixing nut.

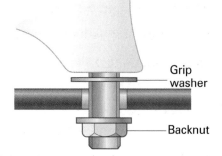

Figure 11.16 Seal to basin tap

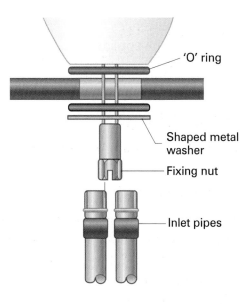

Figure 11.17 Monobloc tap connections

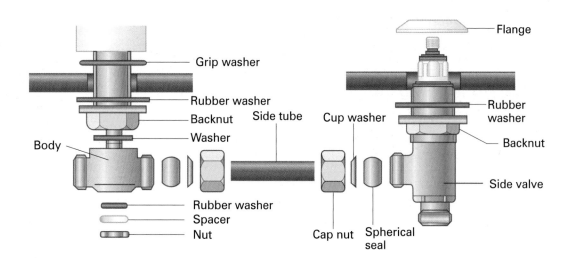

Figure 11.18 Three tap-hole basin

Tap and spout assembly – three-hole

Obviously, these are installed in three-hole basins.

- Remove all the parts that go above the basin
- Fit the spout with its washers, as in figure 11.18
- Loosely connect the tubes between the body and side valves
- Raise the assembly into position from below, making sure the sealing washers are in place
- Fit the body seal, washer and lock nut loosely, then fit the valve flanges on top of the basin, aligning them with the top of the side valve
- Carefully tighten all the components, and then finish by fitting the valve head gear, any shields, drive inserts and hand wheels.

Other appliances

The same principles apply to installing taps on other appliances. Remember to make sure that you use the washers supplied by the manufacturer.

Waste fittings

For washbasins, these include slotted waste fittings, moulded seals and pop-up wastes.

Slotted waste fittings

These can be sealed using non-setting plumbers' compound or silicone sealant. This is done by moulding a thin strip of compound around the flange of the waste and then placing the waste into the basin, making sure that the slots in the waste line up with the overflow slots of the basin. Clean off excess sealant or compound with a soft dry cloth.

Remember

For all tap installations make sure nuts are not cross-threaded

Safety tip

Never use linseed oil compounds on plastic wastes as they degrade the plastic causing it to fail; it can ruin the bath and shower tray as well!

Mould a thin strip of compound around the flange of the waste

Line up slots and apply more compound or sealant around the bottom of the basin and the thread of the waste

A plastic washer should be pressed into the compound, and the back nut is tightened to complete the joint

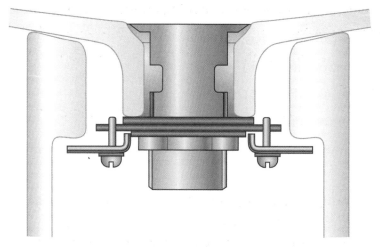

Figure 11.19 Basin/pedestal fixing bracket

In the case of some pedestal basins, a bracket has to be included to fix the basin to the pedestal.

This is done by placing the metal bracket between a rubber washer and the back nut, as shown above. The basin is secured using the two adjustable clips which are tightened against the pre-cast lugs in the pedestal.

Moulded seals

These are supplied for pop-up wastes and combined wastes and overflow, as used on baths, plastic washbasins and sinks. Installation details are shown in the illustration.

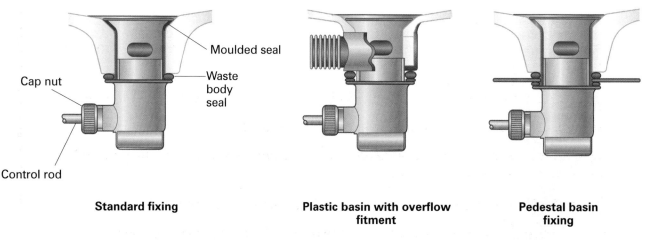

Standard fixing   **Plastic basin with overflow fitment**   **Pedestal basin fixing**

Figure 11.20 Moulded seals

The moulded seal fits between the waste flange and the appliance. For pedestal and plastic basins, a shaped seal and 'O' ring are used together to make the joint for the shaped metal connecting washer, or alternatively the metal pedestal-fixing bracket.

Where the appliance requires an overflow assembly, a rubber ring should be fitted between the overflow flange and the back of the appliance. When fitting this set-up:

- do not cross-thread the overflow grill and the fitting to the rear of the appliance

- do not over-tighten the fitting

- make sure you do not damage the flexible overflow tube when tightening the overflow fitting.

'Pop-up' wastes

Pop-up wastes are available for washbasins and baths. On a bath, the plug unit is raised by twisting the 'circular knob' anti-clockwise, which serves as both an overflow outlet and pop-up waste control. The plug unit is lowered by twisting the control knob clockwise, which allows the plug to drop into the waste by gravity.

Checklist for fixing a washbasin waste

- Fit the waste body into the basin outlet

- Remove cap nut and control rod

- Position the appropriate seals, then couple the pop-up body onto the waste tail thread

- Tighten with the control port correctly aligned with the back of the basin

- Slacken the lock nut on the stem of the pop-up plug and fit into the waste body

- Locate the control rod into the control port so that the end of the rod is in the hole at the end of the plug stem

- Assemble the cap nut

- Check the operation of the plug; it should be flush with the waste flange, and should then lift about 10 mm when operated

- To reduce the lift, the plug is turned clockwise, and to increase the lift anti-clockwise

- Remove the cap nut and control rod and take out the plug

- Tighten the lock nut against the strainer stem of the plug, and then refit the plug and control rod, and tighten the cap nut

- Fit the lift rod through the fitting and locate with the eyelet on the shackle

- Adjust so that the left knob is just clear of the fitting when the plug is fully open

- Tighten the screws on the shackle and check for smooth operation of the assembly.

WC suite

Our installation includes a close-coupled wash-down WC pan and siphonic action cistern.

Dressing the cistern

This will include the siphon, overflow fitting and float-operated valve. The cistern will have holes for either side inlet and outlet, or bottom inlet and outlet. Most cistern components will be plastic, and should be installed using the washers provided.

The shaped rubber sealing for the siphon, and flat rubber washers for the float valve and overflow union, should go inside the cistern.

Fit the siphon first, making sure that the metal bracket for the close-coupled connection is in place, and that a rubber washer has been placed between the bracket and the underside of the cistern. The back nut completes the assembly. When tightening the back nut, make sure the bracket is kept in the correct position for the bolts and locating holes in the pan.

Fit the float valve and overflow.

Assemble the lever and linkage mechanism and check that it works correctly before installation.

Fit the cistern to the pan.

Before securing the cistern to the pan, make sure the soft foam-rubber close-coupling washer is fitted over the siphon nut. Then carefully place in position over the pan. Locate the fixing bolts into the holes in the pan, and tighten the cistern to the pan using the rubber and metal washers and the bolts supplied. Test the joints between the pan and soil pipe. Make the overflow and cold water connections.

Baths

The cradle frame or brackets and legs are usually fitted before the taps and waste, but as part of the dressing process. Bracket and feet assemblies will vary depending on the manufacture, but a typical assembly detail is shown in the illustration. This is sometimes called a cradle frame.

First, place the bath upside down on a dustsheet over a clean, flat surface. Each leg is positioned in the location spigots as shown in figure 11.21. They should be pushed in until the centre section is flush with the baseboard, making sure the sides of the brackets are plumb. Then the centre leg and bracket are located in the centre of the baseboard.

All the brackets are screwed in position using the wood screws provided.

You can then screw the brackets to the location spigots using self-tapping screws.

Safety tip

When tightening any back nuts:

- do not over-tighten
- do not cross-thread.

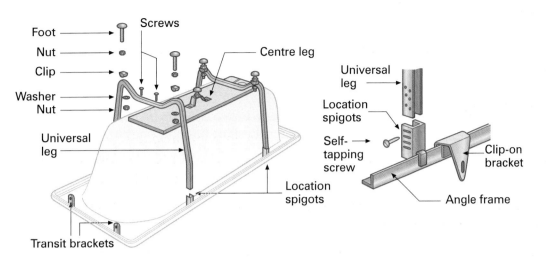

Figure 11.21 Bath fixings

The feet are assembled into the brackets. Stand the bath on its edge and fit the clip-on wall brackets and handles. The taps and waste are installed as described previously.

It is handy at this stage, to adjust the feet while the bath is on its edge. This can be done by using a tape and getting the feet to an approximate measurement required and then 'fine tuning' the feet to the exact measurement using a straight edge and tape. Final adjustments can be carried out with the bath in place.

Other appliances

The waste requirements for shower trays are straightforward and are fitted using the washers supplied with $1\frac{1}{2}''$ waste fitting. On fire-clay trays, non-setting plumbing compound is applied in the same way as with the standard slotted washbasin waste installation.

The waste fitting for a bidet is a $1\frac{1}{4}''$ threaded unit, either slotted or solid depending on whether the bidet is fitted to an overflow; pop-up wastes are also available. Taps can either be pillar or monobloc for over-the-rim applications. A typical arrangement is illustrated for a douche-type (under-rim) bidet.

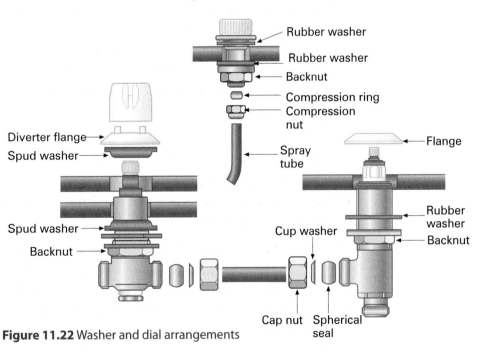

Figure 11.22 Washer and dial arrangements

The installation instructions are similar to those of the tap-and-spout assembly of the washbasin outlined above.

The installation can now begin

There are no set rules for the sequence used to install the appliances. Plumbers will work out the best sequence based on the position of the appliances and the size and layout of the bathroom. They will base their decision on what they think will be the easiest and quickest way.

On jobs where you are replacing a suite and the dwelling has only one WC, the first appliance you install will be the WC suite, to keep any inconvenience to a minimum.

It is not very often that you will install a bath and shower tray in the same room, unless it is a large property. Most baths have shower units installed over the appliance, so a shower tray may be used in an en-suite room, particularly where space is tight, or the occupier has difficulty in using a bath due to age or illness.

Generally speaking it is preferable to install the bath first.

Installing the bath

- Place the bath on its feet and in position
- Check the measurement from the top edge of the bath and or bath panel to the floor. It should be the manufacturer's recommended distance
- Check the bath is level across its length and width. Adjustment to level and height can be made via the feet
- Once the bath is level and at the correct height, tighten the locking nut on each foot
- Mark the position of the fixing holes for the feet and the wall fixings. If the wall is plastered, and you have been instructed to let the bath into the plaster, mark it off for the full length of the bath. Take out the bath
- If the floor is wood, drill pilot holes to receive the wood screws for the feet. This will make fixing much easier. If the floor is solid, drill and plug the floor. Drill holes for the wall fixings
- Chase out the plaster for the wall brackets and, if letting the bath into the plaster, chase out the plaster where you marked the line of the bath: this enables the bath to be let into the wall and provides a really good watertight seal once tiled
- Refit the bath and screw the wall and floor brackets into position. If fixing to timber floors, make sure the length of the screws will not penetrate the underside of the floor
- Service and waste connections are then made. It is best to pre-fabricate the tap connections if using soldered fittings, as this avoids using the blowtorch under the bath. The trap is connected and waste pipe extended to the outside (if there is an external waste system)
- If the bath is fixed flush to the plaster, fill the bath to one-third full, so that the weight causes the bath to settle, and then seal the joint between bath and wall with silicone sealant – after removing the protective film from the bath's surface.

Remember

Bath panels should only be fitted once testing has been completed

The method of fixing bath panels varies depending on the manufacturer, so their fixing instructions should be followed, but generally speaking the following should serve as a guide.

- Check the measurement from the underside of the roll edge of the bath and the floor. Check for any adjustments needed in the event of the floor being out of level

- If the panel has to be cut, firmly support the panel and cut it using a fine-tooth saw, and finish off with a file

- In some cases, panel support frames or bath leg clips are supplied, and these should be fitted

- Use a level or plumb line to mark out for a fixing batten at floor level; once you have made an allowance for any 'kicking space' on the panel, mark the position for the batten. If self-adhesive pads are to be used, an allowance for their thickness should also be made for the position of the batten (four pads should be sufficient)

- Alternatively screws can be used. Mirror screws provide an attractive finish, and some manufacturers provide screw-head caps that match the colour of the bath panel. Pre-drill the panel to take the screws. The panel can then be positioned by inserting the top edge of the panel between the bath roll and the timber frame of the bath, and secured at the base with either the self-adhesive pads or the screws.

Installing the washbasin

Our sample bathroom specification includes a pedestal basin. The taps and waste are in, and during dressing, the metal bracket that will hold the basin to the pedestal has also been fitted.

- Loosely attach the clamps using the pins and washers. Do not tighten them at this stage

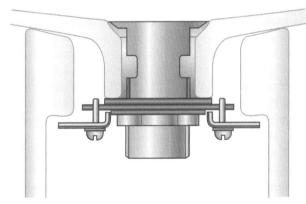

Figure 11.23 Basin fixing bracket

Did you know?

It is often a good idea with pedestal installations to fit the trap to the waste fitting before the basin is secured to the pedestal; this is easier than when the basin is on the pedestal

- Run sealing compound along the top edge of the pedestal; this will provide a better finish to the front

- Place the basin on the pedestal. Adjust the clamps so that they rest against the pre-cast shoulders of the pedestal

- Check the basin for level, make any final adjustments; carefully tighten the pins and then clean off any excess sealing compound.

The basin in this example has moulded screw holes at the back edges of the bowl for fixing it to the wall:

- Offer the basin up to the wall; make sure it is level and in the correct position (you might have to pack the pedestal to achieve level). Make a mark on the wall through the holes

- Remove the basin and pedestal and place them where they will not get damaged, preferably laying them down on a dust sheet

- The moulded holes in the basin are angled. You need to estimate this angle and drill the hole accordingly. If you drill it level, as you screw the basin to the wall the edge of the basin will crack off adjacent to the hole!

- Reposition the basin and pedestal and screw back to the wall using brass or alloy screws and soft washers. This will make the basin easier to remove for any maintenance work or replacement

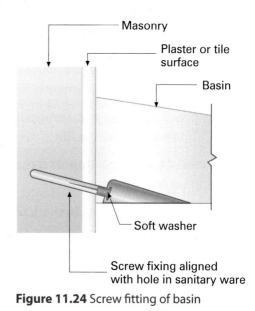

Figure 11.24 Screw fitting of basin

- Some pedestals are supplied with fixing holes in the base. If so, the floor should be marked at the same time as the basin. On wooden floors, pilot holes should be drilled before refitting. On solid floors, plugs are placed in the drill holes prior to repositioning the pedestal

- Some basins are supplied without fixing holes at the back of the bowl, but have moulding holes on the underside edge of the bowl. Brackets are available, such as the one in figure 11.23

- Fit the bracket to the basin and tighten so that it holds firmly in the hole

- Offer the basin to the wall; then adjust the brackets so that they are flush with the wall. Tighten the fixing bolt into the basin, and then mark the position of the bracket holes

- The pipework installation can be prefabricated so that it fits neatly behind the basin and the pedestal. Make sure you have fitted fibre washers before tightening up tap connections for connection to pillar taps. Monobloc taps are usually supplied with compression fittings. Mark off the position of the joints between the pipework stabs and prefabricated pipe at the same time as you make the fixing holes.

Installing the WC suite

All the components have been installed in the cistern during dressing, and the cistern has been fixed to the WC pan. The installation process is very similar to the washbasin:

- Offer the cistern to the wall, check the level of the cistern and pan and mark the holes through the back of the cistern. At the same time, mark the fixing holes through the base of the pan. If a soil pipe is already in, either remove the pan connector or ease the pan outlet into the pan connector

Remember

Do not fully tighten at this stage because you will need to adjust the bracket

Remember

Don't forget to use pilot holes on wooden floors

- Carefully take out the close-coupled suite, drill and plug the holes
- Refit and screw back the cistern; screw down the WC pan. Again, with the cistern use soft washers, and brass or alloy screws. Some manufacturers supply screws with caps that match the colour of the suite
- Some plumbers, once they have positioned the pan, drill through the fixing holes rather than removing the close-coupled suite to do the drilling. This is acceptable, but great care should be taken not to damage the suite with the drill, and to make sure the plug is fully located in the fixing holes
- We have assumed that the hole for the overflow has already been prepared. If not, this should be marked out and drilled before the suite is fixed
- Most float valves are plastic, so, when connecting the pipework with capillary fittings, do not solder the joint when it is connected to the plastic thread because it will melt and distort
- Make sure a fibre washer is in place before tightening the tap connector
- Fit the overflow pipe, which is usually made from plastic.

Safety tip

Close-coupled suites are heavy. Seek assistance if possible when handling the suite

Remember

Do not over-tighten or you will split the plastic. Watch out for cross-threading, too

Shower tray and enclosures

The waste is fitted to the tray during dressing. It will help with future cleaning and maintenance if a removable waste is fitted.

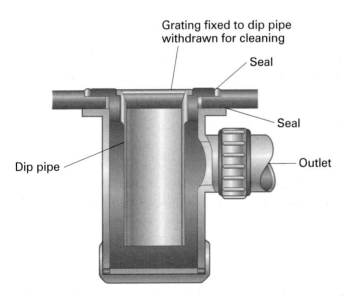

Grating fixed to dip pipe withdrawn for cleaning

Seal

Seal

Dip pipe

Outlet

Figure 11.25 Typical shower tray trap

The most important thing to ensure when fitting the tray is that it is level in all directions. If it is not, you will face big problems when fitting the enclosure.

Fixing to timber floors

- First, check whether the trap and waste pipework can be accessed under the floor. If not, you may have to raise the tray in a similar way to an installation on a solid floor

- If the pipework can be accessed, make sure the floor is clean and dry and coat the underside of the tray with a recommended sealant; alternatively the tray can be bedded on the tile cement

- Position the tray and press firmly into place. Make sure the tray is level in all directions

- Once the bed is dry or cured, the tray edges can be sealed to the adjoining walls.

Fixing to solid floors

On a solid-floor installation, it is likely that you will have to raise the shower tray to gain access to the trap and waste pipe. This is done by building a subframe out of external grade plywood and timber battens. The thickness of the battens will be governed by the amount of clearance you need, but they should be placed at 250 mm intervals. Once you have fixed the subframe, install the tray as you would for a timber floor.

Fitting enclosures

Shower enclosures come in all shapes and sizes, and the best advice here is to follow the manufacturer's instructions carefully. However, here are a few pointers:

- Check the walls are plumb. Most manufacturers' products make allowances for the walls being slightly out of plumb. On older properties, if the wall is seriously out of plumb, some plastering work may be required before you start

- It is almost certain that the frame of the enclosure will be fixed to a tiled surface, and this will involve drilling the tile

- Using the manufacturer's measurements for the frame, mark out for the fixing holes. The best way is to mark the wall the full length of the frame, making sure it is plumb. Then offer the frame to the line and mark out for the fixing holes

- The remainder of the process will involve fitting the panels and the door. Make sure the doors open and close correctly. The installation should be completed by fitting any sealing strips supplied with the kit, and applying a recommended sealant as per the manufacturer's instructions.

Installing the bidet

There is little that is different about the installation of the bidet, other than that it is screwed to the floor using brass or alloy screws. Marking and connection to supply pipework follows a similar pattern to the other appliances.

That completes the installation of our suite.

Did you know?

Acrylic trays often have panels to allow access to the underneath; they do not need to be raised

---

Safety tip

To avoid the drill sliding over the smooth surface of the tile and causing damage (and possible injury to you), stick a piece of clear tape on the hole markings. This will provide a key for the drill bit as you start to drill the hole. You should manually rotate the chuck to break through the glaze and start the drill without hammer action

---

Other types of appliances

Installing a washbasin using brackets

There is a range of brackets used for fixing washbasins – mainly in industrial or commercial situations.

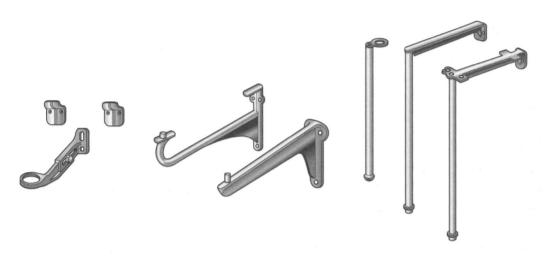

When selecting a bracket for the job, you should:

- make sure the bracket is the correct pattern for the basin
- be confident that it will support the weight of the basin.

The normal fixing height of a basin is 800 mm from the floor to the front rim.

It is a good idea when marking out for the brackets to:

- First, mark the centre line on the wall where the basin is going to be fixed
- Place the basin on the floor with its rim 800 mm from a wall
- Put the brackets in position on the basin and measure the distance to the fixing holes from the wall, and at the same time, the centres of the fixing holes between the two brackets. Halve that distance, and that will be the measurement from the centre line marked on the wall to the centre of each bracket
- Transfer the measurements to the wall where the brackets are to be fixed. Position the brackets on the centre lines and mark the holes
- Drill and fix the brackets, then place the basin in position and check height and level.

Alternatively, if you are working with a mate, mark out the centre line for the basin and the rim height. Then, one person holds the basin in position, checking that it is in line and level. The other person puts the brackets in position and marks the holes.

Countertop basins

Usually, countertop basins are supplied with a cutting template. You must follow this carefully when cutting out the hole. Use a jig-saw blade with fine teeth and a downwards cutting action, as this will reduce any damage to the finish of the worktop.

Once you have cut the hole, seal the exposed edges with a waterproof varnish to prevent any water/moisture damage to the chipboard.

The joint between the lips of the bowl must be sealed using a recommended sealant, and the bowl is then secured using the brackets supplied with the basin.

The edge of the hole for an under-countertop basin is finished off with a strip of laminate in the same material and colour as the countertop.

Wall urinals

These are usually installed at a height of 610 mm from floor level to the front lip of the bowl. Quite often, when a number are installed, one or possibly two will be installed at a height of about 510 mm for children or smaller adults. Dress the urinal bowls first, fitting the inlet spreader and waste.

Urinals are fixed on the type of bracket shown in the illustration, and the measurements can be set out in a similar way to those of washbasin brackets.

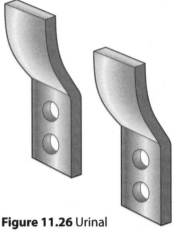

Figure 11.26 Urinal bowl fixing brackets

Slab urinal installation

Slab urinals are found mostly in public buildings. In the main they are manufactured in fire clay, and can be supplied to any length on one or more walls, either in one piece or smaller slabs. The floor channel is manufactured so that the internal surface is laid to a fall. This means the actual channel block can be installed level.

Specialist sanitary appliances

WC macerator unit

The macerator unit is installed behind the WC pan. It collects waste from the WC and reduces it to a 'liquid' state so that it can be pumped via a 22 mm internal bore pipe into a discharge stack. There are a few things to remember about macerators.

- A drain-off valve should be fitted at the base of a vertical rise

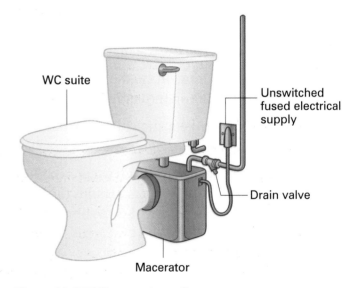

WC suite

Unswitched fused electrical supply

Drain valve

Macerator

Figure 11.27 WC macerator unit

- The macerator can only be fitted when there is another conventional WC in the dwelling
- The electrical supply to the unit must be via a fused spur connection
- The unit is capable of discharging its contents to a maximum of 4 m vertically upwards and 50 m horizontally but not at the same time
- Horizontal pipework must be laid to a fall of 6 mm per metre minimum
- Waste pipework is typically of 22 mm diameter.

Food waste-disposal unit

These units are usually installed under kitchen sinks to dispose of refuse such as vegetable peelings, tea bags, left-over food etc. They require a larger waste outlet than normal, and manufacturers produce sink tops to suit this. The unit should:

- be connected to the electrical supply via a fused spur outlet
- be connected to a waste trap – this must be a tubular P trap as for all kitchen sinks
- be supplied with a special tool to free the grinding blades should they become blocked (keep those fingers clear!)
- have some form of cut-out device that will turn the unit off should it jam.

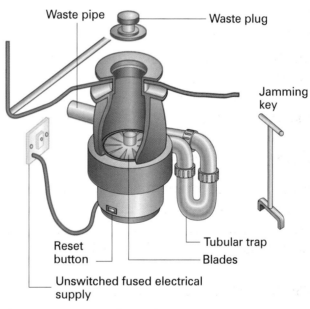

Figure 11.28 Waste-disposal unit

Traps

By the end of this section you should be able to:

- state the purpose of a trap
- state the diameter and depth of the traps used on above ground discharge systems
- identify the various types of traps used on above ground discharge systems
- explain the reason for the natural loss of trap seal.

Traps are used on above ground discharge system pipework and appliances. They are mainly manufactured in plastic (polypropylene to BS3943), although they are also available in brass for use on copper pipework, where a more robust installation is required, and can be chromium plated to provide a pleasing appearance.

The purpose of a trap is to retain a 'plug' of water to prevent foul air from the sanitation and drainage pipework entering the room. The depth of this 'plug' of water will depend on where it is to be used, but BS EN 12056 Part 2 says that, 'Any trap connected to a discharge pipe of 50 mm or less discharging into a main stack should have a seal of 75 mm.'

Where a trap is 50 mm or above, a trap seal of 50 mm is required. This is because the size of the pipe means it is unlikely to discharge at full bore. This is one of two causes of loss of trap seal.

A selection of plastic traps

If the discharge pipe from the trap runs into a gully or hopper head, a seal of 38 mm is allowed. This is because the gap between the gully and the pipe provides an air break should the trap lose its seal, thus no smells can enter the building. The size of the trap is governed by the size of the waste it is connected to. This table gives the minimum size of the waste fitting and trap.

Find out

Why do you think it is necessary to install a trap?

Type of appliance	Waste fitting size (in)	Discharge pipe and trap size (mm)
Sinks, showers, baths washing machines and trough urinals	$1\frac{1}{2}$	40
Washbasins, bidets, drinking fountains and urinals	$1\frac{1}{4}$	32
Stall and slab urinals	$2\frac{1}{2}$	65

Figure 11.29 Waste fitting and trap sizes

The depth of the trap seal is measured as shown in figure 11.30:

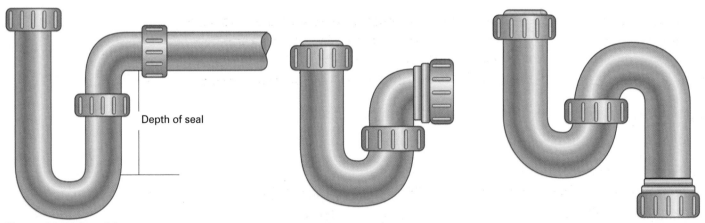

Depth of seal

Figure 11.30 P and S traps

You can see why they are called P or S traps from their shape (note that they are turned here to show how they fit into the appliance). They can be obtained in tubular design, or with a joint connection which allows a few more options when installing pipework and fittings. The traps can also include a cleaning eye for cleaning and maintenance.

P traps are often used where the waste pipe is installed directly through a wall from the appliance and into a drain or directly into a stack. The S trap is used where the pipe has to go vertically from the trap through a floor or into another horizontal waste pipe from another appliance. P traps (and bottle traps) can be converted to S traps using swivel elbows like the one illustrated. S traps can be a real problem as the fall of pipe from the basin is generally too steep, which can lead to trap seal loss and obnoxious smells entering the building.

Figure 11.31 Swivel elbow

Bottle traps

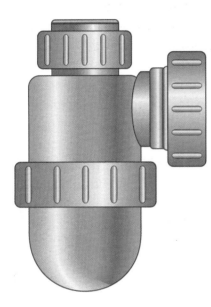

Figure 11.32 Bottle trap

Figure 11.33 Cross-section of bottle trap

Inlet

Outlet

Dip pipe

Effective depth of seal

Base unscrews

These are often used because of their neat appearance and they are easier to install in tight situations. This illustration shows what the trap looks like inside and how the depth of seal is measured. They should be avoided on sinks, where they can be prone to blockage due to food solid deposits collecting in the bottom of the trap.

Tubular swivel traps

These are particularly useful on appliance replacement jobs, as they allow more options when connecting to an existing waste pipe without using extra fittings or altering the pipework. On new jobs they are often used on sinks with multiple bowls, again because of their multi-position options.

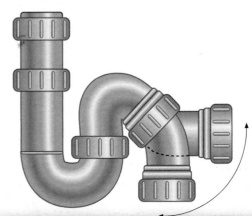

Figure 11.34 Tubular swivel trap

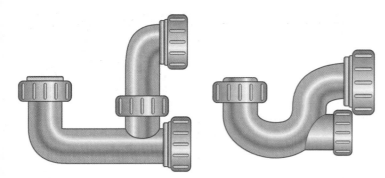

Figure 11.35 Low-level bath taps

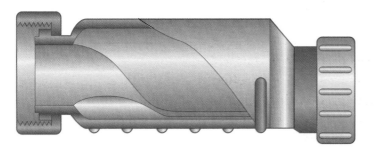

Figure 11.36 Straight-through trap valve

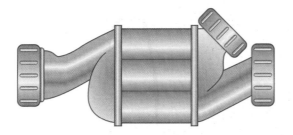

Figure 11.37 Straight-through trap

Low-level bath traps

These are designed so that they can fit into tight spaces under baths and shower trays. The seal on these will only be 38 mm, which means they cannot be connected directly into the soil and vent stack.

Straight-through trap

These are used as an alternative to an 'S' trap where space is limited. They are also easier to hide behind pedestal basins. The main problem with this design is the two tight bends which slow down the flow of water figure 11.37.

An alternative valve works on the simple principle of using an internal membrane as a seal, figure 11.36. The membrane allows water to flow through it when the water is released, then closes to prevent foul air from entering the building. The valve can be used on systems meeting BSEN 12056 – Part 2. It is ideal for fitting behind pedestals and under baths and showers, and is supplied with a range of adaptors so that it can be used in various situations. The valve has the potential to revolutionise the installation of above ground systems, the requirements of which will be covered in the next section. System-design procedures for the straight-through trap will be covered fully at Level 3.

Running traps

You might see these used in public toilets where one running trap is used for a range of untrapped washbasins. On domestic installations it could be used where a P or S trap arrangement is not possible. Running traps are sometimes used with a washing machine waste outlet or for dishwashers, although specialist traps are available for these appliances.

Figure 11.38 Running traps

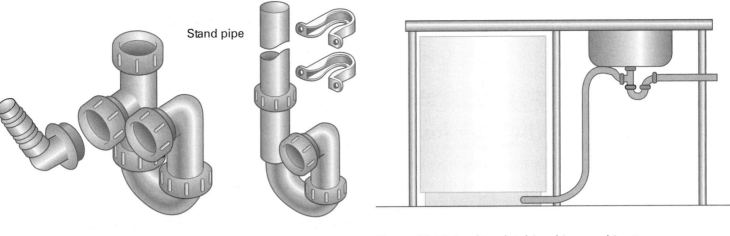

Figure 11.39 Combined sink trap fittings

Figure 11.40 Combined sink/washing-machine trap

Resealing and anti-siphon traps

If an above ground discharge system is designed and installed correctly, loss of trap seal should be prevented. We will look at some of the reasons for loss of trap seals next, but these particular traps are designed to prevent seal loss due to the effects of siphonage (the Hepworth valve provides a very similar function although it is less complicated). These types of traps could be specified or fitted in situations where normal installation requirements cannot be met.

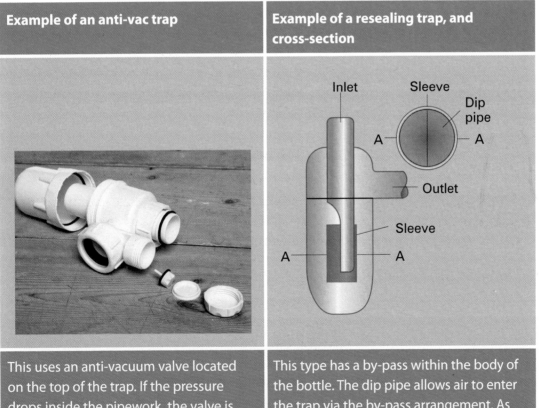

Example of an anti-vac trap	Example of a resealing trap, and cross-section
This uses an anti-vacuum valve located on the top of the trap. If the pressure drops inside the pipework, the valve is activated, allowing air to enter the system, to equalise the pressure	This type has a by-pass within the body of the bottle. The dip pipe allows air to enter the trap via the by-pass arrangement. As the seal is lost due to siphonage, air is allowed into the trap, breaking the siphonic effect

Figure 11.41 Resealing and anti-siphon traps

Access

Whatever type of trap is fitted, it is important that you can get to the trap for cleaning. As you have seen, some traps have cleaning eyes; others can be split at their swivel joints to enable a section of the trap to be removed.

Why traps lose their seal

The main reason for traps losing their seal is bad pipework design and poor installation, which will be dealt with in the next section. However, a trap may lose its seal through natural causes:

Cause	Effect
Capillary action Loss of seal Water is drawn along the strands of the cloth and down the discharge pipe	Only occurs in 'S' trap arrangements, and is not a regular occurrence. If a thread of material becomes lodged as shown in the illustration, water could be drawn from the trap by the effect of capillary attraction.
Momentum	The seal of a trap is removed by force of water. This can happen if you pour a bucket of water into a sink or toilet.
'Wavering'	Wind pressure across the top of the soil and vent pipe, particularly in exposed locations, causes the water in the trap to produce a wave movement and wash over the weir of the trap. This does not happen often, and can be avoided by fitting a 90° bend or a cowl to the top of the vent pipe.
Evaporation	The most common form of natural seal loss. Occurs in very warm ,dry weather when the water in the trap simply evaporates. It is unlikely to occur in traps with a 75 mm seal.

Figure 11.42 Why traps lose their seal

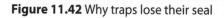

Sanitary systems

By the end of this section you should be able to:

- describe the pipework installation requirements of a:
  - primary ventilated stack system
  - ventilated discharge branch system
  - secondary modified ventilated stack system
  - stub stack system.

- explain the problems caused by poor design or installation.

So far we have looked at the sanitary ware used in above ground discharge installations and the various types of traps used to prevent foul air entering the building. We now turn to the pipework systems that connect the appliances to the drainage system.

At one time – and you may still come across these on maintenance jobs – pipework systems were two-pipe: the pipework from the WC was separate to that of the washbasin, bath and sink. The wastewater only joined together once it entered the drain, as shown.

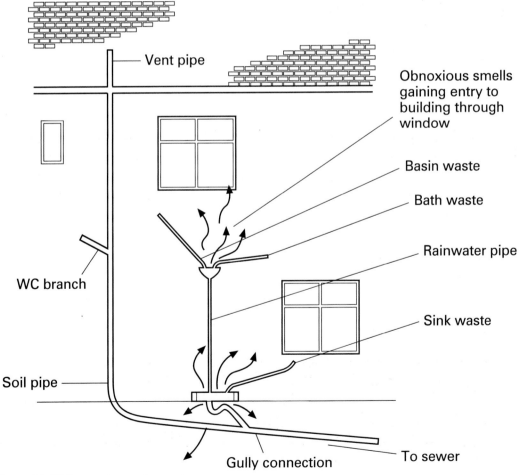

Figure 11.43 External two-pipe system

Today, all pipework systems used on domestic dwellings are one-pipe systems, of which there are four types:

- primary ventilated stack system
- ventilated discharge branch system
- secondary modified ventilated stack system
- stub stack system.

Soil and vent pipe systems can be installed either inside or outside the building. Whichever system is used, the design and installation must comply with Building Regulations Part H1. The general requirements of Part H1 are that the foul water system:

- conveys the flow of water to a foul water outfall (a foul or combined sewer, a cesspool, septic tank or settlement tank)
- minimises the risk of blockage or leakage
- prevents foul air from the drainage system from entering a building under working conditions
- is ventilated
- is accessible for cleaning blockages.

The other important document regarding sanitary pipework is BSEN 12056 Part 2 2000. This gives guidance on the minimum standards of work and materials to be used.

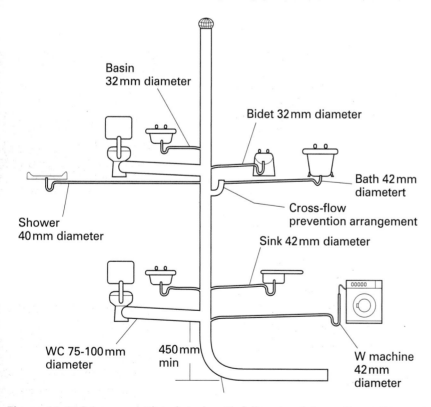

Figure 11.44 Primary ventilated stack, with full range of domestic appliances

Basin 32 mm diameter

Bidet 32 mm diameter

Bath 42 mm diametert

Cross-flow prevention arrangement

Shower 40 mm diameter

Sink 42 mm diameter

WC 75-100 mm diameter

450 mm min

W machine 42 mm diameter

The primary ventilated stack system is the most commonly used. The ventilated discharge branch and secondary modified ventilated stack systems should be avoided wherever possible.

The main aim of the design principles we are going to cover is preventing the removal of the trap seal and allowing obnoxious smells to enter the building.

Primary ventilated stack system

Sometimes called the single-stack system, this type of system is often specified for domestic dwellings because it costs less in terms of materials and installation time, as it does not need a separate ventilating pipe like other one-pipe systems. This is good for the business and the environment.

Building Regulations Part H1 sets out a number of rules about the design of this system. The rules governing pipe diameter and the minimum depth from the lowest connection above the invert of the drain of 450 mm are shown in figure 11.44.

In addition, there are limitations to the maximum lengths of the branch connections and their gradients. These are shown in the table.

	Pipe size	Maximum length	Slope
Basin	32 mm	1.7 m	18–20 mm fall per metre run
Bath	40 mm	3.0 m	19–90 mm/m
Shower	50 mm*	4.0 m	18–90 mm/m
WC	100 mm	6.0 m	18 mm/m min

Figure 11.45   *Normally 40 mm

In this system it is necessary to have the appliances grouped closely together. There is some flexibility, however: for example, if you did install a shower with a 50 mm waste it could be located up to 4 m away from the stack as opposed to 3 m if using 40 mm pipe. The size of the branch pipes should always be at least the same diameter as the trap.

Branch connections

The location of a branch pipe in a stack should not cause cross-flow into another branch pipe. **Cross-flow** can be prevented by working to the details shown in the illustration – again the problem we are trying to prevent is trap seal loss.

It is, however, permissible to have connections from two WCs in opposing positions.

You might find on some installations that it is easier to run the kitchen-sink waste pipe into a gully rather than pipe to the stack. This is permissible as long as the pipe end finishes between the grating or sealing plate and the top of the water seal.

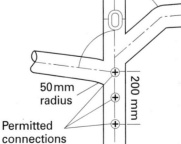

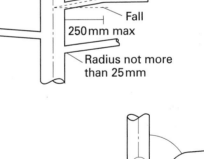

Figure 11.46 Cross-flow

Where the details described earlier cannot be met, separate ventilation will be needed; examples of system types follow. On some larger domestic properties this is not always possible. The alternative is to install separate ventilation pipework. This can be done in two ways:

Definition

Cross-flow – occurs when two branches are located opposite each other

Remember

The use of resealing traps is an option when separate ventilation is required – you covered this in the trap session

1　Ventilating each appliance into a second stack – the ventilated discharge branch system

2　Directly ventilating the waste stack – secondary ventilated stack system

The trap sizes and seals apply as described in the session on traps. The illustration shows what this looks like.

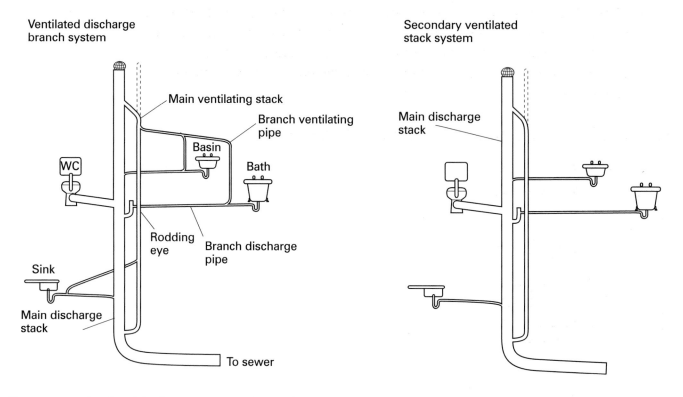

Figure 11.47 Ventilated discharge system and secondary stack system

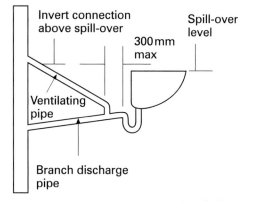

Figure 11.48 Branch ventilating pipes

Branch ventilating pipes

There are some factors to consider:

- The branch vent pipe must not be connected to the discharge stack below the spill-over level of the highest fitting served

- The minimum size of vent pipe to a single appliance should be 25 mm. If it is longer than a 15 m run or serves more than one appliance, then it must be 32 mm minimum

- The main venting stack should be at least 75 mm. This also applies to the 'dry part' of the primary vented stack system.

Stub stacks

These reduce the amount of ventilating pipework on a ventilated discharge stack. They also mean that on internal stacks the need for roof weatherings is avoided. The highest waste connection allowed is 2 m above the invert of the drain, and 1.5 m for the WC. The length of the branch drain from the stack is 6 m for a single appliance connection and 12 m if there is more than one.

These dimensions can be exceeded when an air admittance valve is used to permit air to enter the system. The valve, which operates on negative pressure (the open valve lets air in) and positive pressure (the closed valve contains the smells), can be located in the roof space or the pipe-boxing arrangement to the WC, subject to certain requirements.

General discharge stack requirements

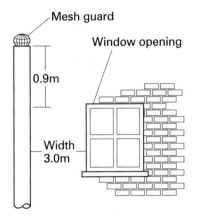

Figure 11.49 An external stack must be terminated as shown

Figure 11.50 A terminal should be fitted to prevent the possibility of birds nesting

Figure 11.51 A vent cowl could be fitted where the stack is sited in exposed windy conditions

Access

All stacks should have access for cleaning and clearing blockages. Rodding points and access fittings should be placed to give access to any length of pipe that cannot be reached from any other part of the system. All systems pipework should be easy to get to in case of repair.

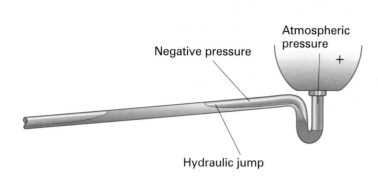

Figure 11.52 Self-siphonage

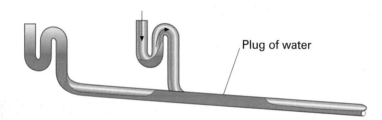

Figure 11.53 Induced siphonage

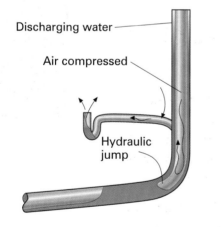

Figure 11.54 Compression

Trap seal loss

In the section on traps we looked at some of the natural causes of trap seal loss. Trap seals can also be affected by poor practice and by not following the regulations – the detail of which we have just covered.

There are three main causes of loss of trap seal caused by poor practice:

1 self-siphonage

2 induced siphonage

3 compression.

Self-siphonage

This is most common in washbasins as their shape allows water to escape quickly. As the water discharges, a plug of water is formed creating a partial vacuum (negative pressure) in the pipe between the water plug and the basin. This is enough to siphon the water out of the trap. Ensuring that the waste pipe is within the lengths allowed and to the correct fall, or that it is ventilated, should prevent self-siphonage. Resealing traps would also avoid this problem.

Induced siphonage

Caused by discharge of water from an appliance that is connected to the same waste pipe as other appliances. As the water plug flows past the second appliance connection, negative pressure is created between the pipe and appliance which siphons the water out of the trap. This arrangement is not acceptable on a primary ventilated stack. In figure 11.53, fitting a branch ventilating pipe between the two traps would solve the problem (as would fitting a resealing trap).

Compression

As the water is discharged from an appliance into the main stack (usually a WC at first-floor level), it compresses at the base of the stack, causing back pressure which can be enough to force the water out of the trap, thus losing the seal.

The use of large-radius bends and minimum 450 mm length between the invert of the drain and lowest branch pipe are defined in the regulations in order to prevent this.

Pipework installation, access and materials

By the end of this section you should be able to:

- explain how to prepare for above ground drainage (AGD) pipework installations
- describe how to install and joint above ground drainage pipework, including:
  - types of materials used
  - jointing methods
  - range of fittings
  - fixing details
- describe how to connect to a below-ground drainage system.

This section overlaps with some of the previous work you have covered, such as materials and methods of jointing. Here, however, we will show how the methods and materials apply to above ground drainage systems. We've also covered preparing for installation in the common plumbing process section, so this will be a bit of revision.

Preparing to install above ground drainage system pipework

You may have to drill or cut your own pipework holes for external discharge pipework. If not, make sure the bricklayer knows in advance where they should be and when they should be cut. On some larger housing jobs, the holes may be left by the bricklayer as the brick/blockwork proceeds, and then made up afterwards.

The installation and jointing of AGDS pipework

Types of material

For domestic plumbing, plastic is generally the material used for AGD systems. For push-fit waste systems, polypropylene to BS5254 is used. For solvent-welded systems PVC and ABS plastics are used. These meet the requirements of BS5255. Compression fittings are available which are suitable for both types of pipe. Waste pipe is available in 32 mm, 40 mm and 50 mm diameters, in 3 m lengths.

The most common plastic for soil and vent pipes and fittings is UPVC to BS4514, which uses ring-sealed push-fit joints. Pipe diameters for domestic use include 82 mm and 110 mm (this is the outside diameter of the pipe), and they are available in lengths of 2.5 m, 3 m and 4 m, either as pipe ends or single-socket pipes.

Range of fittings

Waste systems

These come in three different types: push fit, solvent welded and compression. There is an overlap with the design of the fittings, so we will just illustrate the main features here.

Remember

You can refer back to the generic systems knowledge section in Chapter 5. The documentation relevant to the AGD pipework is Building Regulations H1 and BSEN 12056.

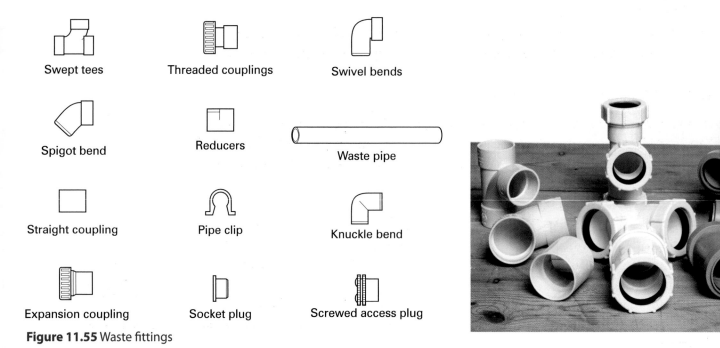

Swept tees

Threaded couplings

Swivel bends

Spigot bend

Reducers

Waste pipe

Straight coupling

Pipe clip

Knuckle bend

Expansion coupling

Socket plug

Screwed access plug

Figure 11.55 Waste fittings

Soil and vent systems

Below is a selection of soil and vent fittings.

Single socket

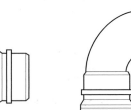

Double socket bend

Metal pipe clips

Plastic pipe clips

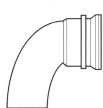

Double socket

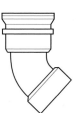

Single socket bend

Offset bend

Socket/spigot bend

Figure 11.56 Soil pipe fittings

Access fittings

We said in the last section that providing access to clean pipework and to clear any blockages is a requirement of the regulations. Access fittings are designed to allow this.

Joints to soil and vent pipes

There are a number of options here. As you have seen from the illustrations, some soil and vent fittings are supplied with a boss access already in place, ready to install the waste pipe. Others have blocked ends in the pipe which can be drilled out and fitted with a rubber boss adaptor.

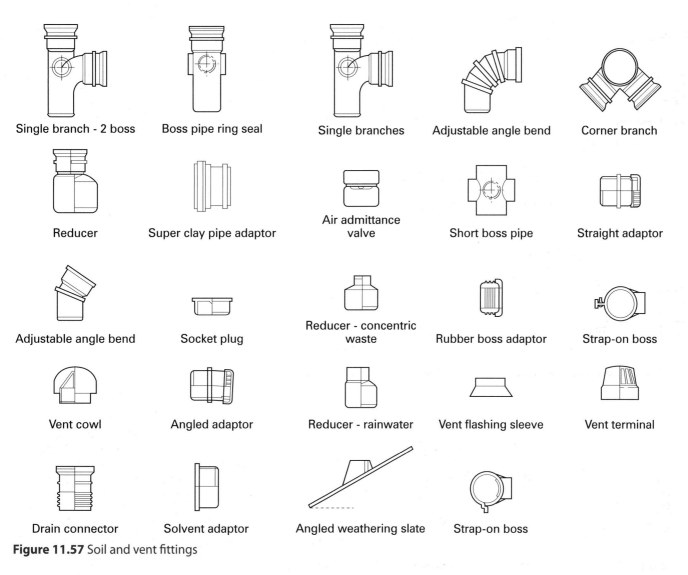

| Single branch - 2 boss | Boss pipe ring seal | Single branches | Adjustable angle bend | Corner branch |

| Reducer | Super clay pipe adaptor | Air admittance valve | Short boss pipe | Straight adaptor |

| Adjustable angle bend | Socket plug | Reducer - concentric waste | Rubber boss adaptor | Strap-on boss |

| Vent cowl | Angled adaptor | Reducer - rainwater | Vent flashing sleeve | Vent terminal |

| Drain connector | Solvent adaptor | Angled weathering slate | Strap-on boss |

Figure 11.57 Soil and vent fittings

Boss pipe fittings

Sometimes, these fixed connections may not suit the installation requirements. In these cases you can drill your own fixing position in the soil and vent pipe and then fit a strap boss. The best way of doing this is to use a hole saw. Make sure it is the correct diameter for the strap boss.

Once drilled, the strap boss is fixed using gap-filling cement.

Fixing details

Plastic waste pipes should be clipped at the spaces given in the table.

Pipe size	Horizontal	Vertical
32 mm	0.8 m	1.5–1.7 m
40 mm	0.9 m	1.8 m
50 mm	1.0 m	2.1 m

Figure 11.58 Waste-pipe sizes and clipping distances

Soil and vent pipes are normally fixed vertically every two metres.

Connections to drains

Plumbers very rarely work on below-ground drainage systems, other than on very large projects. This tends to be a job done by ground workers on large housing contracts.

However, you need to know the basics of the systems and how to connect the soil and vent pipe to the drain, because that *is* a plumber's job. Essentially, you must ensure that **foul drainage** cannot enter the surface water system.

This is because with certain types of system, **surface water** is discharged cheaply via separate pipework to streams etc. – we cannot afford foul-water drainage to be discharged to such pipework, as it could be a potential health hazard.

There is a simple rule here: always know what you are discharging into – then you can't go wrong.

There are three main types of drainage system:

1 combined system

2 separate system

3 partially separate system.

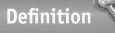

Definition

Foul drainage – anything discharged from a sanitary appliance – WC, bath, basin, sink etc.

Definition

Surface water – the term used for water collected via the rainwater system

svp = soil vent pipe
rwp = rainwater pipe
fwg = foulwater gully

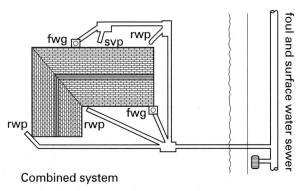
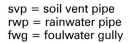

Combined system

Figure 11.59 Combined drainage system

The combined system

In this system, the water from the sanitary appliances (foul water) and the rainwater all go into one sewer.

The separate system

Here, the foul water runs into a separate sewer, and the rainwater likewise into a sewer for surface water.

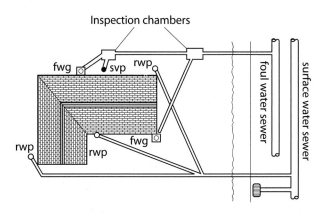

Figure 11.60 Separate drainage system

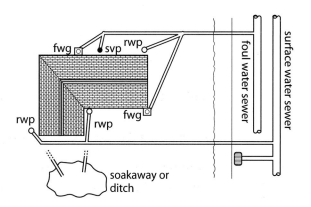

Figure 11.61 Partially separate system

Partially separate system

This system still uses two pipes, but some of the surface water is discharged into a watercourse, soakaway or drainage ditch.

The advantages and disadvantages of each system are shown in the table below.

System	Advantages	Disadvantages
Combined	• Cheap and easy to install • Gets a good flush-out during periods of heavy rain	• More costly to treat the water at the sewage works • At times of heavy rain, inadequately sized drains could overflow • All gullies have to be trapped
Separate	• No need for water treatment of surface water • No need for trapped gullies on surface water drains	• Danger of cross-connections – foul to surface water
Partially separate systems	• Greater flexibility with the system design	• Danger of cross-connection

Figure 11.62 Advantages and disadvantages of different drainage systems

Connecting the soil and vent pipe

On a new job, say on a large building contract, this information will be on the drawings and specification. When you go into the property, it is obvious because all the outlets to the drain will be in place for the soil and vent pipe or downstairs WC.

On a replacement job, the connection will be the same as the existing soil and vent pipe, so you will use that.

The soil and vent pipe are likely to be connected to a drainage pipe. Fittings are made to connect the two. Some drainpipe terminations, particularly on older systems, are finished with a collar. Drain connectors, such as the one illustrated, are inserted into the collar and a joint is made using a sealing compound. This is finished off with a sand and cement joint of a 1 to 4 mix.

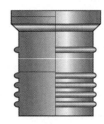
Figure 11.63 Drain connector

Soundness testing

By the end of this section you should be able to:

- describe the procedures for soundness testing of above ground discharge systems (AGDS)
- explain the charging procedures for newly installed AGDS or extensions to existing systems.

Sanitary appliances discharge large volumes of waste water, so it is important that systems are watertight. The test should be carried out in accordance with BSEN 12056, which you'll look at soon. You also need to know what to do to make sure people are aware that testing is going on, and when it is safe for them to use the appliances.

Soundness testing of AGDS

Figure 11.64 shows a test kit. Look at it and then read the testing checklist.

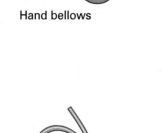
Hand bellows

Test nipple

Water gauge hose

Assembled plug

Hollow plug

Plug testing Y piece

Figure 11.64 Test kit

Testing checklist

- First, seal the system using hollow drain plugs. Fix a test nipple in the one to be used for the test. The bottom test plug can be inserted through the access cover at the base of the stack. If this proves awkward, a testing bag can be inserted and inflated in position, as shown in the illustration.

- Fill the traps with water, and cover the test plugs to make sure they are fully airtight

- The rubber hose, bellows and hand pump are connected to one end. Air is pumped into the system to give a water head of 38 mm. Once this is reached, the plug cock is turned off and the test continued for three minutes.

- Where a pressure drop is found, all the joints should be tested using leak detection fluid

- Smoke tests are also useful to locate leaks, but are not advisable because of the chance of affecting the plastic pipework and rubber seals.

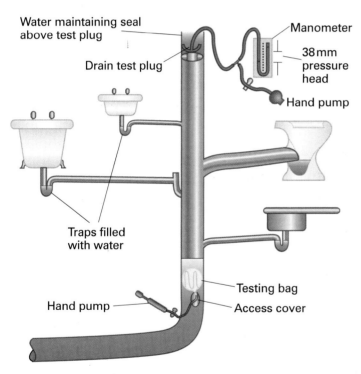

Figure 11.65 Different tests that can be done on a system

Air test in action

Charging procedures

Once the testing equipment has been removed, all the appliances should be filled to their overflow levels and water released; the WC should be flushed at the same time.

Once you have done that, check that the trap seals are not less than 25 mm. This is done as shown in the diagram.

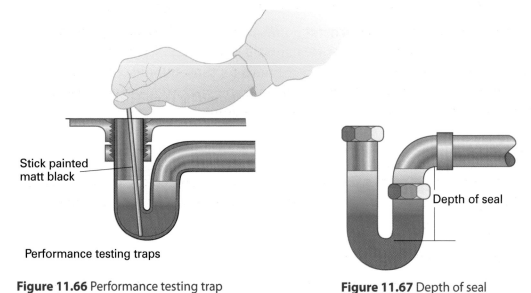

Stick painted matt black

Performance testing traps

Figure 11.66 Performance testing trap

Depth of seal

Figure 11.67 Depth of seal

Remember the depth of seal is measured as shown here.

On the job: Testing an installation

Jenni was asked to test and charge the installation shown here.

1 Label the diagram

2 How should she go about the job? Include the equipment she should use, and how she should make sure the trap seal is adequate.

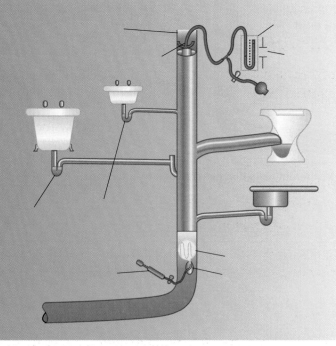

Rainwater and guttering

By the end of this section you should be able to:

- state the materials used for gutter and rainwater systems

- describe the various types of gutter and rainwater systems

- explain the fixing requirements for gutter and rainwater systems.

Gutters are used to collect rainwater that falls on the roof of a building. It flows down the rainwater pipe and into surface-water drains, combined drainage systems or soakaways. Life without gutters and rainwater pipes would see damage to gardens from water running off the roof, staining on brickwork and water getting into houses, leading to such problems as rising damp.

Materials

Most gutters and rainwater pipework systems installed on new domestic properties tend to be plastic. This is PVC to BSEN 9002 94. It has the advantages of being:

- light
- flexible
- cheap
- low-maintenance.

It is available in white, black, brown and various shades of grey. Because PVC gutters and rainwater pipes have a smooth internal surface, they provide a better flow rate than other materials.

Plastic has a high expansion rate, so it can be affected by the variations of winter and summer temperatures. Allowance for expansion and contraction is made in the fittings, so it is important that manufacturers' fitting and fixing instructions are followed.

Other materials

You are quite likely to come across cast-iron and asbestos cement guttering on maintenance jobs.

Cast iron

This is a very strong material, but it requires constant painting to prevent rusting. Leaking joints also take longer to repair/replace than their plastic counterparts, as it takes time to remove the gutter bolt fixings which tend to be corroded and difficult to unscrew. More often than not they have to be cut off using a hacksaw.

Cast iron may still be specified on jobs where its strength is needed, or on listed buildings where original materials have to be used.

Asbestos cement

Asbestos cement gutters were quite popular before the introduction of PVC. They are no longer available because of the dangers of working with asbestos.

Did you know?

Plastic gutters and rainwater pipes do not require painting, making them very low maintenance

Guttering fittings

If you are called to a maintenance job, you must follow the procedures as outlined in Chapter 2 on Health and Safety. Asbestos guttering should not be repaired, but removed and replaced with plastic – with all safety procedures being followed.

Gutter and rainwater systems

Gutters are specified by their shape in cross-section.

Standard half round

Square

Ogee

Figure 11.68 Types of guttering

Rainwater pipes are either square or round. Guttering is supplied in 2 m and 4 m lengths, and can be obtained in a range of sizes from 75 mm to 150 mm; 112 mm is in common use for domestic dwellings, with rainwater pipes in lengths of 2.5 m, 4 m and 5.5 m lengths. For most domestic jobs, 65 mm square and 68 mm external diameters are used.

Fitting rainwater systems

On new work, the gutter will be fixed to **fascia** boards using a fascia bracket. The fascia is the piece of wood (or plastic) that is fixed to the rafter, as shown in figure 11.69. A fascia bracket is also shown.

Union bracket

Angle

Running outlet

External stop end

Internal stop end

Cast iron ring

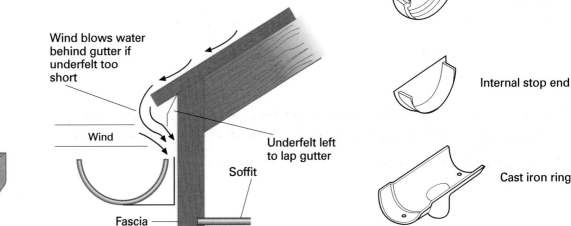

Wind blows water behind gutter if underfelt too short

Wind

Underfelt left to lap gutter

Soffit

Fascia

Fascia bracket

Figure 11.69 Fascia fixed to rafter

Figure 11.70 Fascia brackets

On some older properties, fascia boards may not be fitted. On these jobs, metal rafter brackets like the ones shown in the following illustrations can be used for metal and plastic gutters.

Step-by-step fitting guide for plastic guttering fixed to fascia boards

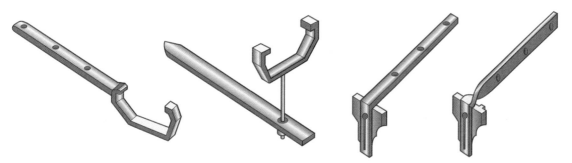

Figure 11.71 Gutter brackets

- The first job is to set out the gutter brackets. Gutters have to be laid to a fall to make sure they drain properly. The fall should be 1 in 600, which for a length of 6 m gutter would be 10 mm – this is hardly noticeable once the gutter is fixed in position.

- Once you have worked out the fall over the total length of the installation, the first bracket is fixed at the highest level on the run, and the last bracket is fixed to give the amount of fall required.

- A string line is fixed between the two brackets. This is used to position the rest of the brackets, which are usually spaced at 1 m centres, unless the manufacturer states otherwise.

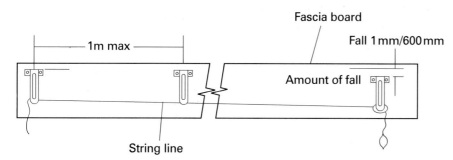

Figure 11.72 Positioning the brackets

- The gutter can then be positioned into the brackets. On the first length, depending on the roof shape, you would fit the stop end (or corner bracket) and union bracket.

- The next joint would be done in position, as shown on page 352.

- Make sure you allow for thermal movement. On a plastic gutter, a raised fixing mark is usually on the inside joint. If expansion marks are not on the gutter joints, you should allow 3 mm for every metre run.

- Once the gutter system is complete, you can start on the downpipe. The brackets are marked out using a plumb line. This is dropped from a masonry nail driven into the nearest mortar joint beneath the running outlet. The brackets are centred with the plumb line and the wall marked through the fixing holes.

- You usually have to install an offset (known as a swan neck), to clear the width of the fascia and **soffit**. The easiest way to do this is to install the top section of downpipe with the first offset on, placing the second offset against the outlet of the running outlet, and taking the measurements in position.

- The fall pipe is finished off at the base with a rainwater shoe or connection to the drain.

- Clip spacings should be every 1.5 m and allowance should be made for thermal movement: 6–8 mm should be left from the end of the pipe to the inside shoulder of the fitting. Clips should also be placed where pipes are jointed.

- All screws used for fixing should be alloy or stainless steel so that they do not rust

- Cast-iron down pipe and gutters should be cut using an angle grinder.

The sizing of gutters will be covered at Level 3.

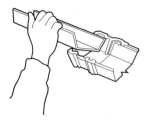

Locate back edge of gutter union bracket

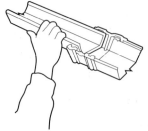

Pull front edge down until level with lip of bracket

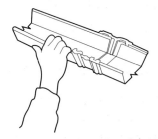

Snap the gutter under the lip of the bracket

Figure 11.73 Fixing the gutter

Figure 11.74 Assembled rainwater pipe

Maintenance and decommissioning

By the end of this section you should be able to:

- describe the maintenance requirements of AGDS

- describe the procedures for decommissioning AGDS.

In previous maintenance sections, we have talked about the importance of regular maintenance and cleaning in order to make sure the system works to maximum efficiency. AGDS are no different, and while these systems do not have as many controls as, say, central heating systems, if they are not looked after they will eventually deteriorate.

As a plumber, you may be called out to a blocked drain or discharge pipe, so we will look at that as well.

Figure 11.75 shows some equipment for unblocking drains, but remember there is much more sophisticated equipment now available.

Basic maintenance and cleaning

In domestic dwellings this is usually restricted to the manual cleaning of pipework. Traps often accumulate hair, soap residue, toothpaste and other objects that are small enough to fall through the grid of the waste hole. Integral overflows are also prone to this, particularly on Belfast sinks. Traps should be cleaned through the access points, if fitted; if not, the trap should be broken at its joints or removed completely.

Overflows can be rodded with wire and flushed. Chemical cleaning agents are also available that can be used for both traps and overflows. When using cleaning agents always follow the manufacturer's instructions.

Access covers to soil and vent pipes should be checked to make sure they are operating properly, and a visual inspection should be made of waste traps and fittings for signs of leakage.

Blocked pipes and drains

Blockages to discharge pipework on sinks, washbasins and baths can often be cleared using a 'force cup' or plunger.

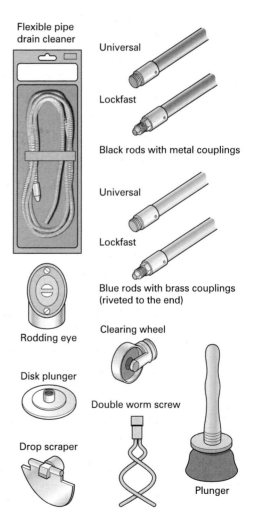

Figure 11.75 Equipment for unblocking pipes and drains

This is done by filling the appliance with water. Make sure you have blocked the overflow with a cloth or similar, then press down repeatedly on the handle of the force cup. This creates a positive pressure on the blockage, and a partial vacuum when it is withdrawn. This is usually enough to remove the blockage. Always check the trap for signs of leakage when you have finished.

Access to blocked drains can be through a drainage inspection chamber or, on newer jobs, through a rodding eye installed for that purpose.

Flexible metal pipe and sink cleaners are also available, which work like 'mini' drain-cleaning rods.

A blocked WC can be cleaned using a drain rod with a rubber plunger. These can also be used to clear blocked drainage pipework. Sections of the rods are joined together until the location of the blockage is reached.

Additional attachments are available. A double worm screw is used for pushing through obstructions or for catching an object (e.g. a cloth) and pulling it back to the operator. A drop scraper is a tool with a hinged steel blade: it is flat when travelling forwards but the blade is lifted when the rods are being pulled back. This can be used for pulling rubbish back to the operator. Clearing wheels help by guiding rods around bends or over obstructions.

Gutters should be checked regularly and cleaned as required. They do collect silt residue off roof tiles, leaves and bird droppings. The silt can build up and becomes a garden for seeds carried by the wind. Plastic gutters can be stripped in lengths between joints, emptied into a bucket and wiped clean. Cast-iron gutters should be painted to prevent rusting.

Leaking joints on cast-iron stacks can be cleaned out and remade using non-setting pipe-jointing compounds. Leaking cast-iron gutter joints can be repaired by removing the old joint, thoroughly cleaning the jointing surfaces of the gutter and fitting, applying jointing compound, and fitting a new gutter bolt.

There are new techniques available for clearing drains of blockages, including jetting techniques and camera systems; these require specialist training to operate and are not covered in this qualification.

Decommissioning systems

Decommissioning AGDS will normally involve stripping out old appliances and pipework to replace with new.

When removing appliances, care should be taken not to damage them.

Removing a cast-iron bath, particularly if it has to be carried downstairs, requires careful handling. Some plumbers break the bath into four pieces for easier removal. A club hammer is the best tool for doing this, and it should be done wearing full-face protection and gloves. Once the appliance has been removed it should be stripped of any scrap metal to be taken for recycling.

Safety tip

Broken vitreous china has edges as sharp as broken glass, so it should be handled wearing thick gloves

Remember

Old pipework systems could be in lead, so you must take the usual precautions when handling this material

On externally mounted pipework, you will have to chop out the mortar between the pipe and the masonry. Take care in doing this: it will mean less making good after you have installed the new pipework.

Taking down cast-iron soil and vent pipes can be dangerous due to their weight, so they need careful handling. It is best to try and take down short sections by partially cutting them with an angle grinder and then tapping the pipework with a hammer, which will cause it to shear; a rope should be tied to each section in turn so that they can be lowered to the floor. Make sure no one is in the area where you are working. Fixing lugs can be broken from the joint, and the nails prized out using a wrecking bar.

Once the stack is removed, make sure the joint to the drain is covered or the joint capped to prevent anyone tripping on it, prevent debris from entering the drain or prevent obnoxious smells coming from the drain – a range of blanking plugs for decommissioning work is invaluable.

Safety tip

Historically, many soil and vent pipes were made in asbestos – always follow the handling rules for asbestos

Remember your personal protective equipment – always wear a hard hat for these jobs!

FAQs

Why is so much fuss made over correct installation of sanitation systems?

It is perhaps the most widely abused system for incorrect installation as people often believe that waste pipes are not very important anyway. In reality it is very important as when things go wrong it causes major disruption and nuisance not to mention the safety and hygiene problems that can arise.

Do I really need to follow all the installation guidelines and measurements given in the Building Regulations?

Yes, you do, there are regulations that MUST be followed in all cases, anyway, and it is in all of our best interests to have sanitation systems that do not become faulty for obvious reasons.

Why does my loo fail to flush easily? It takes several sharp presses of the lever handle in quick succession to get it to flush.

It is almost certainly a worn diaphragm, the thin plastic washer in the siphon has torn or become displaced, simply renew it for a return to normal operation.

FAQ

Why is such a fuss made over allowing rainwater into a foul drain? It is relatively clean rainwater anyway so why not?

The extra quantity of rainwater will place an unnecessary burden of the sewage system as it may not be designed to take this excess, and treatment of sewage is an expensive process, which we all pay for.

Knowledge check

1. What is the relevant BS Code of Practice, and section, affecting above ground drainage and sanitary appliances?

2. Name two materials that are used in the manufacture of baths.

3. What are the two main types of WC pan design?

4. Where would you be most likely to use a straight-through trap?

5. List four reasons for loss of trap seal by natural causes.

6. List the three main pipework systems used on above ground discharge systems.

7. Why is it important to regularly clean and maintain AGDS fittings and pipework?

8. What items of personal protective equipment would be required when removing a cast-iron stack manually?

9. What is the relevant BS covering AGD pipework installation?

10. What are the main types of plastic used for waste and soil-pipe fittings?

11. State one advantage and one disadvantage of a combined drainage system.

12. What's the most common material used for gutters and rainwater pipes on domestic properties?

chapter 12

Sheet lead work

OVERVIEW

Plumbers on-site are sometimes required to carry out weatherings to chimneys, soil-vent-pipe penetrations through roofs and weatherings between building surfaces such as outbuildings. This is done by fabricating or working sheet lead, either by bossing, or by welding it together using oxy-acetylene equipment.

This session will cover:

- **Basic principles:**
  - types of sheet lead
  - codes and sizes of sheet lead
  - characteristics of sheet lead
  - tools and equipment used for sheet lead installation.

- **Fabrication techniques:**
  - safety factors when working with lead
  - sheet lead bossing
  - sheet lead welding
  - fire safety
  - welding equipment and safety

- **Flashings and small weatherings (Part 1):**
  - installation requirements of a chimney weathering set
  - fixing the components of the chimney weatherings

- **Flashings and small weatherings (Part 2):**
  - lead slate
  - testing sheet weatherings
  - maintenance

Basic principles

At the end of this session you should be able to:

- list the types of sheet lead
- list the codes and sizes of sheet lead
- describe the properties of sheet lead
- list and describe the tools and equipment used for working on sheet lead.

The covering of complete roofs, and the weathering of building details in sheet lead has traditionally been the job of a plumber. Over recent years, much of this work has been undertaken by specialist roofing contractors. However, plumbers still carry out sheet weathering in lead on domestic dwellings. This usually takes the form of:

- chimney flashings
- lead slates
- simple abutment flashings.

Types of sheet lead

Sheet lead is categorised by the way it is manufactured. There are two types:

- Cast sheet lead – this was the original production method for sheet lead, and it is still made this way as a craft operation. Production, by specialist lead-working firms, is in relatively small amounts created by running molten lead over a bed of sand. There is no British Standard for this material, and its sheet sizes and thicknesses vary – it tends to be used on historic monuments.
- Rolled (milled) sheet lead to BS 12588:99 – formed by passing a slab of lead back and forth on a rolling mill between two closing rollers, until it has reduced to the required thickness. The sheet is then cut to a standard width ready for distribution.

Milled sheet lead is used by plumbers for sheet weatherings. It is manufactured to British and European Standards and will not vary in thickness by more than 5 per cent at any given point. The standards also set down requirements for its chemical composition.

Did you know?

Milled sheet lead was first used as an alternative roofing material to cast sheet lead at the turn of the twentieth century

Codes and sizes of sheet lead

British Standards specify the codes and standards for sheet lead.

The table shows the code number, the thickness, weight and colour code of the sheet.

Sheet lead for flashings is supplied in rolls in widths from 150 mm to 600 mm in steps of 30 mm: 150 mm, 180 mm and so on. It is usually supplied in 3 m or 6 m lengths. Larger width and lengths are available. For most flashing applications, codes 3, 4, and 5 are appropriate. Code 4 is considered adequate for forming a chimney back gutter if using lead welding techniques, but if you were to 'boss it', code 5 would be more appropriate. Code 3 is used for lead soakers.

BS12588 Code No	Thickness mm	Weight kg/m²	Colour code
3	1.32	14.97	Green
4	1.80	20.41	Blue
5	2.24	25.45	Red
6	2.65	30.05	Black
7	3.15	35.72	White
8	3.55	40.26	Orange

Codes and sizes of sheet lead

Characteristics of sheet lead

The following characteristics of sheet lead make it ideal as a weathering material:

- malleability
- fatigue and creep resistant
- thermal movement
- durability
- corrosion
- patination
- fire resistant
- 'recyclability'.

Malleability

Lead is the softest of all plumbing materials. It is malleable and easily worked, using hand tools and a process called 'bossing', into the most complicated of shapes, or 'dressed' to fit the many types and shapes of roof tiles.

Fatigue and creep resistant

Fatigue is a loss of strength in the lead due to thermal movement, which eventually leads to cracking. The strength of lead relies on the grain structure of the metal. The chemical composition of lead is governed by BS1178, which effectively controls the grain structure to make the lead sheet more resistant to thermal fatigue without effecting malleability.

Creep describes the tendency of metals to stretch slowly over the course of time. Making sure you correctly size and fix individual pieces will reduce the risk of creep.

Thermal movement

Lead has a high coefficient of expansion at 0.0000297 for 1°C. It is important therefore to include regular expansion joints in lead flashings to allow for expansion and contraction due to changes in temperature. Flashings may also be secured into a wall by lead wedges, thus restricting its movement so expansion joints would be needed here as well.

Did you know?

Before metrication, lead was defined by its imperial weight per square foot, i.e. 3lb, 4lb, etc. These weights are where the present day code numbers originate

Definition

Malleability – the ability to be worked without fracture

Did you know?

Creep does not mean that a piece of lead has slipped down a pitched roof when the fixings have failed!

Expansion joints for flashings are usually in the form of laps. To minimise the thermal movement at each lap, it is important that an individual piece of flashing is no longer than as shown in the table.

Code No.	Thickness mm	Use	Maximum length
3	1.32	Soakers	1 m
4	1.80	Flashings	1.5 m
5	2.25	Flashings	1.5 m

Durability

When specified and fitted correctly, sheet lead will provide maintenance-free weather protection for many years. It is extremely resistant to atmospheric corrosion, and will withstand severe weather conditions – hot and cold.

Corrosion resistant

Sheet lead is resistant to most forms of corrosion found in a roofing situation. However, you need to heed the following precautions:

Abutment flashing

- Mortar – unprotected lead damp proof courses and cavity trays may corrode in the presence of some mortars.

- Lichen growth – the acid run off from lichen or moss on a roof may cause small holes to appear in lead sheet under the drip-off point from tiles or slates. A 'sacrificial' flashing may be fitted to the lead gutter. The other option is to treat the growth with a chemical fungicide.

Patination

With time, lead develops a strongly fixed patina (or sheen) which is silver-grey and won't dissolve. However, in rainy or damp conditions, new lead sheet flashings will produce an initial, uneven white carbonate on the surface. Not only does this look unsightly but the white carbonate can be washed off by rain, causing further staining of materials (e.g. brickwork) below the flashings. The use of patination oil, applied evenly with a cloth as the job progresses, will prevent this.

Fire resistant

Lead is incombustible, but melts at 327°C.

'Recyclability'

This is an important property of any material. Sheet lead is totally recyclable. In the UK scrap lead is recovered from buildings due for demolition by a national network of reclamation merchants. It is then returned to manufacturers, where it is carefully refined for re-use in the rolling mills.

Because of this well established recovery network, the sheet lead industry has a solid reputation for environmental awareness.

Tools and equipment used for sheet lead installation

Your Certificate in Plumbing and NVQ will require you to be assessed in fabricating and fixing sheet-lead components. The illustrations here show what these components look like.

You fabricate these components either by **bossing** or lead welding.

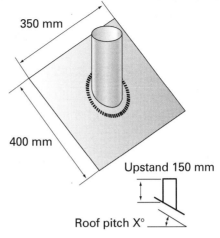

350 mm

400 mm

Upstand 150 mm

Roof pitch X°

Figure 12.1 Full chimney weathering

Cover flashing with lead-welded corners

Back gutter formed by lead welding

Front apron - 1 side formed by bossing, 1 side formed by lead welding

Side flashing lead welded where appropriate, flashing detail determined by centre roofing materials

Figure 12.2 Types of lead weathering

Lead to brick chimney

Lead slate in position

Definition

Bossing – the term for working sheet lead into shape using a range of mallets and dressers

The lead bossing tools are illustrated here. Originally made of boxwood, these tools are now available in durable plastic.

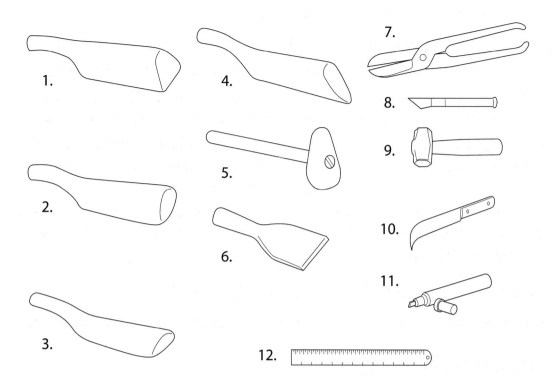

Figure 12.3 Lead bossing tools: 1 Flat dresser; 2 Bossing stick; 3 Bending stick; 4 Setting-in stick; 5 Bossing mallet; 6 Chase wedge; 7 Snips; 8 Plugging chisel – used to chisel out mortar from brickwork to let in the lead work; 9 Club hammer – for use with the plugging chisel; 10 Lead knife; 11 Spirit-based marker pen and straight edge – used for marking and cutting lines without damaging the surface of the lead; 12 Ruler

FAQ

Why is lead work in the qualification? Many plumbing firms don't do it.

The industry required sheet lead to be included so that all trainees become skilled in the basics of a chimney weathering and a lead slate. This means that wherever you go to work in the country employers will know that these skills have been achieved. There will be further courses for those who then wish to expand their range of lead skills in a range of advanced work.

Fabrication techniques

By the end of this section you should be able to:

- state the main safety factors when working with sheet lead and welding equipment
- describe the techniques for fabricating sheet lead using bossing and welding

In this session you will look at what is required to form sheet lead into the various shapes we refer to as flashings. It can be done in two ways: bossing and welding.

Most sheet lead work on larger sites will be done using welding techniques. There will be occasions, however, when you might not have access to welding equipment, and you will need to know how to shape the lead by bossing.

The main components are the pieces that make up the chimney flashings and a lead slate. The lead slate is produced using lead welding techniques, and the chimney flashings a mixture of both.

Safety factors when working with lead

Refer back to the lead safety section in Chapter 2 Health and Safety before continuing.

Sheet lead bossing

We will now look at the way the tools used for bossing are actually used.

Lead at ambient (ordinary surrounding) temperatures is only 300°C below its melting point: compare this to copper – at ambient temperatures it is 1056°C below its melting point. You can therefore imagine that, in many ways, lead behaves at lower temperatures in the same way as harder metals at higher temperatures.

Lead is an outstanding metal for bossing because:

- it is the softest of the common materials
- it is ductile – it will stretch quite a lot before fracturing or splitting
- it does not harden much when it has been worked (work hardening).

Bossing techniques

Bossing is not a skill that can be learnt overnight, but once you have mastered it, you will get a lot of satisfaction from the finished job. The best way to learn it is by watching an experienced lead worker, and then practising.

We are going to look at two techniques for bossing an external corner and an internal corner. Figures 12.4 and 12.5 show what the finished jobs look like.

Did you know?

The term 'bossing' has its origins in the Middle Ages, when 'boss' meant, 'to beat out metals into a raised ornament'.

Definition

Annealing – the process of softening something by heating it. As it has such a low melting point, ambient air temperature has an annealing effect on lead

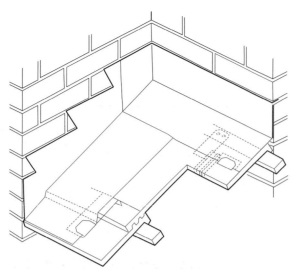

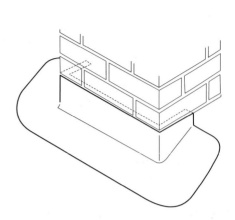

Figure 12.4 Internal corner: This is called an abutment flashing

Figure 12.5 External corner: This is called a front apron

The main aim of bossing is to achieve the required shape without making the lead too thin or too thick.

Internal corner, step by step

Mark out internal corner

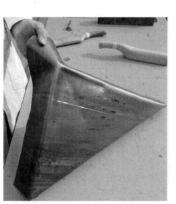

Setting in/starting

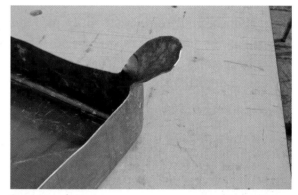

Part complete

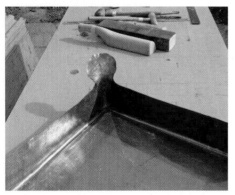

Nearing completion

Trimming away surplus

Final job

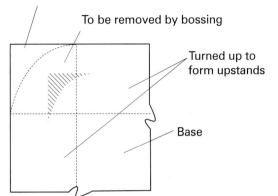

Figure 12.6 Marking the lead

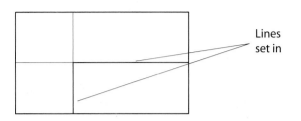

Figure 12.7 Lines to 'set in'

1 Using a spirit-based marker pen, straight edge and tape or ruler, set out the corner. Check the dimensions with a square.

2 Turn the upstands up by 90° using a timber former. A piece of timber about 600 mm long × 100 mm wide × 50 mm thick is ideal.

3 The angles are then 'set in' using the setting in stick. This is done by placing the blade of the setting in stick on the line and tapping it with a mallet. This process fixes the position of the upstands, whilst the bossing is in progress.

4 To get a good square base to the corner and to fix its position, use the mallet to raise a slight groove in the base of the sheet.

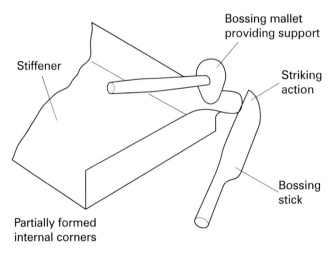

Figure 12.8 Partially finished corner

5 Lead is then worked as shown in figure 12.8, using a bossing mallet on the inside of the corner and a bossing stick.

6 As the process goes on, surplus lead builds up as shown in figure 12.9; this can be trimmed off if it gets in the way.

7 Keep checking that the corner is flush and square.

8 Check the dimensions of the upstands, when finishing off, mark off any excess material and trim off with the snips.

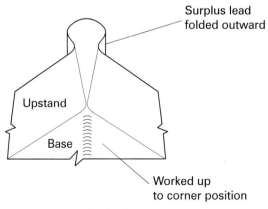

Figure 12.9 Surplus lead

External corner step-by-step

Marked out front apron

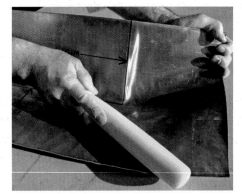

Starting bossing

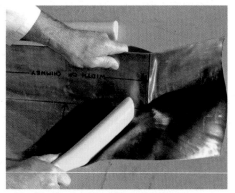

One third completed front apron

Two-thirds completed front apron

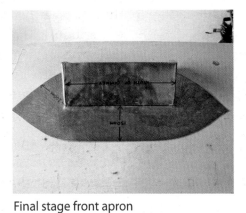

Final stage front apron

Completed front apron in-situ

The principle is the same as for the external corner, that is moving surplus lead around. The marking out is slightly different, however.

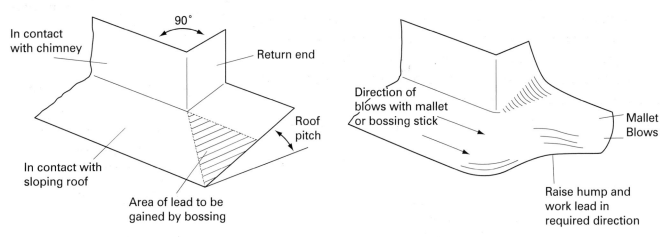

Figure 12.10 External corner – lead gain

Figure 12.11 Bossing an external corner

- Use your marker pen, tape measure, and straight edge as for the internal corner.
- For an external corner, such as the apron on a chimney breast, the lead has to be gained as shown in figure 12.10.

- The first job is to fold the upstands to the angle of the roof pitch.

- Trim the return end to the angle of the chimney breast (usually 90°). It should now look something like figure 12.11.

- As shown, a hump is formed which is worked into the desired position using a dresser, bending or bossing stick.

Finally, a couple of pointers about bossing sheet lead. As the job proceeds and the surplus lead is removed, there's a tendency for it to thicken and crease. If creases are allowed to form they will cause cracking. When bossing an external corner, you have to watch that you do not stretch the lead into position. This will thin the lead, and could cause splitting.

On the job: lead safety

Matt's company specialises in leadwork, particularly on historic buildings, and has just won a contract to replace the lead on the roof of a local church. Matt is given the job of stripping out old sheet lead prior to replacement. The lead is heavily corroded. The vicar is not keen on letting workmen inside the church so won't give Matt's company access to the vestry, which has the only source of running water on the site.

1   What items of PPE should Matt's company provide before Matt starts work?

2   Should Matt ask his employer to insist that the vicar allows the workers access to the running water supply in the vestry? If so, why?

Remember

PPE is your Personal Protective Equipment – use it at all times

Lead welding

Lead welding is a process of joining two pieces of lead by melting the edges of the lead together (called the parent metal) while a filler rod of lead is added. It is called **fusion welding**. The technique can be used to form an internal or external corner, as an alternative to bossing them into shape.

Figures 12.12 and 12.13 show what the finished jobs look like.

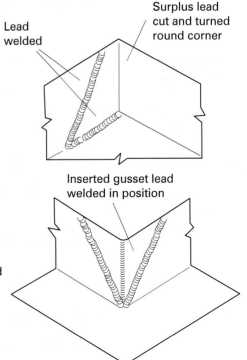

Lead welded

Surplus lead cut and turned round corner

Inserted gusset lead welded in position

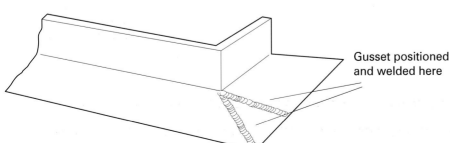

Gusset positioned and welded here

Figure 12.12 External corner

Figure 12.13 External and internal corner

Setting out the external and internal corner for welding

External corner

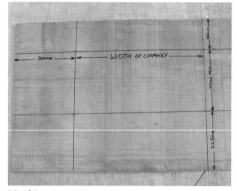

Marking out

Folding

Tacking

Welding

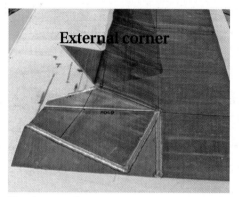

External corner

Nearly complete

Completed welded external corner

External corner

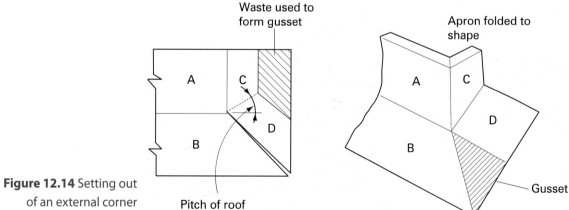

Figure 12.14 Setting out of an external corner

Figure 12.14 shows the setting out of an external corner. It will be used to form part of a front apron.

The apron will be marked out as shown and the joint cut between (B) and (D). The apron is worked into the required position and a gusset cut to fit as shown by the shaded area. This fits flush with the edges of the metal at (B) and (D) and is called a **butt joint**.

Internal corner

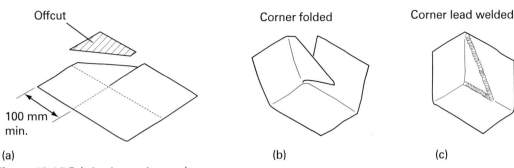

Figure 12.15 Fabricating an internal corner

- Mark out the corner in the same way as the bossed corner, but this time cut the lead as shown in figure 12.15 (a).

- Fold the corner as in figure 12.15 (b)

- This creates an overlap on the two upstands and it is the diagonal lap that will be welded as in figure 12.15 (c). This type of joint is called a **lap joint**.

The welding process

Some plumbers still refer to it this as lead burning, a term that dates from when crude welding techniques were used for jointing lead.

Like lead bossing, this is a very skilful job that will take time to master. Once you have done so, however, it is a very rewarding aspect of the job.

There are two types of lead welded joint:

- butted seam
- **lapped seam**.

Butted seam

Lapped seam

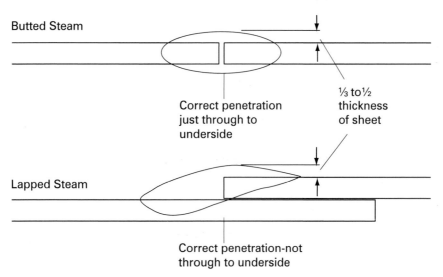

Figure 12.16 Butted seam and lapped seam

Lead welding tools and equipment

You will need:

- Oxy-acetylene welding equipment with a lead-welding torch. These torches have interchangeable heads numbered 1 to 5. In general, nozzle sizes 2 and 3 are used for lead sheet codes 4 and 5. For upright seams, size 1, 2 or 3 nozzles would be used depending upon the lead's thickness.

- snips

- flat dresser

- shave hook

- steel ruler.

Creating the welds

- Clean the surfaces that you are going to weld together

- Mark a width of about 10 mm and, using your metal straight edge as a guide, shave the surface of the metal with the shave hook

- Cut strips to use as filler rods. This is done by cutting a thin strip of lead about 3–5 mm thick and 300 mm long. Again it needs to be shaved clean. Alternatively, lead rods up to 6 mm can be obtained from a supplier.

Now everything is prepared, you can set up the welding equipment.

Now light the blow pipe:

1 Turn on and light the acetylene first.

2 Then feed in the oxygen. A pressure of 0.14 bar (2lb/sq in) for both the oxygen and acetylene is required.

3 You need to achieve a neutral flame to get the best results.

Did you know?

A shave hook is used for cleaning both the edges of the lead sheet to be welded, and the lead filler strips

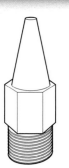

Figure 12.17 Lead welding nozzle

Remember

For a lap joint, don't forget to clean the piece underneath where the faces of the lap joint meet

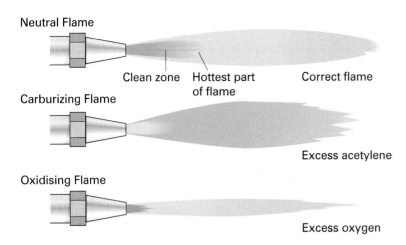

Figure 12.18 Blow-lamp flames

Lead welding – good practice

✓ Flat-butted seam welds should fully penetrate through the thickness of the lead sheet.

✓ Lapped seam welds should penetrate the surface of the lead, but not through to the underside.

✓ The thickness of the seam should be between a third and a half thicker than the sheet.

✓ The width of the weld will depend on the thickness of the lead and the seam pattern. When using code 4, the minimum width of a flat-butted seam should be 10 mm.

Beware of **undercutting** – reducing the thickness of the lead at the side of the weld. This causes a weakness that can result in cracking along the weld line. The main cause of undercutting is holding the flame for too long on the vertical surface.

> **Safety tip**
>
> Whether jointing lead using a butted or lapped seam, you should tack the pieces together. This prevents any movement and keeps the surfaces in close contact

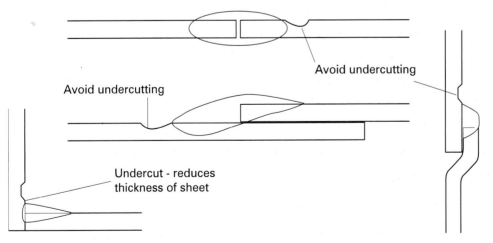

Figure 12.19 Undercutting

Welding flat-butted seams

This seam is used for jointing pieces of flat supported lead sheet as shown.

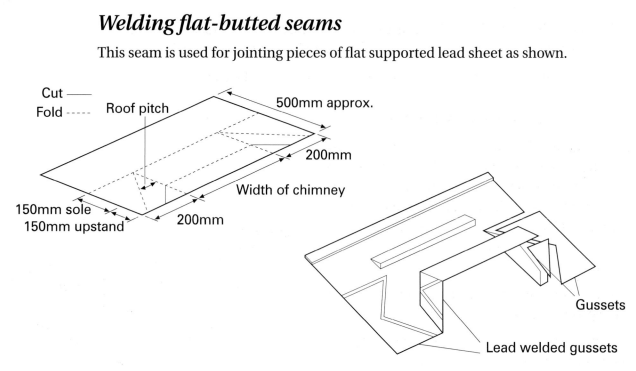

Figure 12.20 Joining flat supported lead sheet using butted seams

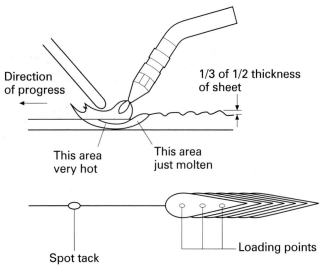

Figure 12.21 Position of blowpipe

The position of the blow pipe and filler rod should be as in figure 12.21.

- The tip of the cone of the flame should be just clear of the molten lead

- Lead is melted off the welding rod into the weld area

- A seam that is between one-third and one-half the thickness of the sheet is built up

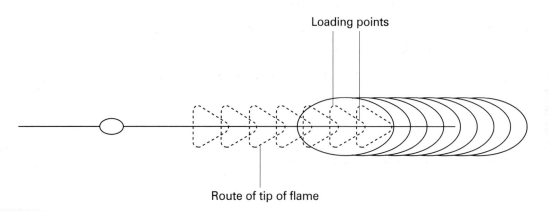

Figure 12.22 The flame is moved forwards to set the pattern of the seam

- The flame is directed into the centre of the seam and is moved forward, either in a straight line or slightly from side to side. This will set the pattern of the seam.

- The weld should just penetrate through the underside of the lead.

Welding a flat-lapped seam

This technique is an alternative to butted seams. It is often preferred when working on-site where there is a risk of fire during the welding process, because the flame will not make contact with the material beneath the lead. Even experienced plumbers will admit that this technique is slightly easier.

The illustration shows a typical example of a welded flat-lapped seam used on lead slate.

The preparation and process of welding a lapped seam is the same as for a butted seam, the only difference is that when welding thicker lead two loadings are used.

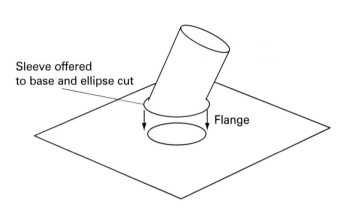

Sleeve offered to base and ellipse cut

Flange

Figure 12.23 Welded flat-lapped seam

Inclined seam on a vertical face

If you can, pre-fabricate sheet lead components before fixing them. There will be occasions, however, when you may have to weld a joint in position. One such situation is an inclined seam on a vertical face.

Although this joint can be made without using filler rods – by using the overlapping edge to make the seam – it is recommended that a filler rod should be used.

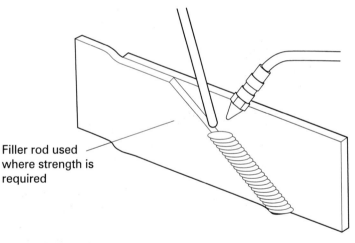

Filler rod used where strength is required

Figure 12.24 Inclined seam

Fire safety when working with lead

If working on-site, always try to use lapped joints. This reduces the risk of fire when using a butt joint against combustible surfaces.

On occasions, butt joints may be unavoidable. When this is the case, you should:

- wet the timber area beneath the weld

OR

- place a non-combustible material beneath the weld.

Remember

Pre-fabricate lead components before fixing wherever possible

Welding equipment and safety

The equipment

Below is a diagram of a typical welding kit.

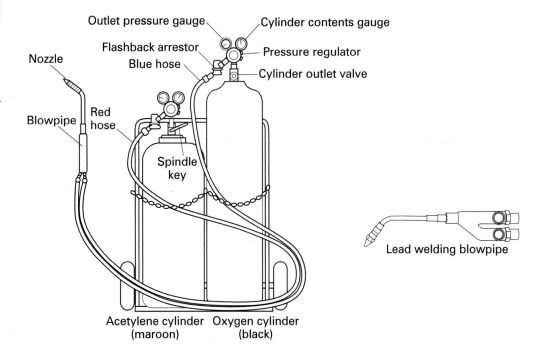

Figure 12.25 Welding kit

FAQ

It seems absurd that we buy a roll of lead and then cut it into short lengths only to weld it together again, surely we could fix it in one long length in the first place.

Not at all, lead expands and contracts with every temperature change so there must be maximum lengths and an overlap time to allow for essential free movement to take place. If it was fixed in long lengths it would cause metal fatigue and cracking will occur.

When using a welding kit, some basic safety procedures *must* be observed.

Safety checklist

1. Gas cylinders should be stored in a ventilated area on a firm base. If possible store oxygen and acetylene separately. Empty and full bottles should also be stored separately.

2. Acetylene gas bottles should be stored upright to prevent leakage of liquid.

3. Oxygen cylinders are highly pressurised. They should be stored and handled carefully to prevent falling. If the valve is sheared, the bottle will shoot forward with great force.

4. Keep the oxygen cylinder away from oil or grease as these materials will ignite in contact with oxygen under pressure.

5. Check the condition of the hoses and fittings. If they are punctured or damaged replace them. Do not try to repair or piece them together.

6. Do not allow acetylene to come into contact with copper. This produces an explosive compound.

7. Make sure the area where you are welding is well ventilated.

8. Erect signs or shields to warn and protect people from the process.

9. Always have fire fighting equipment to hand.

10. Wear protective clothing: gloves, overalls, goggles – clear are fine for lead welding.

11. Make sure that hose check valves are fitted to the blowpipe and flashback arrestors to the regulators. This prevents any possible flashback on the hoses and the cylinders.

12. Allow the acetylene to flow from the nozzle for a few seconds before lighting up.

13. In the event of a serious flashback or fire, plunge the nozzle into water, leaving the oxygen running to avoid water entering the blowpipe.

Flashings and small weatherings (Part 1)

At the end of this session you should be able to describe:

- the installation requirements for a chimney weathering set
- the fixing techniques for a chimney weathering set.

In this section we will look at how to apply the knowledge you have gained from the previous sections. The Level 2 Certificate in Plumbing requires you to competently work on installing chimney weatherings and lead slates, so they will be the main focus of this section.

Installation requirements of a chimney weathering set

The job we describe here is the one that you will do in the sheet lead practical for your NVQ in Plumbing. We also show the side chimney flashings details for both slate and tile.

Front apron

This will have a bossed external corner and a lead welded corner.

The lead joint in the brickwork should be at least 75 mm above the surface of the tiles or slates, and an extra 25 mm added to turn into the mortar joint. The side of the apron needs to be at least 100 mm.

The bossed part of the apron

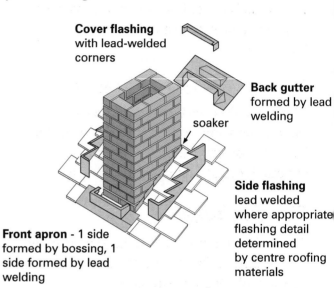

Cover flashing with lead-welded corners

Back gutter formed by lead welding

soaker

Side flashing lead welded where appropriate flashing detail determined by centre roofing materials

Front apron - 1 side formed by bossing, 1 side formed by lead welding

Notes to the drawing
1. All component dimensions to be taken from actual chimney on site
2. Installed components must have minimum laps to guard against water penetration
3. On site components fabricated to suit roof material (traditional slate roof components shown for illustrative purposes)
4. Components to be loosely wedged to pre-raked chimney joints, pointing not required

Figure 12.26 Chimney weathering

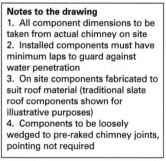

Figure 12.27 Front apron

The lead sheet association recommend that the piece of lead used for the joint apron should be not less than 300 mm wide – 150 mm for the upstand against the chimney plus 150 mm for the apron over the tiles.

The length of the piece will be the width of the chimney plus a minimum of 150 mm for each side. If the roof is covered with deeply contoured tiles increase this to 200 mm. This will form the basis of the dimensions for this example.

The bossed corner is set out and bossed as shown in a previous session on fabrication techniques.

The lead welded part of the apron

Obviously, this will be set out based on the same dimensions as the bossed section. You have already covered these marking out and fabrication techniques above.

Side flashings

This is the first time that we have mentioned side flashings. The illustration shows what they actually look like in real life.

There are two applications for side flashings:

- side flashings using soakers – these are used on roofs covered with slate or double lap plain tiles

- side and cover flashings (combined) – these are used where it is not possible to incorporate soakers, such as over contoured tiles.

Side flashings with soakers

A **soaker** is a piece of lead (code 3). One side is fitted between the slates or tiles, the other is turned up the side of the chimney. These are usually fitted by the roofer as the tiles or slates are fixed. They are then covered by step flashing and, in effect, provide a 'secret' waterproof gutter between the roof and the chimney.

They are fixed by adding 25 mm on to the length of the soaker, which is then turned over the batten and nailed.

The dimensions from which the step flashing is cut is a minimum of 150 mm wide.

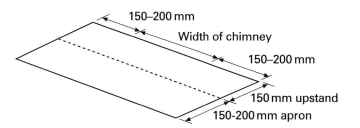

Figure 12.28 Dimensions of the lead sheet

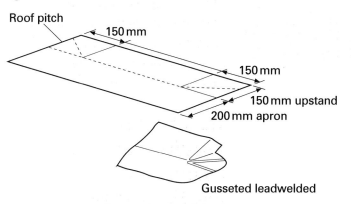

Figure 12.29 Marking out the lead sheet

Side flashings

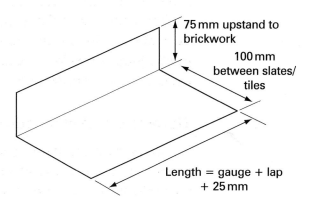

Figure 12.30 Soaker dimensions

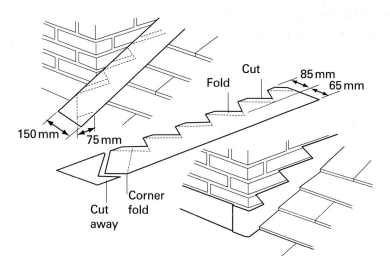

Figure 12.31 Step flashing

Setting out the step flashing

Let's assume a width of 150 mm for the step flashing. The length will be the full length of the side of the chimney. For the overall length you need to add a minimum of 75 mm for turning around the front of the chimney and over the apron.

- A water line of 65 mm is marked on the length of the lead before offering it against the brickwork.

- The lead is placed in position at the side of the chimney and, using a folding ruler, lines are marked that correspond to the bottom of the mortar joint.

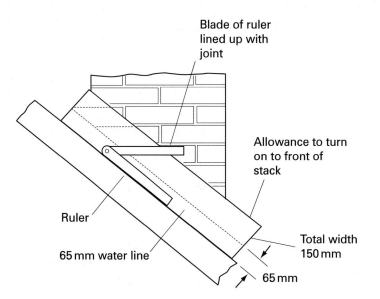

Figure 12.32 Setting out step flashing

- The lead is then marked out and fabricated as shown.

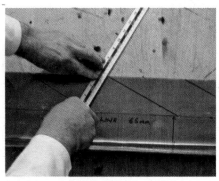

Setting out step flashing

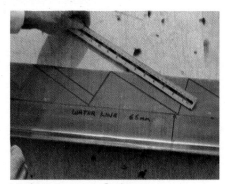

Marking out step flashing

Cutting out step flashing

When installing step flashing on slates or double-lap plain tiles, the allowance for each side of the chimney is 150 mm. If the roof is covered with single lap tiles a measurement of 200 mm is required.

Side and cover flashings

Flashings for both sides and cover require the same techniques as setting out a simple side flashing, but now, because the lead will be extended over the roof covering to a width of not less than 150 mm, the overall width of lead will be greater.

In addition, the piece of lead that is to be turned around the joint of the chimney will have to be fabricated. This can be done by either bossing or lead welding. The details are illustrated here.

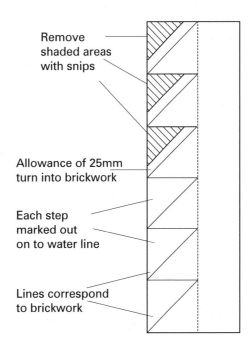

Figure 12.33 Marking lead for step flashing

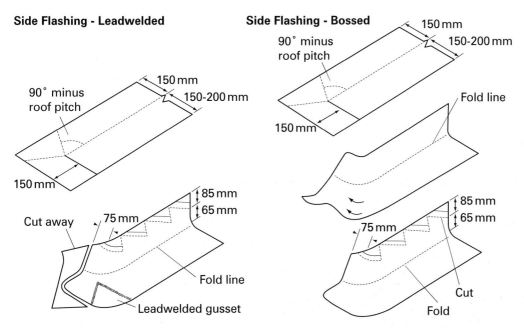

Figure 12.34 Bossing and welding around the joint of a chimney

Back gutter

This forms the final part of the chimney weathering 'jig saw' and a finished gutter is shown in the illustration.

The description here is for a lead welded version, because this is what is covered in the qualification. Again, we have used our own dimensions.

Setting out

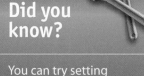

Did you know?

You can try setting out a variety of sheet weathering details with a piece of stiff card

When installing a back gutter on slates or double-lap plain tiles, the allowance for each side of the chimney is 150 mm. If the roof is covered with single lap tiles, a measurement of 200 mm is required.

Once cut, the back gutter is folded into shape and the gussets inserted and welded.

The final part of the back gutter installation is to fit a cover flashing over the upstand of the back gutter.

Cut ——
Fold ----
Roof pitch
500mm approx.
200mm
Width of chimney
150mm sole
150mm upstand
200mm

Gussets
Lead welded gussets

Figure 12.35 Welding a back gutter

Cover flashing

Figure 12.36 Completed chimney weathering with back gutter

A minimum of 100 mm over the width of the chimney is allowed on each side for trimming around the corners, and the bottom of the flashing is left about 5–10 mm from the gutter base.

Fixing the components of the chimney weatherings

Note that we have mentioned leaving 25 mm on each measurement for turning into mortar joints. Hopefully on new work, mortar joints will have been left 'raked out' to a minimum depth of 25 mm. If not, or on maintenance jobs, joints will have to be chipped out with a club hammer and plugging chisel.

Depending on the job you are undertaking, lead fixing wedges may be adequate. These are made using the off-cuts of lead saved while preparing flashings. They are cut into strips about 22 mm wide, rolled up and squashed so that they are thicker than the gap between the lead turn in and the brickwork – about 15 mm. Flatten one edge of the lead to provide a leading edge and place this in the joint. The lead wedge should then be driven in using a wooden chase wedge.

Remember

Make sure all the mortar is removed from the joint

Figure 12.37 Lead wedges

The distance between fixings will depend on the condition of the material you are fixing to. Fix at least one to each side flashing 'step', and for other components, fixings should be placed at 300–450 mm centres.

Once the lead has been fixed, it needs to be pointed with a flexible sealant or silicone.

Flashings and small weatherings (Part 2)

Lead slate in-situ

At the end of this session you should be able to describe:

- the installation requirements for a lead slate
- the fixing techniques for a lead slate
- how to test an installation for water tightness.

Lead slate

The photograph shows what a lead slate looks like in position. They can be formed by bossing, but are much quicker and cheaper to fabricate by lead welding.

This type of flashing is used where a pipe penetrates a roof covering. For most domestic dwellings this is likely to be the plastic soil and vent pipe.

The size of the base will vary depending on the roof covering. A typical slate for a 100 mm pipe will be 400 mm wide. The base should extend 150 mm from the front, and be not less than 100 mm under the slate. The height of the upstand should not be less than 150 mm.

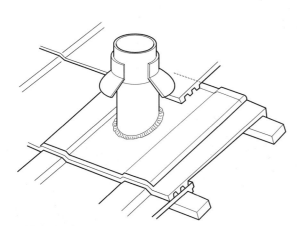

Figure 12.38 Pipe penetrating lead slate

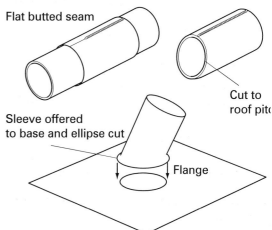

Figure 12.39 Cutting and fitting the upstand

Making the slate

Cut a piece of lead wide enough to give the height of the upstand, and as wide as the circumference of the pipe plus about 5 mm for tolerance. For a 100 mm pipe this is calculated as follows:

3.142×100 mm $= 314$ mm $+ 5$ mm for tolerance

$= 319$ mm (round to 320 mm)

The edges of the upstand are prepared for a butt weld, and then the lead is turned around a rigid pipe and butt welded.

One end of the upstand is cut to the pitch of the roof. This can be done using a bevel and taking the actual angle from the roof. Another method is to 'develop the piece' using a drawing – we will look at this technique now.

Once the upstand is cut to the required angle, the edge is dressed to form a flange. This is then placed on the base, a hole is marked and cut, and the upstand and base are prepared and welded together

Development of the slate piece

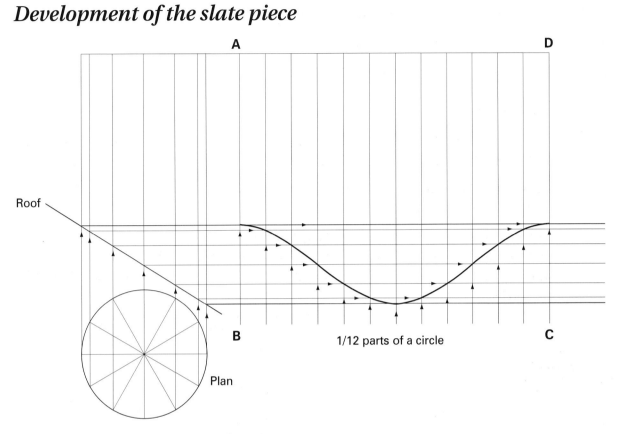

Figure 12.40 Development of slate pieces

- Draw the elevation of the slate piece
- Using a 60° – 30° set square, draw plan and divide circumference into 12 equal parts
- Project points from the plan up on the elevation to give an intersection at the roof line
- Draw a rectangle ABCD, so that AB is a least equal to the height of the slate piece
- Following the arrows, project horizontal lines from the intersection at the roof line and vertical lines from dimension on AB to obtain the intersection marked by a cross
- The true shape of the slate is obtained by joining the intersecting points by hand.

Testing sheet weatherings

When completed, a sheet lead installation needs to be tested to ensure that it is water-tight. The easiest way is to apply water to the completed installation using a hosepipe or bucket. Then check the roof internally for any leaks.

Did you know?

You can try making sleeves from card for a variety of roof pitches and these can then be used as templates for various jobs

Maintenance

Maintenance does not form a large part of the work on the components that we have looked at in this chapter, but you do need to establish the basic procedures for checking systems that have been installed. Areas covered should be based on the following:

- Establish adequate safety procedures for completing work, including safe roof access and working platforms.

- Check internal roof space to establish whether there was any water penetration around the components.

- Check key components for size dimension (establish that the components were installed correctly and in particular could not give rise to capillary attraction).

- Check roof covering adjacent to components – repair/rectify as necessary.

- Check individual roof components for soundness, no splits/cracks etc. – repair/rectify as necessary.

- Check component fixings to ensure soundness of fixing – re-fix (where required and re-pointed), check to establish whether additional fixings were necessary in the event of wind lift.

- Test components for leakage on completion of maintenance activity.

FAQs

What is the white staining often seen on a roof or brickwork below any chimney/leadwork?

This is white carbonate which can form on untreated lead once it is exposed to the weather. It happens as a result of not applying a patination oil to all surfaces of the lead as work progresses. This white carbonate is very unsightly and detracts from a good quality job.

Why does the correct thickness of lead used on jobs matter?

The recommendations for thicknesses of lead are based on sound historical practice and research, You must follow these LSA guidelines if you hope to achieve a good long-lasting job.

Knowledge check

1   State the two types of sheet lead production.

2   State the three codes (number and colour) of lead that are most likely to be used on flashings and where they would be used.

3   List three of the characteristics applicable to sheet lead.

4   What regulation sets out measures covering working with lead safely?

5   How can harmful lead particles enter your body?

6   State two methods of protection when handling lead.

7   What is the main aim of bossing?

8   State two precautions that should be taken when carrying out a butt weld on-site over a timber surface.

9   Excluding soakers, what are the three parts of a chimney flashing set called?

10 What is a soaker, and what is it used for?

chapter 13

Environmental awareness

OVERVIEW

Environmental awareness is an understanding of why it is important to conserve energy, dispose of waste properly and prevent wastage of materials. Environmental awareness is currently very topical; the UK government is keen to promote initiatives which help to conserve energy and reduce waste.

Building Regulations have been improved and updated to ensure that buildings are more energy efficient and that building materials are more effectively used.

In this section you will look at some of the ways you can help the environment in your work as a plumber.

- **The importance of energy conservation**
  - legislation

- **Applying environmental awareness**
  - basic methods of improving efficiency
  - other methods of improving efficiency
  - methods of reducing waste

- **Environmental hazards**
  - waste disposal

- **Customer advice**

You will have realised by now, particularly from the hot and cold water and central heating chapters, just how important energy conservation is to the plumbing industry. What you do when choosing materials and components for systems and how well you install them will have an affect on the environment.

Water Regulations, British Standard specifications and Building Regulations are in place for the public health and welfare of the consumer. They also ensure that working practices, system design and the use of materials are of the highest standards in order to maximise energy conservation.

In this chapter we will look at the legislation and some of the organisations that are involved with energy and our industry.

The importance of energy conservation

By the end of this section you should be able to:

- explain the importance of energy conservation

- state the main requirements of environmental legislation

- explain how the plumbing industry can contribute to energy conservation.

Energy conservation is aimed at reducing the CO_2 emissions into the atmosphere, which has a detrimental affect on the ozone layer, leading to global warming. This needs to be checked, and government has tackled this by bringing in legislation to make buildings better insulated and heating appliances more efficient. Outside of the building industry, vehicles are now taxed on the basis of the amount of CO_2 they produce when burning fuel.

Energy generation (by power stations) is a major source of CO_2 emissions and there are a number of initiatives aimed at producing 'clean' energy, including:

- Wind farms – large windmill type structures that produce electrical power as the blades rotate.

- Solar-powered hot water heating systems – large panels with a system of pipes located behind a glass panel; the water in the pipes is heated by the sun and transferred to the hot water storage cylinder.

- Solar photovoltaics – photo electric cells set in panels use the sun's power to produce electricity.

- Combined heat and power (CHP) – not limited to any one fuel, this uses the heat produced as a by-product of the electricity generation process that is normally lost to the environment. CHP can increase the overall efficiency of fuel use by as much as 70–90 per cent, compared with 35–52 per cent for normal electricity generation.

Solar panels are increasingly being used as alternative energy sources

The government has also brought in measures to minimise waste, as this is also harmful to the environment, using energy to transport it and treat it. Recycling is a way of reducing waste by returning excess and used materials back into the production cycle. A typical example is the scrap sheet lead mentioned in Chapter 12.

Legislation

Part L1 Building Regulations

Approved document L1: Conservation of fuel and power

The Building Regulations 2000 deals with the conservation of fuel and power in an approved document entitled 'Approved Document L1: Conservation of fuel and power' which came into effect on 1 April 2002. It has major implications for the plumbing industry.

It was brought in because it was obvious that big improvements were required to improve the energy efficiency of existing buildings. The following aspects now have to meet the same standards as new buildings:

* glazing

* upgrading of boilers and systems.

Inefficient boilers will require higher levels of insulation for the building.

One of the major requirements of the legislation as far as plumbing is concerned is the certification of heating and hot water systems to show that they've been correctly installed and commissioned, and the provision of operating and maintenance instructions for users.

Why the need for change?

Everyone has probably heard about the greenhouse effect and global warming due to carbon emissions. These new measures aim to

* reduce carbon dioxide emissions from buildings by 25 per cent

* save 1.3 million tonnes of carbon emissions per annum by 2010.

For new dwellings the main changes include:

* big improvements to U values, meaning higher levels of insulation. The lower the U value the lower the heat loss through the building fabric.

* higher standards of insulation for dwellings

* certification of heating and hot water systems to show that they have been correctly installed and commissioned, and that operating instructions have been left for the user

* increased standards of design and workmanship to improve building performance, reduce gaps in insulation, thermal bridging and poor air-tightness (e.g. gaps around windows and door frames, etc.) – it is anticipated that air-tightness testing will eventually become mandatory for dwellings

Definition

The U value – a measure of the thermal transmission of heat through the fabric of a building

- all new buildings to have a SAP (Standard Assessment Procedure)
- minimum levels of boiler efficiency
- minimum requirements for central heating control systems.

The Standard Assessment Procedure (SAP)

SAP is the government's 'Standard Assessment Procedure' for energy rating of dwellings; it is now a compulsory component in Part L1 of the Regulations. Every new house has to have a SAP rating.

What is SAP?

- SAP provides a simple means of estimating the energy efficiency performance of a dwelling. SAP ratings are expressed on a scale of 1–100. The higher the number, the better the rating.

- SAP is also used in the calculation of the carbon index, which can be used to show that dwellings comply with part L1.

- SAP ratings are used to predict heating and hot water costs. These depend on the insulation and air tightness of the house and the efficiency and control of the building's systems and controls.

- The procedure for calculating SAP is shown on SAP worksheets, although in practice most people calculate SAP using a computer program approved by the Building Research Establishment. The SAP calculation is based on:
  - the size of the house
  - the insulation levels
  - ventilation levels
  - heating and hot water systems.

- Once calculated for new dwellings, or new extensions to dwellings, the SAP rating can be submitted for Building Regulations approval and is checked by the local building control department.

- Anyone can provide a SAP rating for the purpose of building control approval and plumbers are often asked by small developers to provide SAP ratings.

What it means to plumbers

The type of heating system and controls installed, and the level of insulation and ventilation influences SAP ratings. Here are a few pointers about SAP:

- Condensing boilers produce higher ratings than traditional boilers and can increase the rating by up to 10 points. An electric heating system using an Economy 7 boiler can reduce the SAP by 11 when compared with a conventional gas fired boiler.

- The Boiler Efficiency Data File is published as part of the Office of the Deputy Prime Minister's (ODPM) Boiler Efficiency Database scheme. It holds data on domestic boilers (gas and oil fired only), current and obsolete, for the purpose of carrying out SAP energy ratings.

Find out

Further information about the Regulations can be obtained at www.construction.detr.gov.uk/const/eep

Did you know?

Sap ratings are similar to the miles per gallon fuel consumption figures for cars

Providing SAP ratings is covered in greater detail at Level 3

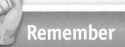

Remember

Boiler interlocks were covered in Chapter 9 on Central-heating systems

- If an existing older boiler is being re-used in a conversion, the plumber must ensure that when the pump switches off, the boiler also switches off. This is normally referred to as a 'boiler interlock'.

- Most new and refurbishment designs will require a minimum of programmer, room thermostat and TRVs to obtain a good rating, with additional points being added for delayed thermostats, energy managers etc.

- Cylinder thermostats are compulsory for a SAP pass.

- Primary pipework insulation is also compulsory to within 1.5 m of the cylinder to prevent losses. The depth of the cylinder insulation also has a significant effect on the rating.

- If the pump is fitted in the heated space, it will have a positive effect on the rating; if it is fitted in the unheated space, it will have a negative effect.

Water Regulations

The Water Regulations, discussed in earlier chapters, are relevant to energy conservation. They are discussed in some detail at Level 3. The purpose of the Water Regulations is to:

- prevent contamination of a water supply

- prevent the waste of water

- prevent the misuse of a water supply

- prevent undue consumption of water

- prevent erroneous measurement (fiddling the meter).

It is clear that two of these areas have a direct impact on energy conservation.

Waste of water

Plumbers see this in the form of dripping taps, leaking joints and overflows on both hot and cold water systems. A dripping hot tap means that the hot water lost is replaced by cold water, which then has to be re-heated, wasting boiler energy. There are around 18 million homes in the UK: if a leaking hot tap lost an average of just $\frac{1}{2}$ litre of water a week in each dwelling, overall 9 million extra litres of water would have to be heated.

A dripping tap represents a waste of water and energy

In the case of cold taps and cistern overflows, wasted water has to be replaced, which on a wider scale means additional treatment and distribution by water companies, using extra electrical and mechanical energy.

Water can also be wasted by burst pipes brought about by freezing – due to a lack of insulation or pipe runs being in exposed conditions.

Undue consumption

Excessive consumption can be caused by bad design, and in particular dead legs. This results in high volumes of cold water being run off before hot water arrives at the tap. Water Regulations also make recommendations about the insulation of hot water pipes and storage vessels, which is another way to reduce heat loss and decrease the load on the boiler.

Applying environmental awareness

By the end of this section you should be able to:

* describe how to improve energy conservation in plumbing systems
* state methods of reducing waste while carrying out plumbing work
* identify potential environmental hazards
* state safe methods of waste disposal
* advise customers on energy efficiency.

We have looked at the legislation that affects what you do as a plumber. Now we will consider how environmental awareness works in practice.

Basic methods of improving efficiency

Central Heating System Specification (CHeSS)

The Central Heating System Specification (CHeSS) was produced by the Building Research Establishment's BRECSU.

BRECSU has been promoting energy efficiency in buildings on behalf of the Government for over 20 years. They have done this as managers of the buildings element of the Energy Efficiency Best Practice Programme, on behalf of the UK's Office of the Deputy Prime Minister (ODPM).

CHeSS was seen as essential because of the difficulties facing the domestic heating installation industry due to a lack of common standards and little understanding of what should be done to improve energy efficiency.

CHeSS gives recommendations for good practice and best practice for the energy efficiency of domestic wet central heating systems, and plumbers should use them as a basis for developing a system specification and costings.

This is an example of the actual guidance from CHeSS for:

- regular boiler and hot water storage systems good practice (HR1)

Details are also provided for other system types.

CHeSS HR1 – Recommended good practice (2000)

CHeSS HR2 – Recommended best practice (2000) **System description**

Domestic Wet Central Heating System with regular boiler and separate hot water store **System description**

Domestic Wet Central Heating System with regular boiler and separate hot water store **Boiler**

A regular boiler (not a Combi) which has a SEDBUK efficiency of at least:
- 78% if fuelled by natural gas (bands A to D)
- 80% if fuelled by LPG (bands A to C, and some from band D
- 83% if fuelled by oil (bands A and B, and some from band C)

Boiler

A regular boiler (not a Combi) which has a SEDBUK efficiency of at least:

- 82% if fuelled by natural gas (bands A to C)
- 84% if fuelled by LPG (bands A and B, and some from band C)
- 85% if fuelled by oil (bands a and B, and some from band C)
  Hot water store EITHER

A hot cylinder, whose heat exchanger and insulation properties both meet or exceed those of the relevant British Standards

OR

A thermal (primary) storage system, whose insulation properties meet or exceed those specified in Reference 4. Hot water store

High performance hot water cylinder (see note 6)

OR

High performance thermal primary storage system (see Note 7)
Controls

Full programmer

Room thermostat

Cylinder thermostat

Boiler interlock

TRVs on all radiators, except in rooms with a room thermostat

Automatic bypass valve controls

Programmable room thermostat, with additional timing capacity for hot water

Cylinder thermostat

Boiler interlock

TRVs on all radiators, except in rooms with a room thermostat

Automatic bypass valve

Guidance notes:
Notes applicable to CHeSS HR2, HC2, HR1, HC1 (year 2000)

1. The specifications list only the main components of a heating system affecting energy efficiency. Other components will be required, such as radiators, circulating pumps, cisterns, and motorised valves. All components must be selected and sized correctly

2. Heating systems should be designed and installed in accordance with relevant safety regulations, manufacturers' instructions (Refs 5,6,7), and British Standards (Refs 9, 11). More detailed advice on domestic wet central heating systems is given in the government's Energy Efficiency Best Practice Programme Good Practice Guides (refs 1, 2) and reference 14

Figure 13.1 CHeSS guidance table

Guidance notes:

Notes applicable to CHeSS HR2, HC2, HR1, HC1 (year 2000)

1 The specifications list only the main components of a heating system affecting energy efficiency. Other components will be required, such as radiators, circulating pumps, cisterns and motorised valves. All components must be selected and sized correctly.

2 Heating systems should be designed and installed in accordance with relevant safety regulations, manufacturers' instructions, and British Standards. More detailed advice on domestic wet central heating systems is given in the government's Energy Efficiency Best Practice Programme Good Practice Guides.

3 A regular boiler does not have the capability to provide domestic hot water directly, though it may do so indirectly via a separate hot water store.

A combination (combi) boiler does have the capability to provide domestic hot water directly, and some models contain an internal hot water store.

A combined primary storage unit (CPSU) is a boiler with a burner that heats a thermal store directly. Each of these may be either a condensing or a non-condensing boiler: condensing boilers are always more efficient.

4 SEDBUK is a measure of the seasonal efficiency of a boiler installed in typical domestic conditions in the UK. The SEDBUK efficiency of most current and obsolete boilers can be found on the website www.boilers.org.uk.

Although SEDBUK is expressed as a percentage, an A–G scale of percentage bands has also been defined in the table at figure 13.2:

Band	SEDBUK range
A	90% AND ABOVE
B	86–90%
C	82–86%
D	78–82%
E	74–78%
F	70–74%
G	Below 70%

Figure 13.2 SEDBUK ranges

5 A hot water cylinder may be regarded as meeting or exceeding the insulation requirements of the relevant British Standard if the manufacturer confirms that the standing heat loss is not greater than 1 watt per litre for the popular 117 litre cylinder

6 A high performance hot water cylinder may either be vented or unvented. The manufacturer must confirm that the heat exchanger and insulation properties exceed the requirements of the relevant British Standards.

7 A high performance thermal (primary) storage system must have insulation properties exceeding by at least 15 per cent those given in the WMA Performance Specification for Thermal Stores, and comply with the Specification in other respects.

8 Systems with regular boilers must have separately controlled circuits to the hot water cylinder and radiators, and both circuits must have pumped circulation. Large properties should be divided into zones not exceeding 150m² floor area, so that heating in each zone can be timed independently.

9 A time switch is an electrical switch operated by a clock to control either space heating or hot water, or both together:
   - a full programmer allows the time settings for space heating and hot water to be fully independent
   - a room thermostat measures the air temperature within the building and switches on and off the space heating; a single target temperature may be set by the user
   - a programmable room thermostat is a combined time switch and room thermostat which allows the user to set different periods with different target temperatures for space heating, usually in a daily or weekly cycle; some models also allow time control of hot water, so can replace a full programmer
   - a cylinder thermostat measures the temperature of the hot water cylinder and switches on and off the water heating
   - a TRV (thermostatic radiator valve) has an air temperature sensor which is used to control the heat output from the radiator by adjusting water flow.

10 Boiler interlock is not a physical device but an arrangement of the system controls (room thermostats, programmable room thermostats, cylinder thermostats, programmers and time switches) so as to ensure that the boiler does not fire when there is no demand for heat. In a system with a combi boiler, it can be achieved by fitting a room thermostat

In a system with a regular boiler, it can be achieved by correct wiring interconnection of the room thermostat, cylinder thermostat and motorised valve (s). It may also be achieved by more advanced controls, such as a boiler energy manager. TRVs alone are not sufficient for a boiler interlock.

11 An automatic bypass valve controls water flow in accordance with the water pressure across it, and is used to maintain a minimum flow rate through the boiler and to limit circulation pressure when alternative water paths are closed.

A bypass circuit must be installed if the boiler manufacturer requires one, or has specified that a minimum flow rate has to be maintained while the boiler is firing. The installed bypass circuit must then include an automatic bypass valve (not a fixed position valve).

All boiler manufacturers provide details of efficiency ratings, and you will find that they meet the SEDBUK rating.

Other methods of improving efficiency

We have covered the following procedures earlier in the book, but they are particularly relevant to efficiency:

- All systems pipework should be insulated and installed to comply with Water Regulations.

- Systems and components should be serviced and maintained regularly to:
  - ensure they are working to design specification
  - they are not wasting water.

- Systems should be designed to:
  - keep dead legs as short as possible
  - avoid over capacity, i.e. heating high volumes of hot water storage that may not be required.

Methods of reducing waste when plumbing

It is important that you get into the habit of keeping waste material to a minimum when you are carrying out plumbing work. Here is a checklist.

Checklist

1. If it is economically possible, try to repair an appliance or component rather than replace it. New components have to be made, and the manufacturing process uses energy. Old appliances have to be transported and disposed of (although replacing an inefficient boiler would be a good move).

2. Take time to carefully measure and set out pipework for bends. This will reduce the amount of wasted pipe when cutting it to length.

3. When using capillary integral solder ring fittings do not use additional wire solder on the joint.

4. Treat screws and other fixings like money – too often they are left all over a job.

5. Be extra careful when fitting sanitary ware, a broken one has to be disposed of and replaced.

6. This also applies to the storage of sanitary ware and other damageable goods.

7. Take care of your tools. Defective tools have to be replaced – and so do stolen ones!

8. Think about how you use water when you are working on a system. Have a thorough check around all the fittings to make sure they have all been soldered. You do not want to be filling and draining systems unnecessarily.

9. Do not overdo the use of jointing compounds, fluxes and other materials.

10. Take care when taking floorboards up; avoid having to replace damaged boards with new ones.

On the job:

Jordan is a second year apprentice working under the supervision of Anthony, who has been with the company quite a few years. They have nearly finished installing a heating system in an occupied terraced house. Jordan decides to use some initiative and begins to insulate the hot water primaries in the airing cupboard.

"Don't bother with that" shouts Anthony, "Waste of time and effort." Jordan is about to tell Anthony about the new regulations and the importance of energy efficiency he learned about at college when Anthony continues, "You apprentices are all the same, get on with something worthwhile."

The job is completed (minus insulation) and while Jordan is putting the tools in the van the customer thanks them and gives a £10 tip each for doing such a good job. Jordan feels guilty because the job is not as complete as it should be.

1   What should Jordan do now?

2   What would you do?

3   Is the insulation important?

Environmental hazards

The main environmental hazard as far as plumbing is concerned is asbestos, dealt with in Chapter 2 Health and Safety. A popular pastime on some of the larger sites is burning waste material. All sorts of things get thrown into a site fire: plastics, paint cans and so on. This activity is not environmentally friendly and is usually against site rules. Particular attention is drawn to the burning of plastics and polystyrene which give off dangerous fumes.

Waste disposal

There are special arrangements for the disposal of some types of asbestos. On larger sites, skips are usually provided for waste. These, once full, are taken to licensed sites for disposal. Do not overfill the skip, and do not put in items that are dangerous, e.g. flammable material.

Dispose of construction waste responsibly

On smaller jobs you might still use a skip, for the disposal of an old bathroom suite for example. If you try to take 'industrial waste' to your local tip you are likely to be turned away or have to pay for its disposal. Remember that the disposal of refrigerators is covered by legislation and they must be dealt with at specialist sites. Do not be tempted to let the customer just dump them in the skip.

Plumbers work with:

- copper
- brass
- lead
- cast iron and low carbon steel.

All of these have a value at the scrap merchant – think about recycling before you think about throwing away.

Customer advice

It is good business practice, as well as a requirement of the L1 Building Regulations: Approved document (central heating systems), to provide information to the users on the control and operation of a new system installation.

You will probably find that a high percentage of callouts are from customers who do not know how to work their system controls properly. Here is a list of what we think should be done for the customer:

- Provide a company folder containing:
  - emergency contact number for burst pipes, etc.
  - general advice line number – you might be able to sort a problem out over the phone
  - customer guidance leaflets that most manufacturers produce, also *go through them with the customer*.
- Label the various fittings or components that the customer may need to use in an emergency, e.g. stop valve, gate valve, etc.
- Show them, or tell them, where the service valves are, and what they do.
- Walk them around the system: show them the various components; tell them what they can touch and what they must not touch.
- Explain what the components do in simple terms, e.g. 'This is the cylinder thermostat, and, when the water in this cylinder gets to that pre-set temperature, it shuts off the heat to the cylinder; you don't have to alter that setting'.
- When you are finished, give your customer the chance to ask questions.

Remember

'Fly tipping' – the disposal of waste in places other than a registered commercial or private site – is illegal.

Recycling is an energy efficient and cost-effective way of getting rid of scrap waste

FAQ

Why does avoiding dead legs in hot-water systems have an impact on the environment?

In two ways: firstly by saving water which is run off to waste while waiting for hot water to arrive, and secondly by saving energy on heating the water in the first place. The hot water that is left in the pipe when we close the tap cools down in a very short time and is wasted.

Surely a little bit of insulation to hot water pipes in an airing cupboard will not make that much difference to the environment?

Well, every little does help. If we were to multiply all the heat losses on a national basis it would be considerable, so you see, it will make a difference.

Knowledge check

1   Why do you think it's important to conserve energy?

2   What area of the Building Regulations covers energy conservation?

3   What does SAP stand for?
Standard Assessment Procedure
Statutory Assessment Procedure
Standard Administration Procedure
Standard Assessment Processes

4   State three examples of how the waste of water would affect energy conservation.

5   What does CHeSS stand for?

6   Explain briefly what CHeSS is.

7   Band D in the SEDBUK range refers to a boiler with an efficiency of:
Below 70 %
70–82 %
78–82 %
82–96 %

8   State three methods other than CHeSS of improving energy efficiency in dwellings.

9   State four methods of reducing waste when carrying out plumbing work.

Glossary

abutment flashing used where a roof needs weathering against a brick wall

adhesion the force of attraction between molecules and another material they are in contact with, e.g. water in a glass, tyres on a road

allen key an hexagonal tool designed to fit into socketed screws to allow them to be turned

alloy a type of metal made from two or more other metals

alternating current type of electrical current as used in domestic properties

ampere a flow of one coulomb in one second

annealing the process of softening something using heat, e.g. copper

anode positively charged electrode

APHC Association of Plumbing and Heating Contractors

Apprenticeship in Plumbing formal agreement between trainee and employer to confirm period of training and expected outcomes

atmospheric pressure the pressure in the environment created by the force of gravity

atom smallest particle of a chemical element

bending spring spring used either internally or externally to aid the manual bending of copper tubing over the knee

bi-metallic strip strip made of two different metals bonded together which have different coefficients of expansion, thereby causing the strip to bend when it is heated

bonding term used to describe electrical earthing arrangement

bossing working lead sheeting into shape using a range of mallets and dressers

BSP British Standard Pipe

butt joint method of joining sheet lead where the edges are welded up against each other

capillary action the process by which a liquid is drawn up the surface of a solid material

catalyst a substance that increases the rate of a chemical reaction while remaining unchanged itself

cathode negatively charged electrode

CE mark mark denoting that an appliance conforms to European Standards

centralised hot-water system stored heated water ready to be supplied to a number of outlets

ceramics fired pottery

chamfering smoothing the cut edge of a pipe by slightly bevelling or angling it

chasing out the process of cutting a recess into plaster or blockwork for a socket or cable

check valve a device used to protect the water supply from any contaminants that may get into it

circuit breakers a safety device that can be set to interrupt an electric current in certain situations

cohesion the tendency for molecules to stick together

compression ring the ring used to form a pressure-tight seal in a compression joint

conduction the transfer of heat energy through a material

conductor a material that allows the transfer of electricity

conduit a preformed PVC steel channel that covers cable runs and protects cables against damage

convection the transfer of heat by means of the movement of a locally heated fluid substance (usually air or water)

corrosion degradation of a material caused by air, water, acids, alkalis, chemicals or electrolytic action

coulomb a unit of measurement of electrons

cross-flow occurs when two branches are located opposite each other allowing water flowing from branch to affect the pressure in the other branch

crutch head the handle of a stop tap

current flow of electrically charged particles

curtilage an area attached to a dwelling house and forming one enclosure within it

deformation when a material will not return to its original stage following the application of force

density the relative lightness or heaviness of solid materials

differential relating to the difference between two opposing forces

direct current electrical term given to current flow as from a battery

earth to run an electric current safely away to earth, or ground

earth equipotential bonding an earthing system that allows a fault or electrical leak on any part of a system to go to earth, rendering the system safe

earth fault loop total earth path around the installation and back to the transformer

electrical conductivity how well or poorly a material will conduct electricity

electricity the flow of electrons through a conductor

electrolyte a solution or molten substance that conducts electricity

electromagnetism the magnetic force produced by an electrical current

electro-motive force producing an electric current

electron elementary particle with a negative electric charge

fascia a piece of wood or plastic fixed to the rafters just below the roof allowing a fixing point for guttering

ferrous metals that contain iron

ferrule a metal fitment used for mains connection

first fix fixing of pipework to floor joists, boards and sheets, stud partitions, roof trusses and timbers etc., sometimes before a building is watertight

flexible push-fit plumbing systems term given to system of pipes and fittings using push-fit joints

fluid a substance such as a gas or liquid that lacks definite shape and is capable of flowing

foul drainage anything discharged from a sanitary appliance – WC, bath, basin, sink etc.

fuse a small unit containing a strip of wire that will melt when the current exceeds a certain level, thereby breaking the circuit

fusion welding the process of joining two pieces of lead by melting the two edges of the lead together (called the parent metal) while a filler rod of lead is added. Also the process of jointing some waste and soil pipework

gas air-like, not solid or liquid

glass hard, brittle, usually transparent substance made by fusing sand (silica) with soda and lime

gravity force of attraction dependent on the relative mass of the bodies in question

gravity circulation movement of water around a property which relies on the fact that cold water is heavier than hot water: gravity therefore exerts a stronger pull on the cold water, allowing the hot water to rise through the system

gripper fixings laid underneath carpets, usually at the edges, to hold the carpet firmly in place

hand bender small portable machine for bending copper tube

hertz the number of cycles of a.c. that are produced every second

immersion heater an electric element fitted inside the hot water storage vessel, controlled by a switch and thermostat

impounding reservoir a man-made reservoir

installation fitting a new system

instantaneous hot-water system heats water on demand

insulator an item to stop transfer of electricity, also used to describe material for reducing flow of heat

JIB for PMES Joint Industry Board for PMES

kinetic movement

lap joint method of joining sheet lead, where the sheets initially overlap

liquid offering no resistance to change of shape, not solid or gaseous

localised hot-water system heats water at the point at which it is required

main earthing terminal as found at electrical point of entry/main distribution board

maintenance carrying out any necessary repairs to worn or broken parts; keeping appliances and systems in good working order

making good repairing and finishing off the brickwork, blockwork and concrete

malleability the ability to be worked without fracture

mandatory units the ones candidates *must* do

manipulative joint a joint in which you work or form the end of the tube

marring the migration of polymers between polystyrene and a PVC cable sheath, potentially leading to damage to the PVC cable

micro-businesses either sole traders or businesses employing 1–4 people

milestones key targets that you need to achieve at specified times within your apprenticeship

Mohs' scale hardness scale, developed in 1812 and named after the German mineralogist Frederich Mohs (1773–1839)

neutron elementary particle that has no electric charge

nucleus the central part and the main mass of an atom

observed assessment carrying out a set task with an assessor in attendance to confirm competence

olive the ring that is compressed onto a pipe in a compression fitting

open vented a system that has a vent pipe always open to atmosphere

orbit circulate around another body

overflow pipe discharges excess water safely

packing-gland nut nut used to compress packing to make valve spindles watertight

patina a protective barrier which forms on metals

P.A.T. tests maintenance records of all portable equipment to ensure it is in safe working order

permeation microscopic particles of oxygen entering the water supply through the external wall of the pipe

plant large, usually heavy machinery used on building sites, such as cranes, excavators, specialised earthmoving equipment, forklift trucks and power access equipment

polybutylene a product of the polymerisation of ethers

potential difference energy used up by one coulomb between two points in a circuit

properties physical and working attributes of materials

proton elementary particle with a positive electric charge

proving unit Low-voltage, inverted d.c. testing device

radial circuit starts at the consumer unit and connects to every point on the circuit, terminating at the last one

radiation the transfer of heat from a hot body to a cooler one without the presence of a material medium (other than air), by means of 'heat' waves

residual current device a highly sensitive unit that measures changes in the electrical current between different electrical conductors in a system and automatically disconnects the circuit if a small change occurs

resistance the degree to which a material or device resists the flow of an electric current

ring circuit connects the consumer unit and every point on the circuit, and returns to the consumer unit

rungs the horizontal 'steps' of a ladder

sacrificial anode a metal that is relatively more easily destroyed such as magnesium: it will be destroyed by corrosion first, therefore giving longer life to other metals

secondary circulation a method of installing hot-water pipework whereby a return pipe is fitted near the top of the cylinder, to maximise energy efficiency

second fix the installation of bathrooms, radiators, staircases, doors, kitchen units, architraves and skirtings after a building has been made watertight

Sector Skills Council links with government and looks after the training and development needs of their specific area e.g. SummitSkills

SEDBUK rating a measure of the seasonal efficiency of a boiler installed in typical domestic conditions in the UK

service regular checking of a system to ensure it is still working properly

simulated assessment assessment that takes place at the centre

soaker a piece of lead fitted between the slates or tiles and then turned up and fitted to the side of the chimney

soffit the underhang of the roof, above the fascia

solid having stable shape, not liquid or gaseous

solvent power ability to dissolve other substances

sparge pipe a flush pipe mounted on the face of a urinal drilled with a number of holes, used to wash the face of the slab and the channel

specific heat capacity the amount of heat required to raise 1kg of a material by 1 degree Celsius

sterilisation purification by boiling or dosing the water supply with chlorine or chlorine-ammonia mix

stand bender a larger type of bending machine that stands on legs/tripod

'stat' a widely used abbreviation for 'thermostat'

stillson type of wrench for use on steel pipe and fittings

stocks and dies stocks are the handle of the tool, dies are the actual cutter

storage hot-water system stored heated water ready to be supplied to one or more outlets

stratification the formation of layers of water in a hot water storage vessel

stiles the upright bars of a ladder

supplementary evidence additional proof that you carried out the work

SummitSkills the sector skills industry organisation dedicated to improving skills in building services, such as plumbing, by training and development

surface tension the way in which water molecules cling together to form a 'skin'

surface water the term used for water collected via the rainwater system

swaging tool a specialist tool for 'opening out' the end of a pipe for manipulative fittings

swarf debris produced from cutting action

termination end point

thermoplastic a plastic that softens when heated

thermal heat

thermal conductivity how well or poorly a material will conduct heat

thermal insulators wooden, ceramic and plastic materials that are poor conductors of heat

thermosetting plastic a plastic that keeps it shape after initial forming

throating when a pipe's diameter has been reduced at the bend by the roller, guide and former being too tight

TN-S system electrical installation system in which exposed metalwork is connected to the main earthing terminal and in which there is a separate earth conductor

TRVS thermostatic radiator valves

undercutting reducing the thickness of lead sheeting where two pieces are welded together

venturi a device that causes a differential pressure in a pipe to activate a switch such as on a gas water heater

vitreous china ceramic produced from a solution of clay and water, called slip, or casting clay which contains ball clay, china clay, sand, fixing agent and water

voltage electro-motive force, measured in volts

warning pipe alerts occupants to a problem with float valves

water cycle water's journey from the sky to your tap

water governor a pressure-reducing valve

water mains the network of pipes that supply wholesome water to domestic and commercial properties

water table the natural level of water under the earth

water undertaker the legal term for the water companies that supply domestic water

wholesome water good-quality water provided for human consumption

workplace evidence record formal written evidence of your work carried out on site

Index